Obesity Epidemic and the Environment

Obesity Epidemic and the Environment

Latin America and the Caribbean Region

Matheus Koengkan

Faculty of Economics
University of Coimbra, Coimbra, Portugal
Center for Advanced Studies in Management and Economics of the Universidade de Evora (CEFAGE-UE) Evora, Portugal

José Alberto Fuinhas

Faculty of Economics
University of Coimbra, Coimbra, Portugal
Centre for Business and Economics Research (CeBER) Coimbra, Portugal

Aida Isabel Pereira Tavares

ISEG, Lisbon School of Economics and Management, Lisbon, Portugal
CEISUC, Centre of Studies and Research in Health of the University of Coimbra, Coimbra, Portugal

Nuno Miguel Barateiro Gonçalves Silva

Assistant Professor
University of Coimbra, Coimbra, Portugal
Centre for Business and Economics Research (CeBER)
Coimbra, Portugal

ELSEVIER

ACADEMIC PRESS
An imprint of Elsevier

Academic Press is an imprint of Elsevier
125 London Wall, London EC2Y 5AS, United Kingdom
525 B Street, Suite 1650, San Diego, CA 92101, United States
50 Hampshire Street, 5th Floor, Cambridge, MA 02139, United States
The Boulevard, Langford Lane, Kidlington, Oxford OX5 1GB, United Kingdom

Notices

Knowledge and best practice in this field are constantly changing. As new research and experience broaden our understanding, changes in research methods, professional practices, or medical treatment may become necessary.

Practitioners and researchers must always rely on their own experience and knowledge in evaluating and using any information, methods, compounds, or experiments described herein. In using such information or methods they should be mindful of their own safety and the safety of others, including parties for whom they have a professional responsibility.

To the fullest extent of the law, neither the Publisher nor the authors, contributors, or editors, assume any liability for any injury and/or damage to persons or property as a matter of products liability, negligence or otherwise, or from any use or operation of any methods, products, instructions, or ideas contained in the material herein.

ISBN: 978-0-323-99339-5

For information on all Academic Press publications visit our website at https://www.elsevier.com/books-and-journals

Publisher: Stacy Masucci
Acquisitions Editor: Elizabeth Brown
Editorial Project Manager: Veronica III Santos
Production Project Manager: Niranjan Bhaskaran
Cover Designer: Miles Hitchen

Typeset by TNQ Technologies

Contents

Biographies

Matheus Koengkan

Matheus Koengkan holds a Ph.D. in economics from the University of Évora. He is a researcher in energy and environmental economics at the Department of Economics, Management, Industrial Engineering and Tourism (DEGEIT), University of Aveiro, Aveiro, Portugal. He has published in international journals, such as *Energy*; *Environmental Economics and Policy Studies*; *GeoJournal*; *Environmental Science and Pollution Research*; and *Environment Systems and Decisions*.

ORCID: https://orcid.org/0000-0002-0062-4476

José Alberto Fuinhas

José Alberto Fuinhas, Ph.D. in economics, is a professor of applied energy economics and intermediate econometrics at the Faculty of Economics of the University of Coimbra, Coimbra, Portugal. He is a researcher in macroeconomics, energy economics, and environmental economics, at the Centre for Business and Economics Research (CeBER), sponsored by the Portuguese Foundation for the Development of Science and Technology. He has published in international journals, such as *Energy, Economic Modelling*; *Energy Policy*; *Energy Economics*; *Renewable and Sustainable Energy Reviews*; *Applied Energy*; *Environmental Science and Policy*; *Environmental Resources and Economics*; and *Energy Sources Part B: Economics, Planning, and Policy*.

ORCID: https://orcid.org/0000-0002-6937-5420

Aida Isabel Pereira Tavares

Aida Isabel Pereira Tavares holds a Ph.D. in economic analysis, awarded by the Autonomous University of Barcelona, Spain, and a Ph.D. in management - decision aiding science, awarded by Faculty of Economics of the University of Coimbra,

Coimbra, Portugal. Her research is focused on different areas of applied health economics and policy, and she has published several articles in international peer-reviewed journals such as *European Journal of Health Economics, International Journal of Health Economics and Management, International Journal for Quality in Health Care, International Journal of Health Planning and Management, BMC Health Services Research,* and *International Journal of Medical Informatics.* Currently, she is an assistant professor at Lisbon School of Economics and Management, University of Lisbon, and she is a researcher in the Centre for Health Studies and Research of the University of Coimbra, Portugal, which is funded by national funds through FCT—Foundation for Science and Technology, I.P., under the Multi-annual Financing of R&D Units 2020–2023.

ORCID: https://orcid.org/0000-0003-3487-1202

Nuno Miguel Barateiro Gonçalves Silva

Nuno Miguel Barateiro Gonçalves Silva, Ph.D. in economics, is a professor of micro-economics, monetary, and financial economics at the Faculty of Economics of the University of Coimbra, Coimbra, Portugal. He is a finance and environmental economics researcher at the Centre for Business and Economics Research (CeBER), sponsored by the Portuguese Foundation for the Development of Science and Technology. He has published in international journals, such as the *Journal of Economics Dynamics and Control* and *Environmental Science and Pollution Research.*

ORCID: https://orcid.org/0000-0002-5687-3818

Context of the obesity problem in the Latin American region

1

1.1 Introduction

Overweight and obesity are defined as abnormal or excessive fat accumulation that may impair health. Overweight and obesity can be measured by the body mass index (BMI), a simple weight-for-height index commonly used to classify overweight and obesity in adults (WHO, 2020).

The World Health Organization (WHO) uses this index and defines it in two categories using cut-off points. For example, an individual with BMI between 25.0 and 30.0 is determined to be 'overweight', while a BMI greater than 30.0 is considered 'obese'. In 2016, about 39% (2.0 billion) of the world's adult population older than 18 years (38% of men and 40% of women) were overweight, and 13% (620 million, 11% of men and 15% of women) were obese in 2016 (International Food Policy Research Institute, 2016a, b, pp. 1–7; World Health Organization, 2020). The global prevalence of overweight and obesity has more than doubled between 1975 and 2017, raising serious concerns in public health practitioners and governments.

Overweight and obesity are linked to higher premature deaths worldwide than the problem of underweight, meaning that more people are obese than underweight in the world—this occurs in every region except some parts of sub-Saharan Africa and Asia (WHO, 2020). In this context, there is a link between overweight and obesity and the increased level of avoidable and preventable mortality because overweight and obesity are risk factors for several of the world's leading causes of death, such as diabetes, several types of cancer and cardiovascular diseases. Obesity does not directly cause any of these health conditions, but it can increase their likelihood of occurring. Indeed, in 1990, the obesity problem was related to 2.16 million deaths worldwide and in 2017, reached a value of 4.72 million (see Fig. 1.1 below).

For this reason, obesity is considered one of the world's most significant health problems. It was considered a problem of high-income countries, but nowadays, it is an emerging and prevalent problem in low- and middle-income countries also, particularly in urban settings. In general, the increase of overweight and obesity problems in low- and middle-income countries is related to economic development, globalisation and urbanisation, leading to significant changes in people's diet and physical activity. The determinants for the overweight and obesity problem will be better explained in Chapter 2 of this book.

Overweight and obesity are therefore rising in low- and middle-income countries. More than 300 million adults in 2016 were overweight in the Latin America (LA) region, while more than 100 million were obese. Obesity has become a public health issue in the region, where around 57% (302 million) of the region's adult population are overweight, and 19% (100.8 million) are obese (Pan American Health

Obesity Epidemic and the Environment. https://doi.org/10.1016/B978-0-323-99339-5.00003-0

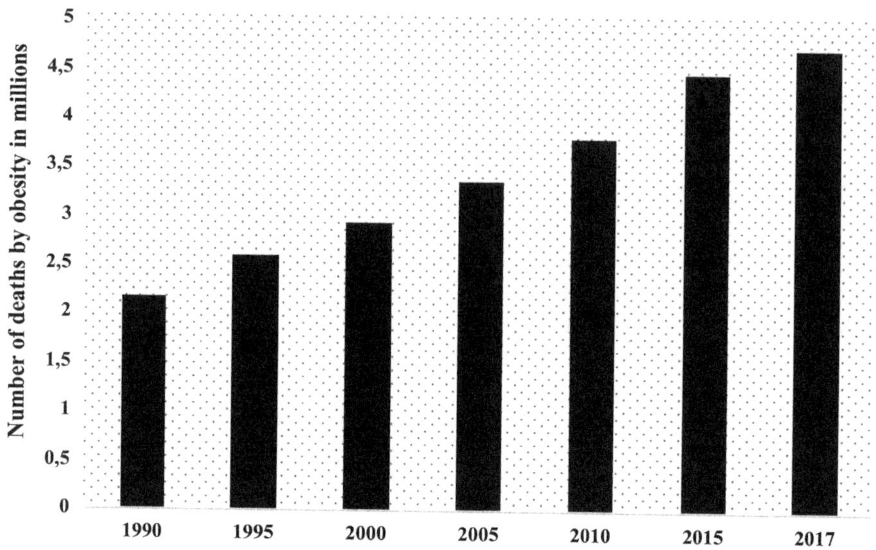

Figure 1.1 The number of deaths caused by obesity worldwide between 1990 and 2017. The total annual number of deaths caused by obesity in millions, measured across all age groups and both sexes.
The authors created this figure with data from the Our World in Data database. (2021). *Obesity*. https://ourworldindata.org/obesity.

Organization, 2011). Exhibit 1.1 discusses the increase of obesity and overweight in the Latin America and Caribbean (LAC) region.

This substantial number of obese people may be reflected in the number of deaths caused (in)directly by obesity. For instance, in 1990, obesity caused 163,664 deaths, and in 2017, it caused 433,431 deaths in the LA region (see Fig. 1.2 below).

An exponential increase in obesity in this region, which affects 19% of the population, has almost doubled the global level by 13%. This problem not only has a high economic cost but also has environmental costs, and it threatens the lives of hundreds of thousands.

The increase in the overweight and obesity problem in the LA region began in the 1980s and accelerated from the 2000s. This increase is related to the economic gains caused by liberalisation, privatisation, foreign investment and infrastructure investment. These processes are derived from the structural and stabilisation programmes imposed on LA countries by the International Monetary Fund (IMF) (Koengkan et al., 2018; Koengkan, Santiago, & Fuinhas, 2019; Koengkan, Santiago, Fuinhas, & Marques, 2019). These neoliberal programmes of adjustment consisted of the complete opening of the economies to international trade and capital, deregulation of the economy, privatisation, reduction of public expenditures, creation of conditions for foreign investment and the reduction of the role of the government in the economy (Koengkan, Fuinhas, & Fuinhas, 2021, pp. 1–7; Koengkan, Fuinhas, & Silva, 2021; Santiago et al., 2020).

Exhibit 1.1 Obesity and overweight populations in the LAC region

More than 300 million adults were overweight in the LAC region, and of these, more than 100 million were obese in 2014. Obesity and overweight are defined as abnormal or excessive accumulation of fat that may impair health (Garcia-Garcia, 2021). Worldwide, 39% (2.0 billion) of the adult population (38% of men and 40% of women) were overweight, and close to 13% (600 million, 11% of men and 15% of women) were obese in 2014. The global prevalence of obesity more than doubled between 1975 and 2014) (International Food Policy Research Institute, 2016a, b, pp. 1–7; NCD-RisC, 2016; World Health Organization (WHO) 2020).

Obesity has become a significant health challenge in the LAC region. Around 57% (302 million) of the adult population in the LAC region (54% men and 70% of women) are overweight, while 19% (100.8 million) are obese (15% in men and 24% in women) (Garcia-Garcia, 2021).

In other low- and middle-income countries, the overweight problem impacts 61% of women and 54% of men, and obesity problem affects 24% of women and 15% of men. Indeed, the obesity problem is more prevalent in women than in men. In 14 LAC countries, there is a prevalence among females, more significant than 20%. The highest prevalence of obesity problem in the adult population is found in El Salvador (33%) and Paraguay (30%) for women and in Uruguay (23%) and Chile (22%) for men (Ng et al., 2014).

Moreover, the prevalence of overweight and obesity in children in the LAC region is also high, which impacts 16% of children. It ranges from more than 12% for girls in Chile, Uruguay and Costa Rica, to less than 5% in Bolivia, Ecuador, Peru, Honduras and Guatemala. The highest prevalence of obesity in children is found in Chile (12%) and Mexico (11%) in boys and Uruguay (18%) and Costa Rica (12%) in girls (Ng et al., 2014).

Therefore, overweight and obesity are significant risks for non-communicable diseases like cardiovascular disease (heart disease and stroke) and the leading cause of death (30% of death due to all factors) in the LAC region from diabetes, hypertension and chronic kidney disease (Garcia-Garcia, 2021).

The adoption of these policies occurred between 1980 and 1992, when several countries, for example, Mexico and Costa Rica in 1989, Venezuela in 1990, Uruguay in 199 and Argentina and Brazil in 1992, were going through in-depth processes of financial and trade openness, liberalisation of foreign investment, privatisation of substantial portions of the public sector and reduction of import barriers (Aizenman, 2005, pp. 1–30; Koengkan, 2020; Koengkan & Fuinhas, 2020). Notably, before this adjustment, the region's annual growth rate was approximately 0.35% in 1990. After the imposed 'macroeconomic adjustment', LA's GDP per capita had a yearly compound growth rate of 4.58% in 1994. The 'commodities boom' that occurred between the beginning of the

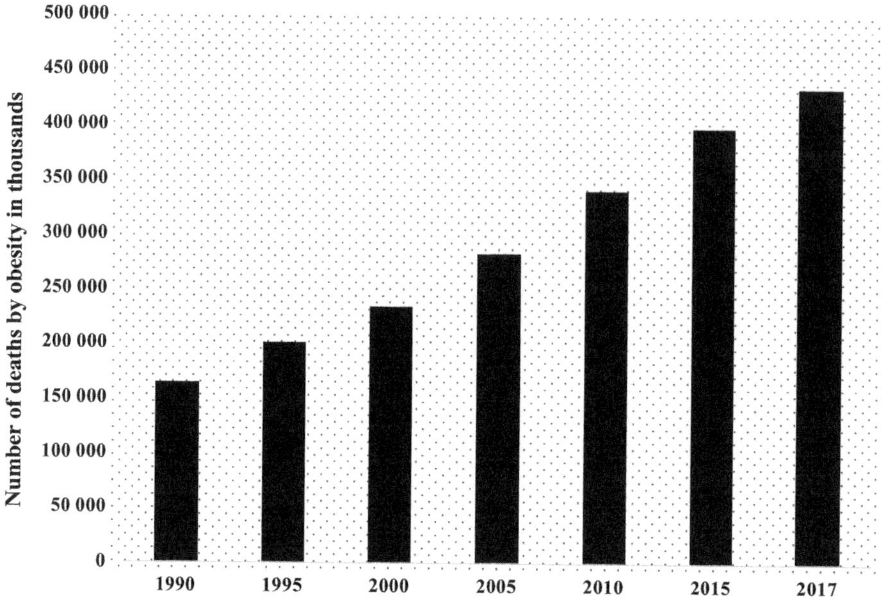

Figure 1.2 The number of deaths caused by obesity in the Latin American region between 1990 and 2017. The total annual number of deaths caused by obesity in thousands, measured across all age groups and both sexes.
The authors created this figure with data from the Our World in Data database. (2021). *Obesity*. https://ourworldindata.org/obesity.

2000s and the end of 2014 also accelerated the process of openness, as well as the economic gains in the region (Koengkan, 2020; Koengkan & Fuinhas, 2020; Koengkan, Fuinhas, & Fuinhas, 2021, pp. 1–7; Koengkan, Fuinhas, & Silva, 2021).

Therefore, this liberalisation process positively changed market production, employment, transportation, home production and leisure and food systems through midstream and downstream processing and wholesale, retail and transportation methods of supermarkets and large processors. Moreover, fast-food chains are fed by modernised procurement systems, and there has been co-evolution among these segments. As a result, urban and even rural LA areas are experiencing a rapid and ubiquitous transformation (Popkin & Reardon, 2018).

This liberalisation process made a reduction of undernourishment possible in the region. Since 2000, the number of LAs suffering from undernourishment has dropped from more than 60 million to 39 million (Radwin, 2018). All these socio-economic and technological factors have influenced the nutrition transition and the reduction of physical activities. Consequently, the negative impact of overweight and obesity has emerged and has significantly increased in the region. The effects of economic development, globalisation and urbanisation on the LA region's overweight or obesity problem will be better explained in Chapters 3 and 4 of this book.

1.2 Share of overweight adults in the LA region

Worldwide, 39% (2.0 billion) of the adult population aged 18 years and older were overweight in 2016, while in 1990, this value was 27%. This means an increase of 48% between 1990 and 2016. As mentioned before, overweight is defined based on the BMI: the threshold value is lower than that of obesity, with a BMI equal to or greater than 25.

In Fig. 1.3 below, we can see the share of overweight adults across various countries and regions. The overall pattern is very closely aligned with the distribution of overweight across the world: the percentage of overweight people tends to be higher in richer countries and lower incomes, but the evolution over time is identical. The portion of overweight is larger than the portion of obesity.

In most high-income countries, around two-thirds of adults are overweight. In the United States, for example, 70% of the adult's population were overweight in 2016, with 67% of people being overweight in the United Kingdom, 65% in the United Arab Emirates and the 59% in the European region. However, in upper-middle-income economies, such as China, overweight affected (34%) of the adult population in 2016, while in the LA region, 57% of the adult population were affected. Moreover, in lower-middle-income economies, for example, the overweight problem was present in 19% of the adult population.

In the LA region, 57% of the adult population was considered overweight in 2016, while this value was 37% in 1990 (see Fig. 1.4 below). This means that there was an increase of 54% between 1990 and 2016.

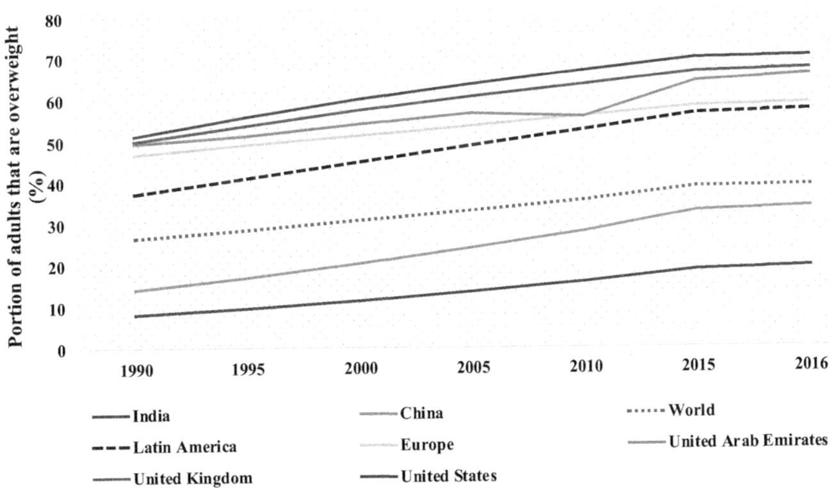

Figure 1.3 Share of adults that are overweight (%) worldwide between 1990 and 2016. Being overweight is defined as having a body mass index (BMI) greater than or equal to 25. BMI is a person's weight in kilograms divided by their height in metres squared.
The authors created this figure with data from the Our World in Data database. (2021). *Obesity*. https://ourworldindata.org/obesity.

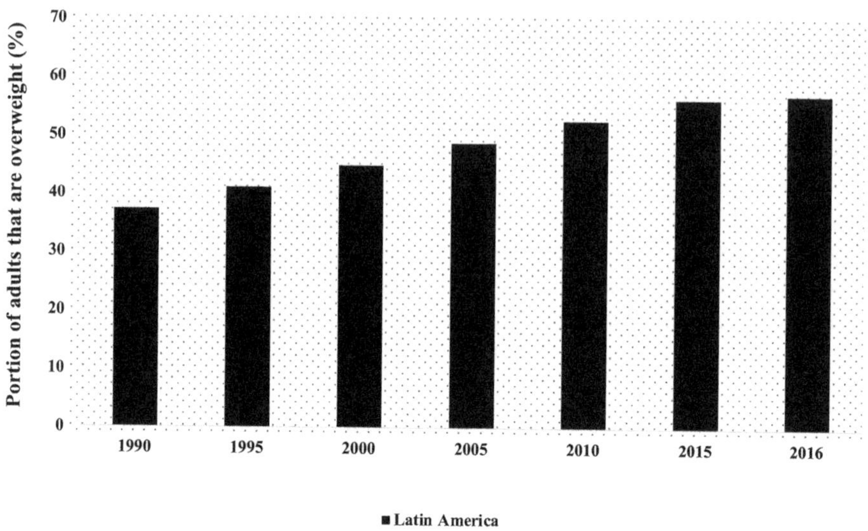

Figure 1.4 Share of adults that are overweight (%) in the Latin American region between 1990 and 2016. Being overweight is defined as having a body mass index (BMI) greater than or equal to 25. BMI is a person's weight in kilograms divided by their height in metres squared. The authors created this figure with data from the Our World in Data database. (2021). *Obesity*. https://ourworldindata.org/obesity.

The increase in the share of overweight adults in the population is related to rapid economic development. Indeed, when we examine the leading LA economies (e.g., Argentina, Brazil, Chile, Colombia, Ecuador, Mexico, Peru and Venezuela, we identify similar growth trends in overweight since the 1990s (see Fig. 1.5 below).

In 1990, the overweight problem affected 49% of the adult population in Argentina, while 46% of people were overweight in Chile, 46% in Venezuela, 45% in Mexico, 39% in Peru, 39% in Colombia, 36% in Ecuador and 35% in Brazil. In 2016, the overweight problem reached the following values: Chile 64%, Mexico 64%, Argentina 63%, Venezuela 63%, Colombia 59%, Brazil 57%, Peru 56% and Ecuador 55%.

In all countries in the region, overweight affects at least half the population, with the highest rates registered in Chile (64%), Mexico (64%), Argentina (63%) and Venezuela (62%). Moreover, the increase rates are well over 1% point per year in many LA countries (e.g., Chile, Mexico, Argentina and Venezuela). In some cases, like Brazil, they appear to be accelerating (Popkin & Reardon, 2018). Over the last 20 years, there has been a rapid increase in overweight problems across the population (Crowley, 2017).

Moreover, women are more affected by overweight and obesity than men. In the LAC region, 38% of men and 40% of women were overweight, and in many countries, such as Chile and Mexico, the figures reach two-thirds of women and over half of men (Popkin & Reardon, 2018). These two countries have registered significant growth in BMI, both for women and men.

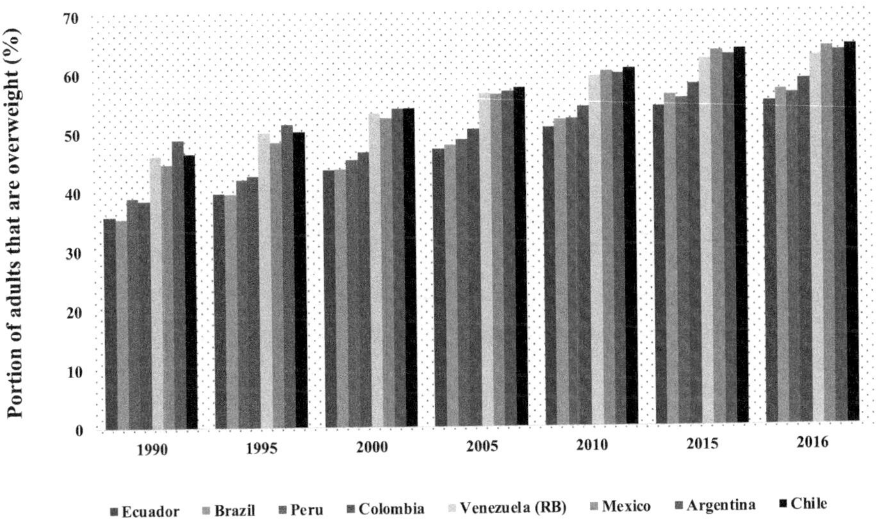

Figure 1.5 Share of overweight adults (%) in leading Latin American economies, between 1990 and 2016. Being overweight is defined as having a body mass index (BMI) greater than or equal to 25. Obesity is characterised by a BMI greater than or equal to 30. BMI is a person's weight in kilograms divided by their height in metres squared.
The authors created this figure with data from the Our World in Data database. (2021). *Obesity*. https://ourworldindata.org/obesity.

1.3 Share of obese adults in the LA region

Globally, 13% of adults aged over 18 years were obese in 2016, while in 1990, this value was 6.8%, that means, an astonishing increase of 91% between 1990 and 2016. In most high-income countries, such as the United States, 36% of adults were obese in 2016, while 32% of people were obese in the United Arab Emirates, 28% in the United Kingdom and 23% in the European region. However, in upper-middle-income economies, such as China, obesity affected 6.2% of the adult population in 2016, while in the LA region, 19% of the adult population were affected. Moreover, in lower-middle-income economies, for example, India, the obesity problem was present in 4% of the adult population (see Fig. 1.6 below).

In the LA region, 19% of the adult population were obese in 2016, while this value was 9% in 1990 (see Fig. 1.7 below), that is an increase of 108% between 1990 and 2016.

The increase in the percentage of adults that are obese is related to rapid economic development, as mentioned before. Indeed, when we examine leading LA economies (e.g., Argentina, Brazil, Chile, Colombia, Ecuador, Mexico, Peru and Venezuela), we find this trend of growth in the obesity problem (see Fig. 1.8 below).

In 1990, the obesity problem affected 17% of the adult population in Argentina, while 17% of people were obese in Chile, 16% in Mexico, 15% in Venezuela, 12% in Colombia, 10% in Peru, 10% in Brazil and 9% in Ecuador. In 2016, the obesity

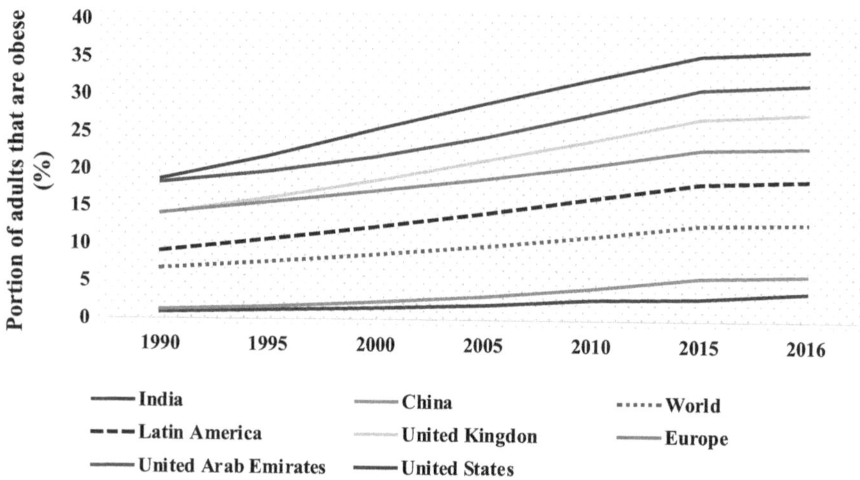

Figure 1.6 Share of adults that are obese (%) worldwide between 1990 and 2016. Obesity is defined as having a body mass index (BMI) equal to or greater than 30. BMI is a person's weight in kilograms divided by their height in metres squared.
The authors created this figure with data from the Our World in Data database. (2021). *Obesity*. https://ourworldindata.org/obesity.

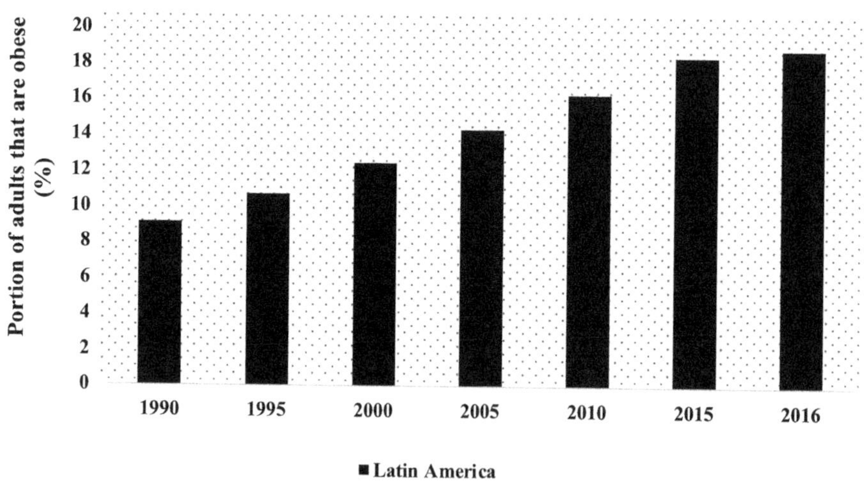

Figure 1.7 Share of adults that are obese (%) in the Latin American region between 1990 and 2016. Being overweight is defined as having a body mass index (BMI) greater than or equal to 25. BMI is a person's weight in kilograms divided by their height in metres squared.
The authors created this figure with data from the Our World in Data database. (2021). *Obesity*. https://ourworldindata.org/obesity.

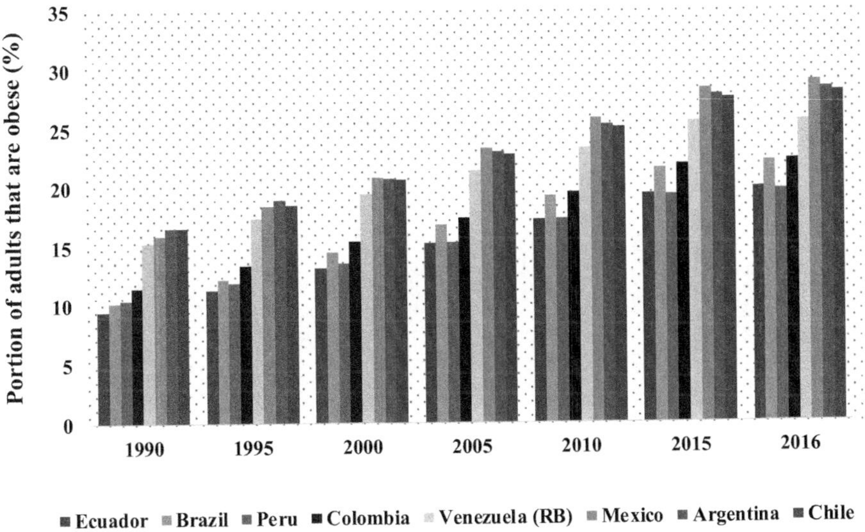

Figure 1.8 Portion of adults that are obese (%) in leading Latin America (LA) economies between 1990 and 2016. Being overweight is defined as having a body mass index (BMI) greater than or equal to 25. Obesity is characterised by a BMI greater than or equal to 30. BMI is a person's weight in kilograms divided by their height in metres squared.
The authors created this figure with data from the Our World in Data database. (2021). *Obesity*. https://ourworldindata.org/obesity.

problem reached the following values: Mexico 30%, Chile 28%, Argentina 28%, Venezuela 26%, Colombia 22%, Brazil 22%, Peru 20% and Ecuador 20%.

In the LA region, the prevalence of obesity in women is similar to that in countries with higher per capita income (Kain et al., 2003). The evolution of BMI in women in Argentina, Chile and Mexico between 1990 and 2016 can be seen in Fig. 1.9 below.

As shown in the figure above, in Argentina, the average BMI in women was 25 in 1990 and reached 28 in 2016 and in Chile; in 1990, the BMI was 26 and attained a value of 28, while in Mexico, the BMI in women was 26 in 1990 and reached 29 in 2016. In contrast, the prevalence of obesity in men is less significant than in women (see Fig. 1.10 below).

As shown in Fig. 1.10, in Argentina, the average BMI in men was 26 in 1990 and reached 28 in 2016, and in Chile, in 1990, the BMI was 26 and attained a value of 28, while in Mexico, the BMI in men was 25 in 1990 and reached 27 in 2016. Moreover, some authors, such as Peña and Bacallao (2000) and Kain et al. (2003), argue that the increase of prevalence of obesity in the LA region arose from poverty and was found mainly in urban areas. Monteiro et al. (2002) investigated the obesity problem in the adult population in Brazil. The authors found that the obesity problem increased in both genders in Brazil's Northeast and Southeast regions, that is, in areas where poverty is observable and where urbanisation has taken off very rapidly. The same authors add that the rate of overweight women was almost double that of men at the

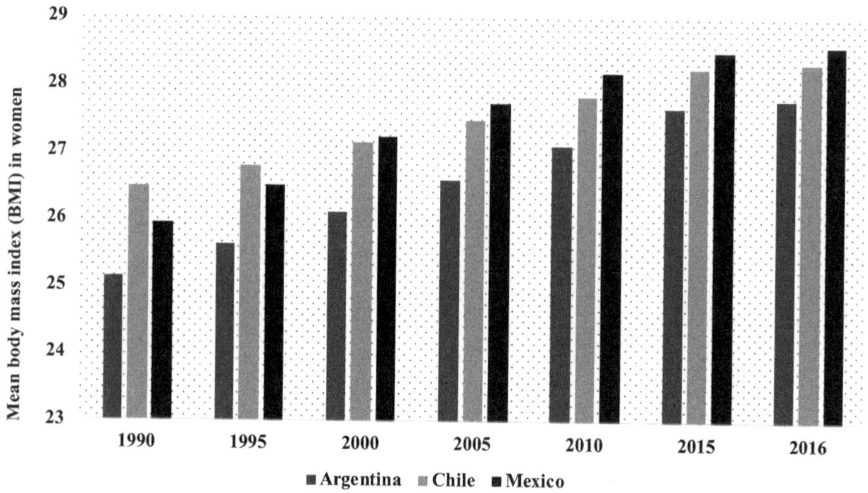

Figure 1.9 Mean body mass index (BMI) in women in Argentina, Chile and Mexico between 1990 and 2016. Being overweight is defined as having a BMI greater than or equal to 25. Obesity is characterised by a BMI greater than or equal to 30. BMI is a person's weight in kilograms divided by their height in metres squared.
The authors created this figure with data from the Our World in Data database. (2021). *Obesity*. https://ourworldindata.org/obesity.

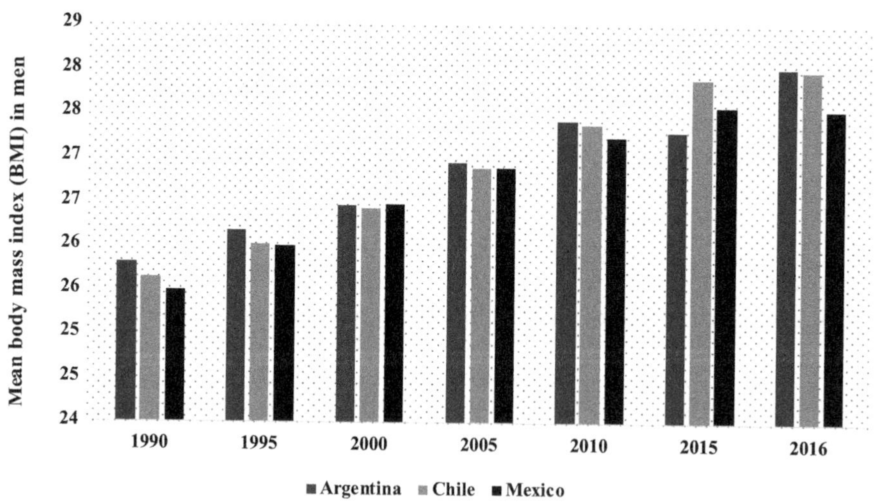

Figure 1.10 Mean body mass index (BMI) in men in Argentina, Chile and Mexico between 1990 and 2016. Being overweight is defined as having a BMI greater than or equal to 25. Obesity is characterised by a BMI greater than or equal to 30. BMI is a person's weight in kilograms divided by their height in metres squared.
The authors created this figure with data from the Our World in Data database. (2021). *Obesity*. https://ourworldindata.org/obesity.

national level and point out that obesity in women increased in the more impoverished regions, while it declined in wealthier ones.

1.4 Share of overweight or obese children in the LA region

The prevalence of overweight or obesity in children is remarkable in the LA region. According to the WHO, the share of overweight or obese children and adolescents aged 5–19 has risen from 4% in 1975 to around 18% in 2016 (Our World in Data database, 2021).

The prevalence of this problem is extremely high in some high-income countries. For example, in the United States, 18% of children were overweight or obese in 1990, while in 2016, this value reached 24%. In Germany, 18% of children were overweight or obese in 1990, while in 2016, this value rose to 26%. In the United Kingdom, 18% of children were overweight or obese in 1990, while in 2016, this value reached 26%. The United Arab Emirates is the worst in this respect, where 17% of children and adolescents were overweight or obese in 1990, and by 2016, this number had risen to of 33% (see Fig. 1.11 below).

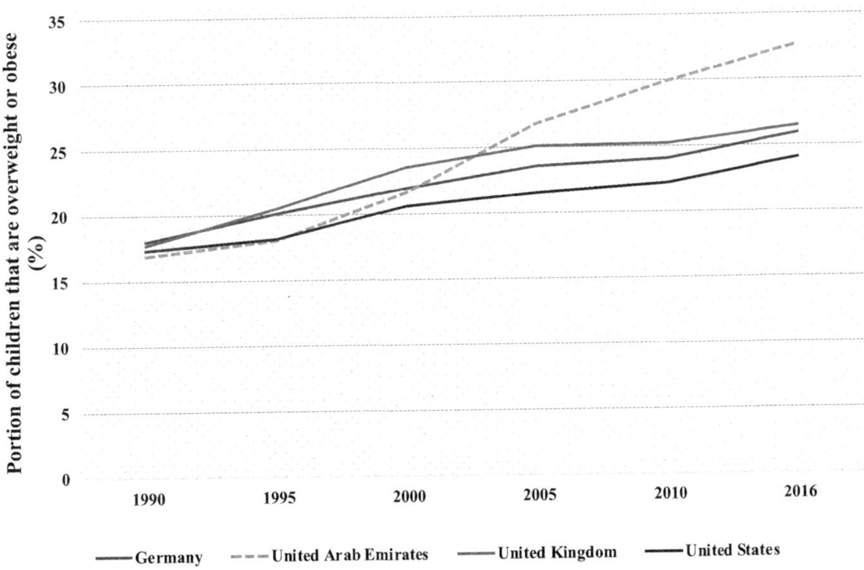

Figure 1.11 Share of children that are overweight or obese (%) in the United States, Germany, the United Kingdom and the United Arab Emirates between 1990 and 2016; portion of children aged 2–4 years old who are defined as overweight or obese; a child is classified as overweight if their weight-for-height is more than two standard deviations from the median of the WHO Child Growth Standards.
The authors created this figure with data from the Our World in Data database. (2021). *Obesity*. https://ourworldindata.org/obesity.

Upper-middle-income economies such as Argentina, Brazil, Bolivia, Chile, Colombia, Ecuador, Mexico and Peru have likewise registered an increase in the prevalence of overweight or obesity in children (see Fig. 1.12 below).

As shown in Fig. 1.12, in Argentina, the prevalence of overweight or obesity in children in 1990 was 10%, and in 2016, this value reached 17%. In Brazil, in 1990, this value was 14% and went up to 33% in 2016. In Bolivia, the prevalence of overweight or obesity in children in 1990 was 23%, while in 2016, this value increased to 29%. In Chile, 24% of children were overweight or obese in 1990, while in 2016, this value reached 45%. Meanwhile, in Colombia, 12% of children were overweight or obese in 1990, while in 2016, this value went 13%. In Ecuador in 1990, 15% were overweight or obese, and in 2016, this value reached 21%. In Mexico, 23% of children were overweight or obese in 1990, while in 2016, this value increased to 24%. Finally, in Peru, 25% of children were overweight or obese in 1990, and in 2016, this value decreased to 23%.

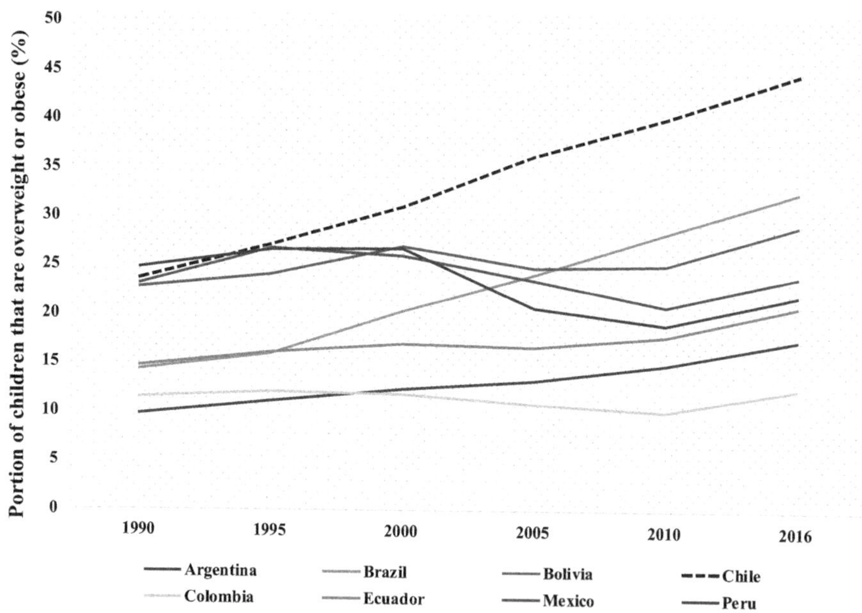

Figure 1.12 Share of children that are overweight or obese in Argentina, Brazil, Bolivia Chile, Colombia, Ecuador, Mexico and Peru between 1990 and 2016; portion of children aged 2–4 years old who are defined as overweight or obese; a child is classified as overweight if their weight-for-height is more than two standard deviations from the median of the WHO Child Growth Standards.

The authors created this figure with data from the Our World in Data database. (2021). *Obesity*. https://ourworldindata.org/obesity.

1.5 Conclusions

This chapter approached the context of the obesity problem in the LA region. The overweight and obesity problem in the LA region began in the 1980s and accelerated from the 2000s. The increase in this problem is related to the economic gains caused by liberalisation and privatisation, the growth of foreign investment and infrastructure investment caused by structural and stabilisation programs imposed on LA countries. In parallel with reducing public expenditures, creating open economies, trade, investment and capital movements have triggered economic growth and fast urbanisation and integration in the global economy.

Therefore, this liberalisation process brought about a positive change in the market production; the aspects of employment, transportation, home production and leisure; the food systems through midstream and downstream processing and wholesale, retail and transportation methods. Therefore, supermarkets, large processors and fast-food chains became supplied by modernised procurement systems. As a result, urban and rural LA areas experiencied a rapid and ubiquitous transformation. This rapid change resulted in an exponential increase in overweight and obesity problems.

This health problem affects women, men and children but in a heterogeneous way. In general, women are the most affected group. In 1990, the overweight problem reached 49% of the adult population in Argentina, while in Chile, 46% of people were overweight, 46% in Venezuela, 45% in Mexico, 39% in Peru, 39% in Colombia, 36% in Ecuador and 35% in Brazil. In 2016, the overweight problem reached the following values: Chile 64%, Mexico 64%, Argentina 63%, Venezuela 63%, Colombia 59%, Brazil 57%, Peru 56% and Ecuador 55%.

In all countries of the region, overweight affects at least half the population, with the highest rates registered in Chile (64%), Mexico (64%), Argentina (63%) and Venezuela (RB) (62%). Moreover, the rate of increase is well over 1% point per year in many LA countries (e.g., Chile, Mexico, Argentina and Venezuela). In some cases, like Brazil, they appear to be accelerating (Popkin & Reardon, 2018). Over the last 20 years, there has been a rapid increase in overweight problems across the population.

Women are more affected by overweight and obesity than men. About 38% of men and 40% of women are overweight, and in many countries, as in the case of Chile and Mexico, the figure reaches two-thirds of women and over half of men.

The obesity problem reached 17% of the adult population in Argentina in 1990, while in Chile, this was 17%, Mexico 16%, Venezuela 15%, Colombia 12%, Peru 10%, Brazil 10% and Ecuador 9%. Then, in 2016, the obesity problem reached the following values: Mexico 30%, Chile 28%, Argentina 28%, Venezuela 26%, Colombia 22%, Brazil 22%, Peru 20% and Ecuador 20%. In the LA region, the prevalence of obesity in women is similar to that in countries with higher per capita income.

The prevalence of overweight or obesity in children is also remarkable in the LA region. According to WHO reports, the portion of overweight or obese children and adolescents aged 5–19 has risen from 4% in 1975 to around 18% in 2016. Indeed, the upper-middle-income economies such as Argentina, Brazil, Bolivia, Chile, Colombia, Ecuador, Mexico and Peru have likewise registered an increase in overweight or obesity in children. In Argentina, the majority of overweight or obesity in children in 1990 was 10%, and in 2016, this value reached 17%. In Brazil, in 1990, this value was 14% and went up to 33% in 2016. In Bolivia, the prevalence of overweight or obesity in children in 1990 was 23%, while in 2016, this value increased to 29%. In Chile, 24% of children were overweight or obese in 1990, while in 2016, this value reached 45%. Meanwhile, in Colombia, 12% of children were overweight or obese in 1990, while in 2016, this value reached 13%. In Ecuador, in 1990, 15% were overweight or obese, and in 2016, this value reached 21%. In Mexico, 23% of children were overweight or obese in 1990, while in 2016, this value increased to 24%. Finally, in Peru, 25% of children were overweight or obese in 1990, and in 2016, this value decreased to 23%.

References

Aizenman, J. (2005). *Financial liberalisations in Latin America in the 1990s: A reassessment.* NBER working paper series, 11145. URL: https://www.nber.org/papers/w11145.

Crowley, E. (2017). *A crisis of overweight and obesity in Latin America and the caribbean.* http://www.ipsnews.net/2017/01/a-crisis-of-overweight-and-obesity-in-latin-america-and-the-caribbean/.

Garcia-Garcia, G. (2021). *Obesity and overweight populations in Latin America.* URL: https://www.thelancet.com/campaigns/kidney/updates/obesity-and-overweight-populations-in-latin-america.

International Food Policy Research Institute. (2016a). *Global nutrition report 2016: From promise to impact: Ending malnutrition by 2030* https://doi.org/10.2499/9780896295841. Washington, D.C.

International Food Policy Research Institute. (2016b). *Global nutrition report 2016: From promise to impact: Ending malnutrition by 2030.* Washington, DC. URL: https://www.ifpri.org/publication/global-nutrition-report-2016-promise-impact-ending-malnutrition-2030.

Kain, J., Vio, F., & Albala, C. (2003). Obesity trends and determinant factors in Latin America. *Cadernos de Saúde Pública, 19*(1), 1–10. https://doi.org/10.1590/S0102-311X2003000700009

Koengkan, M. (2020). Capital stock development and their effects on investment expansion in renewable energy in Latin America and the Caribbean region. *Journal of Sustainable Finance and Investment*, 1–16. https://doi.org/10.1080/20430795.2020.1796100

Koengkan, M., & Fuinhas, J. A. (2020). Exploring the effect of the renewable energy transition on CO_2 emissions of Latin American and Caribbean countries. *International Journal of Sustainable Energy*, 1–25. https://doi.org/10.1080/14786451.2020.1731511

Koengkan, M., Fuinhas, J. A., & Fuinhas, C. (2021). *Does urbanisation process increase the overweight epidemic? The case of Latin America and the caribbean region.* SSRN. https://doi.org/10.2139/ssrn.3826196

Koengkan, M., Fuinhas, J. A., & Marques, A. C. (2018). Does financial openness increase environmental degradation? Fresh evidence from MERCOSUR countries. *Environmental Science and Pollution Research, 25*(30), 30508–30516. https://doi.org/10.1007/s11356-018-3057-0

Koengkan, M., Fuinhas, J. A., & Silva, N. M. B. (2021). Exploring the capacity of renewable energy consumption to reduce outdoor air pollution death rate in Latin America and the

Caribbean region. *Environmental Science and Pollution Research, 28*, 1656–1674. https://doi.org/10.1007/s11356-020-10503-x

Koengkan, M., Santiago, R., & Fuinhas, J. A. (2019). The impact of public capital stock on energy consumption: Empirical evidence from Latin America and the Caribbean region. *International Economics*, 1–20. https://doi.org/10.1016/j.inteco.2019.09.001

Koengkan, M., Santiago, R., Fuinhas, J. A., & Marques, A. C. (2019). Does financial openness cause the intensification of environmental degradation? New evidence from Latin American and Caribbean countries. *Environmental Economic Policy, 21*, 507–532. https://doi.org/10.1007/s10018-019-00240-y

Monteiro, C., Wolney, A., Conde, L., & Popkin, B. (2002). Is obesity replacing or adding to undernutrition? Evidence from different social classes in Brazil. *Public Health Nutrition, 5*, 105–112.

NCD risk factor collaboration (NCD-RisC). (2016). Trends in adult body mass index in 200 countries from 1975 to 2014: A pooled analysis of 1698 population-based measurement studies with 19·2 million participants. *Lancet, 387*, 377–396. URL: https://www.thelancet.com/journals/lancet/article/PIIS0140-6736(16)30054-X/fulltext.

Ng, M., Fleming, T., Robinson, M., Thomsom, B., & Graetz, N. (2014). Global, regional, and national prevalence of overweight and obesity in children and adults during 1980–2013: A systematic analysis for the global burden of disease study 2013. *Lancet, 384*, 766–781. https://doi.org/10.1016/S0140-6736(14)60460-8

Our World in Data database. (2021). *Obesity*. https://ourworldindata.org/obesity.

Pan American Health Organization. (2011). In P. Ordúñez-García, & C. Campillo-Artero (Eds.), *Regional consultation: Priorities for cardiovascular health in the Americas. Key messages for policymakers*. Washington, D.C.: PAHO.

Peña, M., & Bacallao, J. (2000). *Obesity and poverty. A new public health change*. Scientific Publication 576. Washington, DC: Pan American Health Organization.

Popkin, B. M., & Reardon, T. (2018). Obesity and the food system transformation in Latin America. *Public Health/Nutrition, 19*(8), 1–23. https://doi.org/10.1111/obr.12694

Radwin, M. (2018). *Latin America's 'double burden' of malnutrition: Rising obesity and undernourishment*. https://www.worldpoliticsreview.com/articles/26840/latin-america-s-double-burden-of-malnutrition-rising-obesity-and-undernourishment.

Santiago, R., Koengkan, M., Fuinhas, J. A., & Marques, A. C. (2020). The relationship between public capital stock, private capital stock and economic growth in the Latin American and Caribbean countries. *International Review of Economics, 67*, 293–317. https://doi.org/10.1007/s12232-019-00340-x

World Health Organization (WHO). (2020). *Obesity and overweight. Key facts*. http://www.who.int/mediacentre/factsheets/fs311/en/#.

Determinants of obesity in the Latin America and Caribbean region

2

2.1 Introduction

In the Latin America and the Caribbean (LAC) region, adult obesity has tripled since 1975; it affects about 105 million people, but 25% of this population is hungry (UN, 2019). Obesity, a well-known phenomenon worldwide, is powerfully prevalent in Latin America (LA), where it can be said to have reached the proportion of an epidemic health problem (see Chapter 1).

Obesity, despite not being a disease, is a risk factor for several other diseases, and it is framed in a myriad of determining factors. This determinants framework includes individual and genetic, socioeconomic, demographic, urban, environmental and technological aspects (e.g., Brahmbhatt, 2017; McLaren, 2014). These determining factors can be said to work as human body energy savers, which result in the accumulation of body fat, making people overweight and obese. This process of body fat accumulation takes place daily as energy intake is more significant than energy expenditure, and it has become more important from the 1980s on.

The counterfactual example of this trend in the LAC region is Cuba. Due to the collapse of the USSR in 1989, Cuba entered an economic crisis, resulting in a reduced energy intake from 2899 calories to 1863 calories and increased physical activity. The medium-run effect of this crisis was a decline in the prevalence of obesity and a reduction in diabetes and cardiovascular diseases (Franco et al., 2007). The results found in Cuba suggest that policy measures designed to reduce energy accumulation without affecting nutritional sufficiency may positively impact the population health by leading to a decline in diabetes and cardiovascular disease prevalence and mortality.

In this way, well-designed policies need to be based on well-grounded knowledge of the determinants of obesity. So, this chapter aims to describe the main factors associated with overweight and obesity in the LAC region.

2.2 Conceptual model for the determinants of obesity

Obesity and overweight are not illnesses or the absence of health, but they constitute a risk factor for several severe diseases such as diabetes and cardiovascular diseases (see Chapter 1). For this reason, it would not be entirely appropriate to explain the determinants of obesity based on the conceptual model, describing the determinants of health proposed by Dahlgren and Whitehead (1991, pp. 1–69). Instead, we follow

Obesity Epidemic and the Environment. https://doi.org/10.1016/B978-0-323-99339-5.00001-7

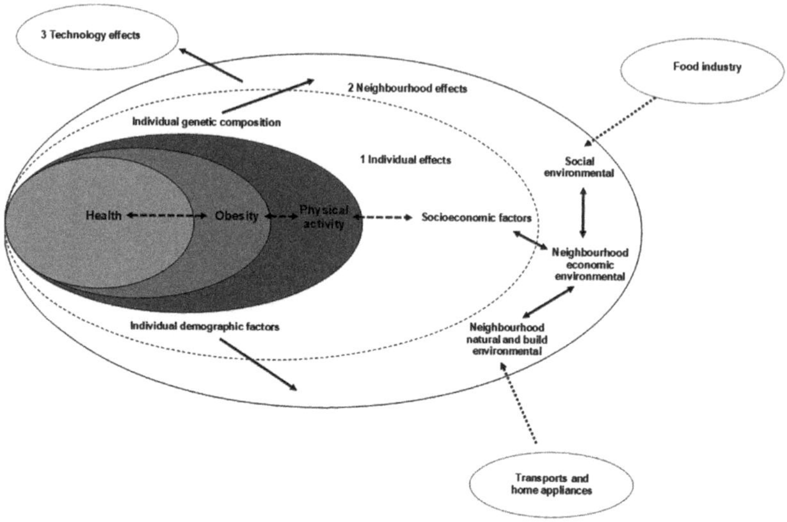

Figure 2.1 A conceptual model for the determinants of obesity.
This figure was adapted from Amarasinghe et al. (2009).

the conceptual model proposed by Amarasinghe et al. (2009), inspired within the spirit of the Dahlgren and Whitehead model, which we have improved with an additional layer of determinants. Fig. 2.1 below shows the conceptual model for the determinants of obesity.

The original model developed by Amarasinghe et al. (2009) begins by placing obesity at the centre, directly linked with health and physical activity. Both health and physical activities are determinants of individual weight, but weight also influences health status and individual choices in physical activity.

The concern for understanding obesity places it at the centre of the model. The determinant factors are found and grouped into three main layers: individual, neighbourhood and technological effects. The model considers three layers of influence and their possible interactions explaining obesity.

In the first layer, we find non-observable determinants, such as genetic features and observable demographic and socioeconomic factors, including age, gender, ethnicity, income, education, civil status and employment status. The second layer is dedicated to the influences arising from the neighbourhood. In this layer, surrounding factors are found; these include the social, economic, natural and how the community is built. Finally, the last layer of determinants was not in the original model. Still, we think it to be relevant as a determinant of obesity, that is, the effects arising from technological improvement. Technological improvements included factors such as those taking place in the agriculture and food industries, transport and home appliances and equipment (e.g., fridges, microwaves, air conditioning, television and video game consoles). All these factors in the three levels of influence interact with each other, contributing to increased weight in people and emergency obesity.

2.2.1 Health and physical activity

Health and physical activity are the two factors most closely associated with obesity and overweight. In general, people who present levels above the average level of adiposity in the body tend to have lower health status and lower levels of physical activity. High levels of adiposity in the body raise several health issues such as cardiovascular and metabolic problems. These may result in heart diseases, hypertension, diabetes and others. Physical activity plays a crucial role in regulating body weight because it is responsible for caloric energy balance, meaning the balance between caloric energy intake and caloric energy expenditure.

The concern with health and physical activity in relation to obesity is a long-standing topic in several international organisations: PAHO (2021a), NCD Alliance (2021), WHO (1998), and across South American governments. Health and physical activity are conditioned by several modern features mainly associated with economic growth, industrialisation, urbanisation and technological improvement. The decline in physical labour, usually associated with rural work, and the change in lifestyles towards a sedentary life are shaping human bodies. Exhibit 2.1 discusses the reduction of physical activity in the LAC region.

According to Popkin (2006), migrations from the countryside to urban centres have created new lifestyles. People have become more sedentary and reduced their consumption of calories. Still, they also have easy access to low-quality food resulting

Exhibit 2.1 The decline of physical activity in the Latin America region

Latin America is the region in the world with the highest rate of people who do not adhere to physical activity to stay healthy. This index of physical activity reached 39% of the total population in 2018. Indeed, the region exceeds the group of Western countries with high incomes. In general, these people do less physical activity than those with low incomes, with 37% and 16% of their populations.

Brazil has the highest rate of sedentary lifestyles in the LAC region, where 47% do not do enough physical activity to stay healthy. Other countries from the region have the same situation, such as **Costa Rica**, **Argentina** and **Colombia**, wherein the share of the population not doing enough physical activity is, respectively, 46%, 41% and 36%, in 2018. However, other countries like **Uruguay**, **Chile** and **Ecuador** present better rates of the population doing exercise—only 22%, 26% and 27% of the people, respectively, do not meet the minimum physical activity requirements.

Therefore, this reduction in physical activity in the LAC region is related to rapid urbanisation, which has caused people to leave places where they should exercise to work, especially in agriculture, to settle in cities where they are unemployed or have jobs in industry, which are much more sedentary and in which they execute repetitive movements (Agência Brasil, 2018).

Table 2.1 The increase in availability of food calories in the South America region.

South America region	1961	1971	1981	1991	2001	2011	Accumulated growth rate (%)
Food supply kcal/person/ day	2329	2468	2624	2658	2797	3027	30.0

Notes: The authors created this table with the database from The Food and Agriculture Organization (FAO) (2021). *Food balances (old methodology and population).* URL: http://www.fao.org/faostat/en/#data/FBSH.

from low-quality calories that increase body fat mass. Indeed, Table 2.1 below shows the increase in food calories availability per capita in the South America region from 1961 to 2011.

As presented in the table above, in the South America region, the caloric consumption was 2329 kcal in 1961 and reached a value of 3027 kcal in 2011. In 50 years, easy access to food is reflected in a 30% increase in food calories per day per capita. This value is higher than 2100 kcal a day recommended by the World Health Organization (WHO).

Indeed, when we address the country individually from the LAC region, we can identify a growth trend in caloric consumption. For example, Brazil had a caloric consumption of 2489.00 kcal a day in 1975, and in 2014, this value reached 3263.00 kcal, which is an increase of 31% in this period. Cuba had a caloric consumption of 2675.00 kcal in 1975 and in 2014 reached a value of 3409.00 kcal, which is an increase of 27%. The Dominican Republic had a caloric consumption of 2083.00 kcal in 1975 and in 2014 reached a value of 2614.00 kcal, which is an increase of 25% in this period. Colombia had a caloric consumption of 2292.00 kcal in 1975 and in 2014, reached a value of 2804.00 kcal, which is an increase of 22%. Chile had 2490.00 kcal in 1975 and reached a value of 2979.00 kcal in 2014, which is an increase of 20%. Costa Rica had a caloric consumption of 2425.00 kcal in 1975 and in 2014 reached a value of 2848.00 kcal, which is an increase of 17% (see Table 2.2 below).

However, in the LAC region, there is a group of countries with low caloric consumption. For example, Bolivia in 1975 had a caloric consumption of 2131.00 kcal a day and in 2014, had a consumption of 2256.00 kcal. In this period, the country registered an increase of 6%. Uruguay had a caloric consumption of 2890.00 kcal in 1975 and in 2014, had a value of 3050.00 kcal, where the country registered an increase of 6%. Haiti had a caloric consumption of 1930.00 kcal in 1975 and reached a value of 2091.00 kcal in 2014, which is an increase of 8% during this period (see Table 2.2 below). Moreover, other countries have registered a decrease in caloric consumption in the region, such as Argentina in 1975, which had a caloric consumption of 3259.00 kcal, and in 2014 had a consumption of 3229.00 kcal, that is a decrease of 1% (see Table 2.2 below).

Moreover, there is evidence that in LA countries, the consumption of fruits and vegetables is lower than the recommended quantity (Hall et al., 2009) although the prevalence of physical inactivity is above the recommended level of physical activity appropriate to maintain good health status (Hallal et al., 2012).

Table 2.2 Daily caloric supply in Latin America and Caribbean (LAC) countries.

	Countries	1975	1980	1990	2000	2010	2014	Accumulated growth rate (%)
Daily caloric supply (per person) kcal/person/day	Argentina	3259.00	3222.00	2911.00	3260.00	3155.00	3229.00	−1
	Bolivia	2131.00	2096.00	1938.00	2076.00	2177.00	2256.00	6
	Belize	2413.00	2652.00	2484.00	2688.00	2839.00	2751.00	14
	Brazil	2489.00	2698.00	2719.00	2880.00	3230.00	3263.00	31
	Chile	2490.00	2682.00	2569.00	2833.00	2915.00	2979.00	20
	Colombia	2292.00	2521.00	2563.00	2779.00	2665.00	2804.00	22
	Costa Rica	2425.00	2565.00	2742.00	2800.00	2848.00	2848.00	17
	Cuba	2675.00	2893.00	2920.00	3031.00	3160.00	3409.00	27
	Dominican Republic	2083.00	2261.00	2069.00	2224.00	2532.00	2614.00	25
	Haiti	1930.00	1967.00	1747.00	1958.00	2169.00	2091.00	8
	Mexico	2715.00	2999.00	2969.00	3037.00	3041.00	3072.00	13
	Uruguay	2890.00	2823.00	2527.00	2811.00	2984.00	3050.00	6
	Venezuela	2391.00	2712.00	2362.00	2454.00	2805.00	2631.00	10

Notes: The authors created this table with the Our World in Data (2021). *Obesity*. URL: https://ourworldindata.org/obesity.

2.2.2 Individual effects

The first individual characteristics to influence weight are genetics. It is the interplay of genetics, epigenetics, metagenomics and the environment that trigger obesity (e.g., Den-Hoed & Loos, 2014; Pérusse et al., 2014; Thaker, 2017; Waterland, 2014). Nevertheless, these are not entirely observed characteristics, and they are not subject to health policies or health measures to be changed. On the other hand, age and gender are individual characteristics that are observable but cannot be changed by health measures, and they do influence the level of fat in the body (Thaker, 2017).

2.2.2.1 Low birthweight and child malnutrition

From womb to tomb, the influence of the nutritional conditions of the mother on the baby and its future is well known. Food deprivation during pregnancy is associated with obesity, and similarly, stunting seems to be responsible for obesity, especially in urban areas (Kain et al., 2014). Child malnutrition is another life situation that has future effects and increases the likelihood of overweight and obesity in adulthood (Ruel, 2000). Moreover, Popkin et al. (2012) pointed to the growing concern with obesity in low- and middle-income countries. According to these authors, the problems associated with obesity are beginning to surpass the issues related to malnutrition. The central point to explain the increase in the incidence of obesity in developing countries is the mismatch between technological progress and human biology. In other words, the capacity to produce more abundant and caloric foodstuff grows faster than the ability of people to adapt to it.

In the LAC region, most countries are near the regional average of 9.5% of preterm births. However, Colombia is the only country in the region significantly above the average of preterm births in 2015, with a value near 15%. Moreover, Brazil is another country in the LAC region with a high level of preterm births and in 2015 registered a value of 11%. Indeed, the lowest level of preterm births was identified in other countries from the LAC region, such as Cuba with nearly 6% of preterm births, Mexico with 7% and Ecuador with 8% in 2014. Fig. 2.2 below gives the preterm birth rate in LAC countries.

Moreover, in the LAC region, these rates are lower than the global rate. According to Howson et al. (2012), there are many ways to reduce the high rate of preterm births in LAC countries, for example, improving obstetric and neonatal care and establishing referral systems with a higher capacity of neonatal care units and staff and equipment.

When we address the low birthweight problem in the LAC region, we find that 10 out of 100 newborns have low weight. In the LAC region, a significant difference exists between the countries related to rates of low weight at birth, where these rates range from low rates in some countries, for example, Cuba with 5.3% and Chile with 6.2% in 2015, to the highest rates in others countries, for example, Guyana with 16%, and Haiti with 23% in 2015. Fig. 2.3 below shows the low birthweight rates in LAC countries.

Despite this, low birthweight in the LAC region decreased an average of 0.4% points in 26 LA countries from 2000 to 2015. However, some countries, such as

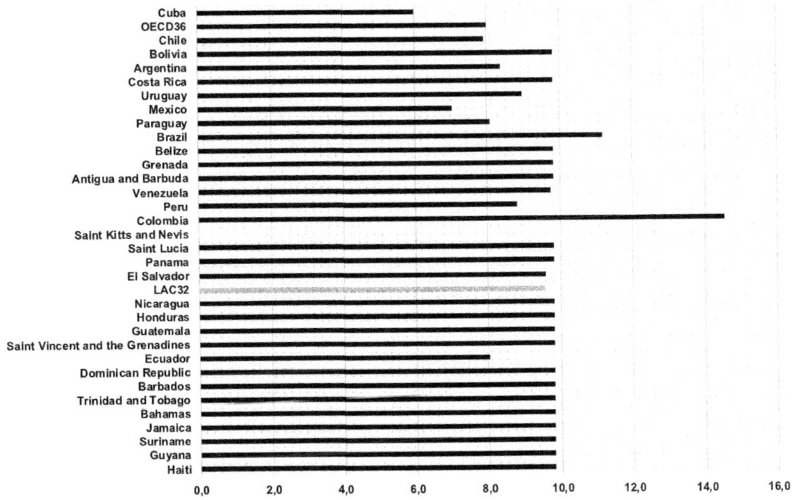

Figure 2.2 The preterm birth rate in Latin America and Caribbean countries (per 100 live births) in 2015.
The authors created this figure with the Organisation for Economic Co-operation and Development (OECD) (2021a). *Health at a glance: Latin America and the Caribbean 2020: Preterm birth and low birth weight*. URL: https://www.oecd-ilibrary.org/sites/53620b68-en/index.html?itemId=/content/component/53620b68-en.

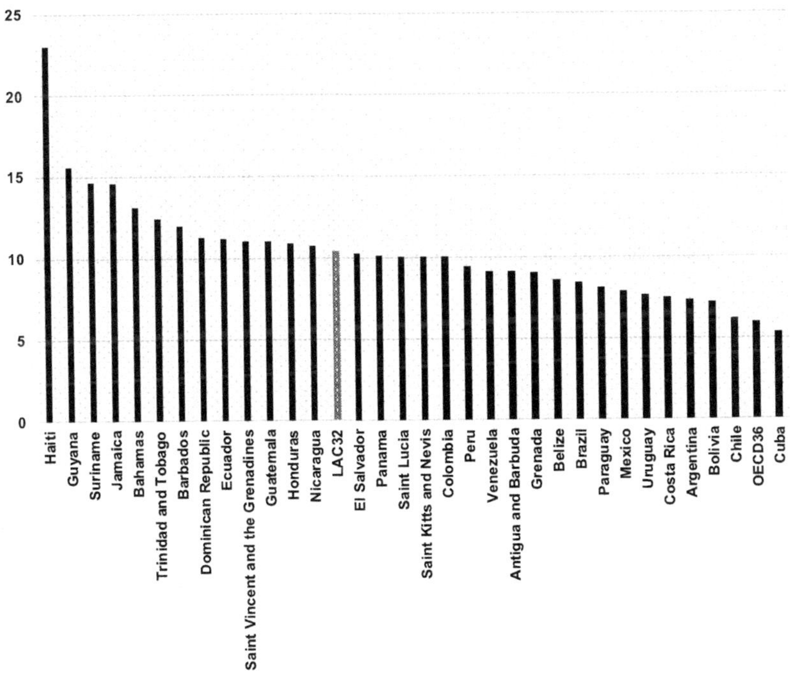

Figure 2.3 The low birthweight rates in Latin America and Caribbean countries (per 100 live births) in 2015.
The authors created this figure with the Organisation for Economic Co-operation and Development (OECD) (2021a). *Health at a glance: Latin America and the Caribbean 2020: Preterm birth and low birth weight*. URL: https://www.oecd-ilibrary.org/sites/53620b68-en/index.html?itemId=/content/component/53620b68-en.

Brazil, Chile, Costa Rica and Venezuela, have registered an increase in low birthweight newborns. At the same time, the most significant reduction happened in Suriname, Guatemala and Honduras, with more a decrease of than 1% point between 2000 and 2015 (see Fig. 2.4 below).

According to OECD (2021a), antenatal care can help women prepare for delivery and understand warning signs during pregnancy and childbirth. Therefore, the high coverage of antenatal care is associated with higher birthweight in LAC countries (see Fig. 2.5 below).

The negative correlation presented above does not apply equally in all countries from the LAC region. Therefore, according to OECD (2021a,b), in 2018, Barbados and Trinidad and Tobago reported that they had 98% and 100% of at least four antenatal care visits. Still, their low birthweight prevalence was 12% and 12.4%, respectively (see Fig. 2.3 above). These values are above the LAC average of 10% (see Fig. 2.3 above). On the other hand, countries like Bolivia, Grenada and Paraguay showed an antenatal care coverage below the 24 countries average of 87% in 2018. However, in these countries, the low birthweight prevalence is 7.2%, 9% and 8.1% (see Fig. 2.3 above). Indeed, these differences between countries are attributed to cultural practices and preferences, such as different approaches to privacy or perceptions about what antenatal and postnatal care entail (OECD, 2021a).

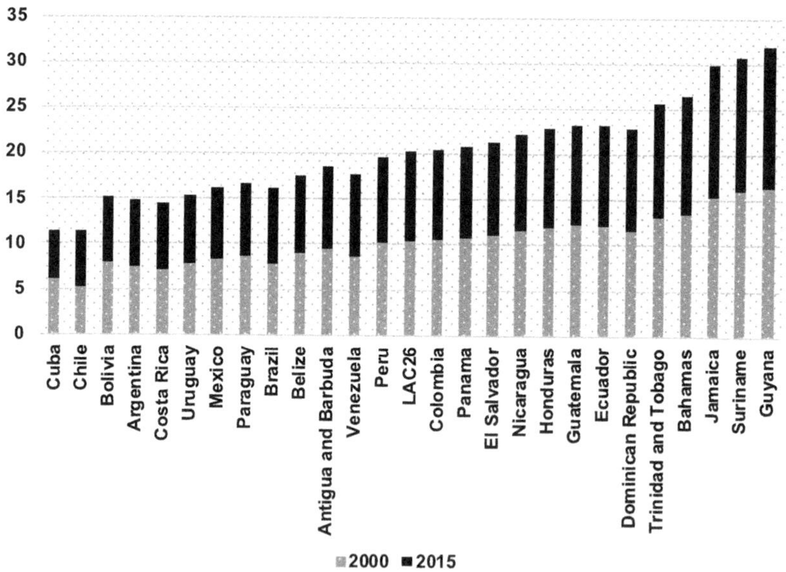

Figure 2.4 The low birthweight rate in Latin America and Caribbean countries (per 1000 live births) between 2000 and 2015.
The authors created this figure with the Organisation for Economic Co-operation and Development (OECD) (2021a). *Health at a glance: Latin America and the Caribbean 2020: Preterm birth and low birth weight.* URL: https://www.oecd-ilibrary.org/sites/53620b68-en/index.html?itemId=/content/component/53620b68-en.

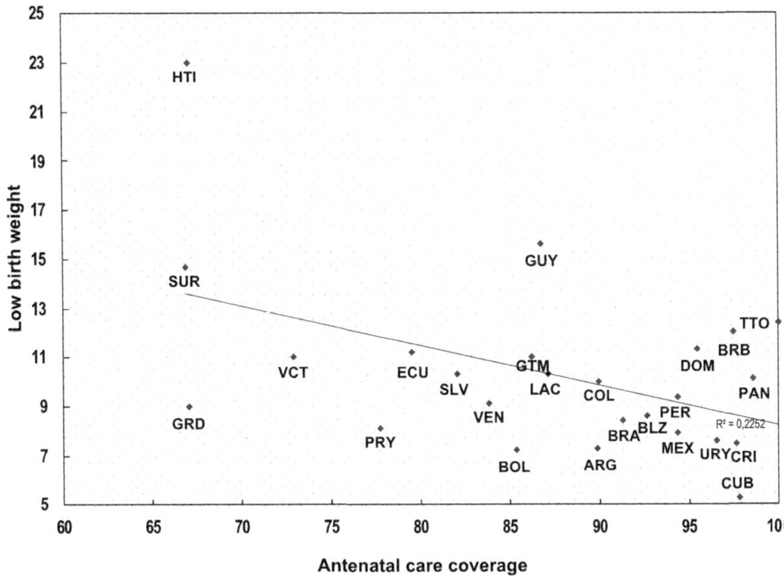

Figure 2.5 Antenatal care coverage and low birth weight in Latin America and Caribbean countries in 2018.
The authors created this figure with the Organisation for Economic Co-operation and Development (OECD) (2021a). *Health at a glance: Latin America and the Caribbean 2020: Preterm birth and low birth weight.* URL: https://www.oecd-ilibrary.org/sites/53620b68-en/index.html?itemId=/content/component/53620b68-en.

Another factor that contributes to the increase in obesity in LAC countries is related to child malnutrition. In the LAC region, child malnutrition is lower than in other world regions, but it is still a significant problem in several countries. Indeed, the region has a rate of 13% of children below the age of 5 with malnutrition (see Fig. 2.6 below).

However, in some countries in the LAC region, this rate is higher—nearly 47% in Guatemala, 24% in Ecuador, 23% in Honduras and 22% in Haiti, while in other countries, we can identify the presence of lower rates of prevalence of stunting among children under the age of 5, for example, Chile with 2%, Saint Lucia 2.5% and Paraguay with 5.6% in 2018 (see Fig. 2.6 above).

Moreover, the wasting rates are also lower than in other regions, with an average of 2.5% among children below 5. Some countries from the LAC region have significantly higher rates, such as Barbados with 7%, Guyana with 6.4% and Uruguay with 6.4%. The lowest rates are observed in Chile with 0.3%, Peru with 0.5%, Guatemala with 0.8% and Colombia with 0.9% (see Fig. 2.7 below).

Therefore, the countries with higher prevalence of stunting tend to have higher than average under-five mortality rates, reflecting that about half of all deaths before the age of 5 can be attributed to malnutrition. This positive correlation is presented in Fig. 2.8 below.

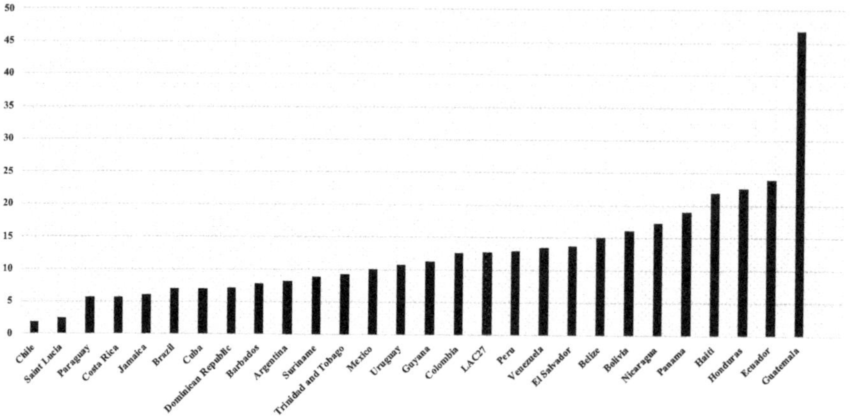

Figure 2.6 Prevalence of stunting among children under the age of five in Latin America and Caribbean countries in 2018.
The authors created this figure with the Organisation for Economic Co-operation and Development (OECD) (2021b). *Health at a glance: Latin America and the Caribbean 2020: Child malnutrition*. URL: https://www.oecd-ilibrary.org/sites/53620b68-en/index.html?itemId=/content/component/53620b68-en.

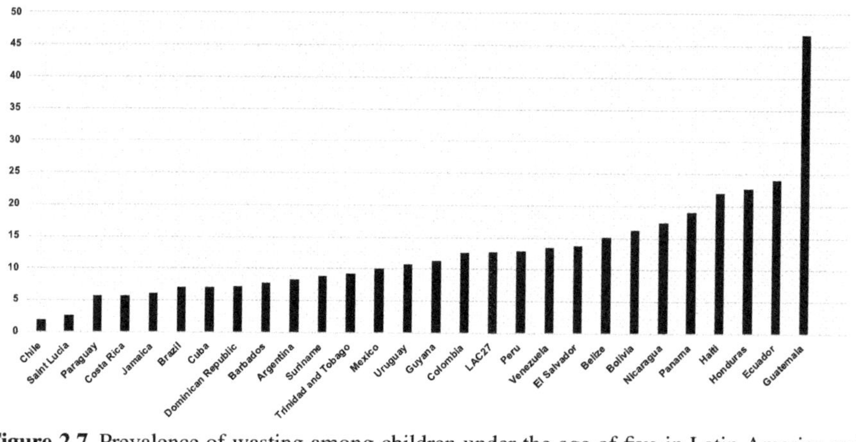

Figure 2.7 Prevalence of wasting among children under the age of five in Latin America and Caribbean countries in 2018.
The authors created this figure with the database from Organisation for Economic Co-operation and Development (OECD) (2021b). *Health at a glance: Latin America and the Caribbean 2020: Child malnutrition*. URL: https://www.oecd-ilibrary.org/sites/53620b68-en/index.html?itemId=/content/component/53620b68-en.

In the LAC, Guatemala deviates significantly from the trend by having a stunting rate almost four times the region and an under-five mortality rate eight points over. This is mainly due to the high poverty rate and significant income distribution inequality in the country. This inequality is translated into the impossibility of

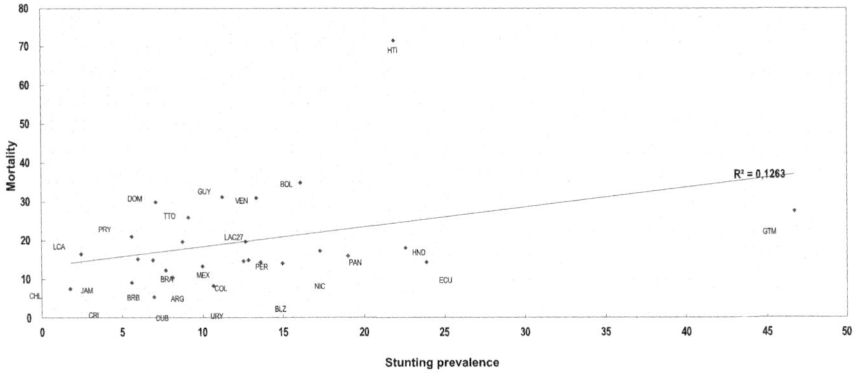

Figure 2.8 Mortality and prevalence of stunting under the age of five (per 1000 live births) in Latin America and Caribbean countries in 2018.
The authors created this figure with the Organisation for Economic Co-operation and Development (OECD) (2021b). *Health at a glance: Latin America and the Caribbean 2020: Child malnutrition.* URL: https://www.oecd-ilibrary.org/sites/53620b68-en/index.html?itemId=/content/component/53620b68-en.

affording a basic food basket for half of the population. Additionally, this country also has to face the effects of natural disasters and climate change that damages food production (OECD, 2021b).

2.2.2.2 Education and income

Income and education are two related characteristics, which are strongly associated with health and obesity (Grossman, 1972). From the perspective of human capital theory and considering the demand for health, people with more education are better endowed with the skills to make better choices for their health and people with high incomes can afford better services and goods for taking care of their health. In this way, education and income contribute to creating higher individual health capital, so health status tends to be better.

Empirical evidence confirms the relevance of income and education as determinant factors of obesity (e.g., Jiwani et al., 2019; Lee et al., 2019). Education and income are strongly correlated, meaning that higher education levels correspond to increased incomes and vice-versa. In the LAC region, government expenditures on education have been growing since the end of the 1990s. This progressive change happened after adopting structural and stabilisation programmes imposed by the International Monetary Fund (IMF) between 1989 and 1992 and the commodities cycle boom between 2002 and 2012 (see Chapter 1), whereby several countries in the LAC region were able to invest heavily in education. For example, in 1998, Argentina had public expenditures on education of 4%, and this value increased to 5.5% in 2016; Brazil in 1998 had public expenditures on education of 5%, and in 2016, this value reached 6.3%. Chile had public expenditures on education of 3.3% in 1998, and in 2017, this value reached 5.4%. In 1998, Colombia had public expenditures on education of 3.9%, and

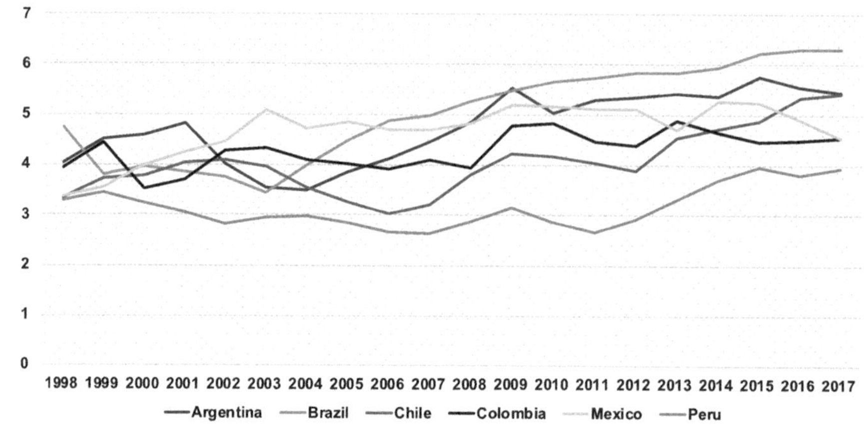

Figure 2.9 Government expenditure on education, total (% of GDP) in Latin American countries between 1998 and 2017.
The authors created this figure with the World Bank Open Data (2021). URL: http://www.worldbank.org/.

this value reached 4.5% in 2017. Mexico had public expenditures on education of 3.4% in 1998, and in 2017, this value reached 4.5% (see Fig. 2.9 below).

Increasing education gives individuals more resources (cognitive, communicative, relational) to make better choices, so more education tends to be related to lower levels of obesity (e.g., Kain et al., 2003; Lee et al., 2019). On the other hand, increasing income has a blurred association with obesity. Most likely, there is no linear relationship between income and obesity. Undoubtedly, poverty and meagre incomes are associated with obesity (e.g., Popkin, 2006; Popkin & Reardon, 2018), and communities with low incomes and high levels of poverty are usually characterised by low levels of physical activity. Therefore, they are also characterised by the high risk of overweight and obesity (e.g., Powell et al., 2004). In these communities, low-level education, poverty, and food insecurity constitute a daily challenge due to the small food expenditure; that is uncertainty, or limited food access is associated with low-quality diets, which end up potentialising the growth of obesity in people (e.g., Amarasinghe et al., 2009; Townsend et al., 2001).

As family income increases and food insecurity disappears, the choices and quantity of food also increase. However, in transitional societies, like those in LA, income tends to be associated with obesity, at least up to a certain income level (Kain et al., 2014).

In the LAC region, an increase in the level income is related, as mentioned before, to the adoption of structural and stabilisation programmes imposed by the International Monetary Fund (IMF) that occurred between 1989 and 1992, as well as by the commodities cycle boom that occurred between 2002 and 2012 (see Chapter 1). For example, in Argentina, the gross domestic product (GDP) per capita based on purchasing power parity (PPP) (Constant 2017 international dollars) was 14,144.76 USD in 1990 and reached a value of 22,063.90 USD in 2019. Brazil had a GDP per capita,

PPP in 1990 of 10,517.67 USD and in 2019, rose to a value of 14,651.61 USD. Mexico in 1990 had a GDP per capita, PPP of 14,998.03 USD, and in 2019 reached a value of 19,765.91 USD. Bolivia in 1990 had a GDP per capita, PPP of 4587.42 USD, and in 2019, this value was 8,724.47 USD (see Fig. 2.10 below).

This phenomenon contributes to diminishing poverty and improving quality of life. Fig. 2.11 shows poverty in the LAC region between 1980 and 2015.

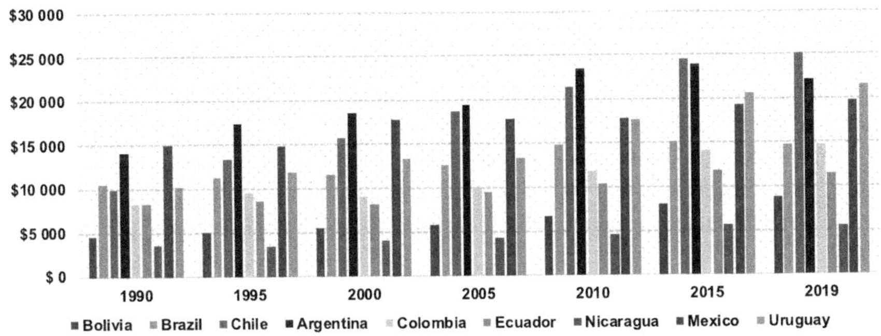

Figure 2.10 Gross domestic product (GDP) per capita, PPP (constant 2017 international $) in LA countries, between 1990 and 2017.
The authors created this figure with the World Bank Open Data (2021). URL: http://www.worldbank.org/.

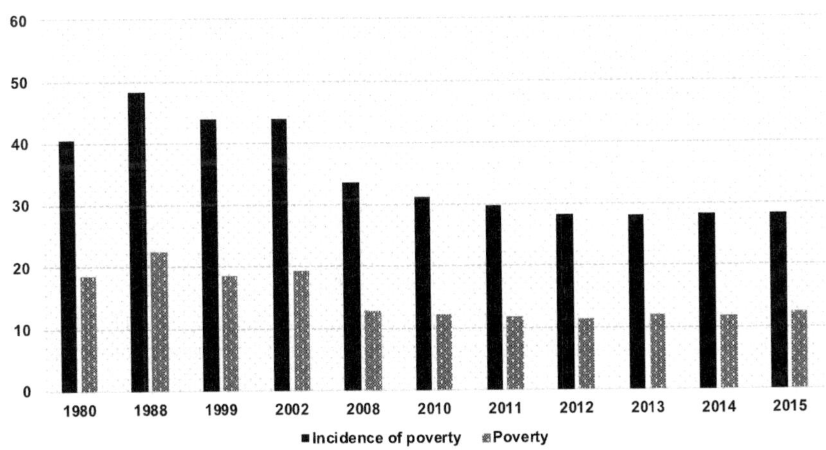

Figure 2.11 Incidence of poverty and poverty (%) in the Latin America and Caribbean region, between 1980 and 2015.
The authors created this figure with the Economic Commission for Latin America and the Caribbean (ECLAC) (2016). *Social panorama of Latin America 2015*. Santiago de Chile: Economic Commission for Latin America and the Caribbean (ECLAC). URL: https://www.cepal.org/en/publications/39964-social-panorama-latin-america-2015.

The incidence of poverty in the LAC region declined from 44% to 28.2% of the population, while poverty dropped from 19% to 12% of the population. In the LAC region, poverty is not evenly distributed among population groups, and it is higher among women, rural dwellers and indigenous and Afro-descendent people (Cecchini, 2017, pp. 1–8). The population from the LAC region changed from a traditional diet based on grains, fruits and vegetables and low in fat, to cheap fast-food and pre-cooked meals, which are extremely high in calories with high fat, high salt and sugar and low fibre, which is hardly healthy. Exhibit 2.2 discusses the decline of poverty in the LAC region between 2002 and 2012.

Moreover, increased participation of women in the labour market is associated with the increase in overweight and obesity in the region. The rise in women' participation in the labour market increased from about 20% in the 1960s to slightly more than 60% at the beginning of 2010 in South America (Bustelo et al., 2020). This phenomenon also contributed to diminishing poverty and improving the quality of life in the LAC region. Nevertheless, on the other hand, working women have less time to dedicate to cooking healthy meals, which favours the change in dietary pattern (Ruel, 2000).

Exhibit 2.2 The decline of poverty in the Latin America region 2002–12

Latin America is the most unequal region in the world. Despite this, poverty and extreme poverty in the LAC region experienced a substantial reduction between 2002 and 2012. The incidence of poverty in the region declined from 44% to 28% of the population, while poverty dropped from 19% to 12% of the population. Indeed, in the LAC region, poverty is not evenly distributed among population groups. It is higher among women, rural dwellers, indigenous, and Afro-descendent people (Cecchini, 2017, pp. 1–8).

However, with the "commodities boom", many commodity-exporting countries, like Bolivia, Brazil, Peru and Venezuela, saw more substantial declines in their poverty and inequality. In contrast, the commodity importers like Honduras, Nicaragua and Panama experienced a minor improvement. Therefore, the boom of commodities allowed commodity exporters to experience a significant boost in trade and economic growth. Moreover, the commodity sector expanded and drew in labour, with rising wages and employment. The demand for more workers also spilled over to other sectors, such as construction and commerce. At the same time, government revenues increased, which supported higher public investment and spurred job creation. All this allowed an increase in employment gains for the lower-skilled workers, as well as the development of income transfer programmes (e.g., *Renta Dignidad* in Bolivia, *Pension 65* in Peru and the conditional cash transfer programme *Bolsa Familia* in Brazil) and so reduced poverty and inequality even more (Balakrishnan & Toscani, 2018).

2.2.3 *Neighbourhood effects: social, economic and built environment*

Urbanisation increased worldwide significantly in a quarter of a century: from about one-third of the population living in urban areas to a share of nearly two-thirds of the world population (Jaitman, 2015). The LAC region is highly urbanised, with four global and massive cities to be found: São Paulo, Buenos Aires, Rio de Janeiro and Caracas. The process of urbanisation in the region contributed to the urban population growth by 30% in the 1940s, 60% in the 1970s and by the 2000s, 70% of people were living in cities. This means that the most significant urban centres, such as Buenos Aires (Argentina), São Paulo, Rio de Janeiro (Brazil) and Caracas (Venezuela), had an increase of 10% in the urban population in specific periods (Martins, 2002, pp. 303–313). Today, 80% of the region's population lives in cities, making LA the world's most urbanised region. In comparison, the European Union is 75% urbanised, and East Asia and the Pacific region are 50% urbanised. In 2050, about 90% of the LA region's population will be living in cities, according to (Arsht, 2014). Fig. 2.12 below shows the growth of the urban population in the LAC region between 1960 and 2019.

As can be seen in the figure above, the urban population has been growing since the 1960s. In 1960, the urban population was 49% of the total population, and in 2019, this value reached 81% of the total population. Indeed, the rapid urbanisation process in the LAC region is related to rapid economic development caused by an increase in economic growth. This increase is associated with the structural and stabilisation programmes imposed on LA countries by the IMF. These adjustment programmes are neoliberal policies that comprised the complete opening of the economies to

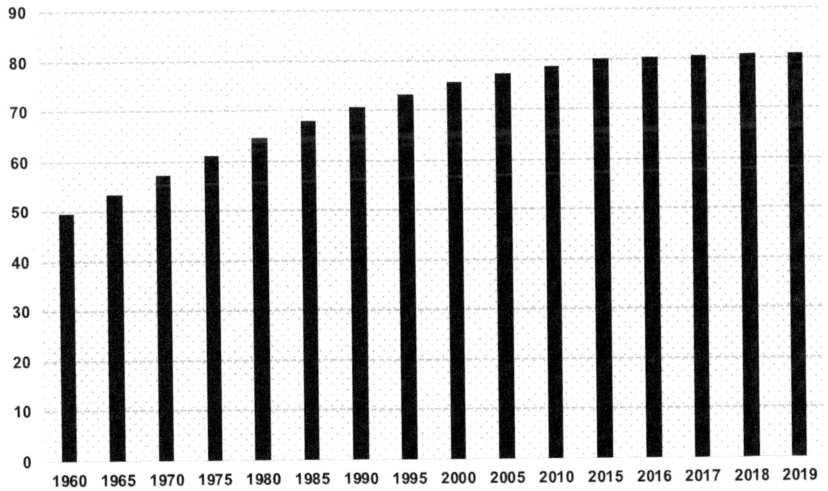

Figure 2.12 Urban population (% of the total population) in the Latin America and Caribbean region, between 1960 and 2019.
The authors created this figure with the World Bank Open Data (2021). URL: http://www.worldbank.org/.

international trade and capital movements, deregulation of the economy, privatisation and reduction of public expenditures, creating appropriate conditions for foreign investment and reducing the role of the government in the economy. Moreover, the "commodities boom" that occurred between the initial of the 2000s and the end of 2014 has also accelerated the process of openness and economic growth in the region (e.g., Koengkan & Fuinhas, 2020a,b).

Indeed, urbanisation caused by economic development, globalisation and technological revolution accelerated changes in the region's diet, which began in the 1980s. This change was also accompanied by declines in physical activity caused by urbanisation, which consequently contributed to the overweight or obesity problem in the region. Moreover, this process also impacted food systems, where the supply and demand changed through processing and wholesale, retail and transportation methods. The impact on the food system generated several supermarkets, fast-food chains and large food processors that are supplied and reinforced by modernised procurement systems and the coevolution among these segments. Together and simultaneously, these changes have created a rapid and omnipresent transformation in rural and urban areas in the LAC region.

There is a broad consensus in the literature that urbanisation leads to a nutritional transition in which a change of eating habits is observed. Therefore, the discussion is mostly around the mechanisms that operate that transition and how that transition can differ depending on different starting situations. Chee et al. (2019) explore how market integration and urbanisation contribute to an increase in overweight people in indigenous populations. In a qualitative case study conducted on the Kichwa people of Ecuador, they conclude that changing children's food preferences with a turning away from traditional foods and poor meal timing associated with urbanisation and market integration, pose an increased obesogenic risk. The authors stress the loss of traditional culture as the pivotal issue in the increase in obesity.

Popkin (1999) explains the inconsistency of the effects of urbanisation on overweight between high-income countries and low- and middle-income countries. In low- and middle-income countries, urbanisation is linked with an increase in access to a wide variety of unhealthy food. In contrast, in rich countries, where access to unhealthy foods is widespread both in urban or rural areas, increases in urbanisation tend to imply a greater level of education, better services, improved urban equipment and broader access to information that tends to reduce the proportion of overweight people (e.g., Böckerman et al., 2017; Goryakin & Suhrcke, 2014).

Moreover, according to PAHO (2021b), the overweight and obesity in urban areas may derive from the strong association between urbanisation and the consumption of ultra-processed food in the LAC region. For instance, countries such as Uruguay (95%) and Argentina (91.5%) have very high urbanisation rates. They account for annual sales per capita of ultra-processed food more significant than 150 kg, whereas in the least urbanised countries, such as Kenya (24.8%) or Vietnam (32.3%), the annual sales per capita of this kind of food is less than 50 kg. Additionally, the employment available for residents in urban areas is less physically demanding than the work pattern in rural areas. This work shift is significantly associated with the increase in the prevalence of overweight and obesity in urban areas (Popkin, 2006).

In the context of the difference between developed and developing countries, Wang et al. (2020) found that urbanisation increases the risk of overweight in China since urbanisation shifts peoples' lifestyles in ways that increase calorie consumption and decrease calorie expenditure. In the same vein of research focused on the relationship between urbanisation and overweight in developing countries, Goryakin and Suhrcke (2014) pointed out a positive relationship between urbanisation and overweight across all country wealth groups but considerably stronger in developing countries. Siervo et al. (2006), in a study focused on the Gambia, showed a positive relation between urbanisation and obesity.

The fast, unplanned growth of cities in this region resulted in very deficient areas of residence for poor people coming from rural areas and may be pointed to as another factor in the increase in overweight and obesity in the LAC region. Outskirts and slums are unhealthy places to live; there is a lack of social services, with overcrowding, poverty, property rights and crime coming together. This concentration of poverty in cities is known as "urbanisation of poverty" (UN, 2003), and in LA countries, about 60% of the poor live in urban areas (e.g., Jaitman, 2015; Popkin, 1998).

Indeed, poor neighbourhoods seem to be designed to promote obesity (e.g., Ewing & Meakins, 2014; Lee et al., 2019; Ruel, 2000). The inexistence of walking lanes, recreational spaces and fresh food markets (Bassett & Perl, 2004), along with all kinds of socioeconomic deprivation, is associated with an increase of 20% in overweight (Van-Lenthe & Mackenbach, 2002).

2.2.4 Technology effects

Technological improvements have taken place since the Industrial Revolution. However, since the mid-20th century, the upgrades have revealed a significant impact on humans' life and health. These technological improvements included medical advances and improvements in the transportation market and home equipment. The technological advancements in these areas resulted in increased quality of life. Still, they have also resulted in the lower level of calories burnt by human bodies and increased people's weight up obesity levels. On the other hand, it should be noticed that technological changes, in particular those that are strictly related to people daily life, use non-ecological energy sources such as those generated from oil and coal.

2.2.4.1 Transport

The technological improvements in transportation, particularly in cars and motorcycles, have contributed to a decrease in walking or movement by bicycle that people used to use. Nowadays, transportation across the city is done using motor vehicles that contribute to air pollution and decreased physical activity. Although evidence is limited, some studies from China show that people who bought cars or motorcycles are more likely to become overweight, particularly in urban areas (Popkin, 2006).

2.2.4.2 Other technological improvements

2.2.4.2.1 Home appliances

The technological improvements in home appliances contributed to the change of dietary patterns. Both fridges and the microwaves are used to preserve and prepare pre-cooked meals, which favour weight gain. On the one hand, these technological improvements have increased women's participation in the labour market and made food more accessible and faster. On the other hand, these technological improvements complement the development of the food industry associated with processed food with low nutritious quality and a highly unhealthy effect.

2.2.4.2.2 Air conditioning, elevators, television, video games and internet

Air conditioning is another well-appreciated home appliance. It certainly has improved the quality of life, but it is known that heat forces the body to use calories to cool. The massive use of these appliances means that calorie saving might have increased the urge to eat in hot weather conditions. In contemporary society, technological developments have entered people's lives without realising how dependent they would become on them, and how they would impact the daily energy balance needed to regulate body weight and energy saving. Things like elevators, television, video games and the internet have become human body energy savers, which means the accumulation of calories and overweight or obesity.

2.2.4.3 Other macro-determinants

Several other macro-determinants create an aggregated environment which is favourable to overweight and obesity. According to Tovar and Must (2014), the process of globalisation is one of these macro-determinants. Therefore, this process has also encouraged changes in almost all aspects of humankind's behaviour, ranging from cultural to social and economic issues (Koengkan & Fuinhas, 2021). Indeed, this process has several dimensions, such as the economic, social and political, that have evolved at different speeds. Economic and social globalisation makes a substantial contribution to the progressive path for overweight and obesity.

According to Fox et al. (2019) and Popkin (1998), one of the main features of economic globalisation is the change in food systems that arises as a natural outcome. The food chain extension, caused by economic globalisation, enables economies of scale in food production. This process enabled a diet rich in high-energy caloric foods with high sugar and salt content. Indeed, this kind of food is less expensive and thus more accessible to lower-income people. Globalisation also offers a ready supply of processed foods by multinational food corporations, fast-food chains and multinational supermarket chains.

Moreover, according to Koengkan and Fuinhas (2021), globalisation has a complementary effect on the demand side of food systems. The socioeconomic dimension of globalisation has increased people's time constraints and consequently reduced home-made food consumption. As a side effect, it increases the consumption of processed food by multinational food corporations, fast-food chains and multinational supermarket chains.

Moreover, economic globalisation also impacts people's energy-caloric expenditure. This impact occurs due to the penetration of new technologies made available by increasing trade and economic liberalisation. These new technologies reduce people's need for physical activity (Koengkan & Fuinhas, 2021). Indeed, innovations in technology have evolved on a path that allows labour-saving behaviours in industrial sectors and that has made home appliances and motorised transportation more accessible (e.g., Bell et al., 2002; Sobal, 2001). Social globalisation promotes individual transportation, communication and other activity-sparing systems (Sobal, 2001).

Additionally, social globalisation also impacts how food is distributed, marketed and consumed. Thus, the process of social globalisation encourages extensive exposure to global eating practices or Westernisation of food consumption. A clear example of Westernisation of food consumption is McDonaldisation or Cocalisation, which promotes food consumption with higher-energy, caloric content, contributing to increased overweight and obesity (e.g., Koengkan & Fuinhas, 2021; Koengkan et al., 2021, pp. 1–9). In short, the overweight and obesity epidemic is stimulated by economic and social globalisation via economic growth that facilitates nutrition transition.

In the LAC region, this process of globalisation began with the trade and financial liberalisation in the 1970s in Chile with the profound shift toward free-market economies during the dictatorship of Pinochet (Ahumada & Andrews, 1998). On the other hand, in many other countries in the region, the implementation of the neoliberal economic model took place during the process of the "Washington consensus", a combination of measures to promote the "macroeconomic adjustment", and the "Brady Plan", as mentioned in Chapter 1. Indeed, this adjustment occurred between 1989 and 1992, where Costa Rica and Mexico in (1989), Venezuela (1990), Uruguay (1991), Argentina (1992) and Brazil (1992) passed schemes of extensive trade and financial liberalisation, with the privatisation of significant portions of the public sector, liberalisation of foreign investment, reduction of import barriers and with the development of economic stabilisation programmes (e.g., Aizenman, 2005; Koengkan, 2020; Vásquez, 1996). In the 1990s, as mentioned earlier, the LAC region grew and became more and more integrated into the global economy. Indeed, in the first half of the 1990s, most LA countries adopted unilateral opening policies, reducing their tariffs and eliminating other trade restrictions.

Moreover, several regional agreements within the framework of *Asociación Latinoamericana de Integración* (ALADI) were strengthened in this period. For example, Mexico joined North American Free Trade Agreement (NAFTA), and the Common Market of the South (Mercosur) was created (Koengkan, 2020). Between 1990 and 1999, LA imports grew at an average rate of 11%, while its exports increased at an average rate of 8.1%, improving its share in world trade (Terra, 2003). Hence, the trade in the LAC region in the 1990s has been characterised by rates of import growth, which are much higher than exports. It should be recalled that imports had been drastically reduced in the wake of the debt crisis that followed the Mexican financial crisis of 1982 (Ventura-Dias et al., 1999). Then, in this period, imports had a vital role in modernising the production process. Modern machinery and better industrial inputs contributed to the technological upgrading of the industrial basis in the region (Ventura-Dias et al., 1999). According to World Bank Open Data (2021), in 1989, trade (% of GDP) in the region was 31.7%, and at the end of 2003, the value rose to 42.8%, an increase of 35.01% between 1989 and 2003.

Financial liberalisation in the LAC region followed the same path as that of trade liberalisation, where the inflow of capital into the LAC region resumption after the Brady Plan in the early 1990s (Koengkan, 2020). The magnitude of the financial liberalisation in the LAC region can be grasped with the index of capital mobility. In the 1980s, the index capital mobility was at 40 and in the 1990s arose to about 75 in a normalising, completely free capital mobility at 100 (Aizenman, 2005). Moreover, the financial liberalisation caused by the Brady Plan promoted the entrance of foreign direct investments (FDIs) flows. FDI inflows grew dramatically between 1990 and 1997. Indeed, developing countries received most of these flows, with LA countries receiving 15% of these inflows in 1990, and this value reaching 38% in 1997. LA was no different; these flows rose from eight billion USD in 1988 to 55.3 billion USD in 1997 (Birch & Halton, 2008). The increase of FDI flows in the region aligns with World Bank Open Data (2021), where in 1989, these flows were 18 billion USD and reached a value of 93 billion USD in 2003.

However, this process of globalisation that began in the 1970s by trade and financial liberalisation intensified with the "commodities boom" that occurred between the 2000s and 2014, when the region was experiencing an average economic growth rate of 7.40% (Santos, 2015). According to Carneiro (2012, pp. 1–47), the cycle of commodity prices in LA economies impacted economic openness or, more precisely, the degree of dependence on external demand vis-à-vis domestic demand or markets. The same author adds that between 1990 and 1993, the degree of economic openness was 28.6, 38.5 between 1998 and 2001 and 44.7 between 2006 and 2009 on a scale of 0 until 100, where 100 represents an open economy. In this period, between 1990 and 2009, the degree of economic openness had a growth of 50.71% (Fuinhas et al., 2021, pp. 19–234).

These two phases in the insertion of LA economies between 1989 and 1992 and between 2004 and 2014 influenced the degree of globalisation of the region's economies (Koengkan, 2020). Fig. 2.13 below shows the degree of globalisation in the LAC region between 1970 and 2018.

As shown in the figure above, the degree of globalisation of LA is measured by the KOF Globalization index (2021). In 1970, this index was 39 on a scale of 0 until 100, where 100 represents a 100% globalised economy. In 2018, this value reached 60. Indeed, the subcomponents of the globalisation index (e.g., economic, social and political) followed the same trend. In 1970, the economic globalisation was 36 and reached a value of 55 in 2018. Social globalisation had a value of 43 in 1970 and a value of 70 in 2019. Furthermore, political globalisation in 1970 had a value of 37 and reached a value of 55 in 2018 (Fuinhas et al., 2021, pp. 19–234).

2.3 The nutrition transition

The nutrition transition is a dynamic process that reflects the change in the diet patterns of people in different countries towards what has been called the "Western diet" and

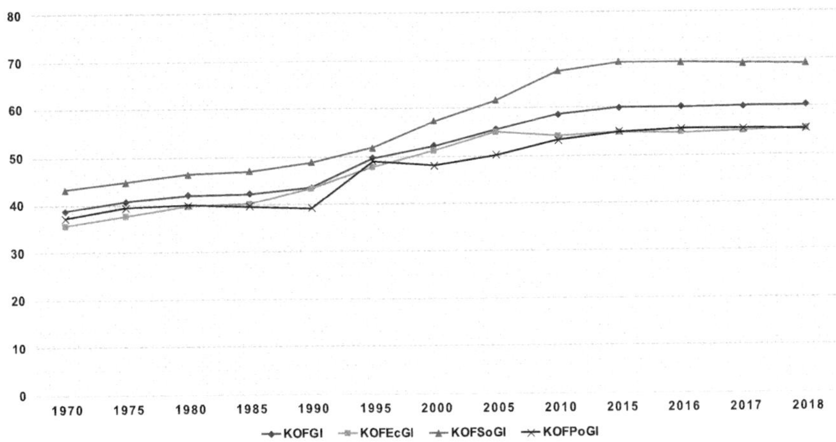

Figure 2.13 Globalisation index (%) in the Latin America and Caribbean region, between 1970 and 2018.
The authors created this figure with the KOF Globalization index (2021). URL: https://www.kof.ethz.ch/en/forecastsand-indicators/indicators/kof-globalisation-index.html.

towards a physically inactive lifestyle. The "Western diet" is characterised by high content of saturated fats, sugar, salt and processed and refined foods. This change results in changes in body composition (high fat mass and low muscle mass), stature and consequently higher levels of morbidity (Popkin & Gordon-Larsen, 2004).

The nutrition transition is taking place as the population around the world faces two other shifts, the demographic transition (the change from high fertility and mortality rates to low fertility and mortality rates) and the epidemiological transition (the change from the high prevalence of contagious diseases to high prevalence of non-communicable diseases). Both demographic and epidemiological transitions are associated with economic growth, industrialisation and urbanisation. The nutrition transition is a phenomenon happening jointly with the demographic and epidemiological transition. While the transition of nutrition in developed countries was a progressive change, it is rising fast in developing countries as LAC countries (Popkin, 2002).

Several patterns or stages may describe the dynamic process of the nutrition transition, and the last three represent the changes occurring in most low- and middle-income countries. Accordingly, in summary, pattern 3 describes the decrease of famine as national income increases; pattern 4 happens when diet and physical activity lead to the emergency of new chronic health problems and increases disability and, finally, pattern 5 takes place when the behavioural change in people reverses the previously negative trend. All these different and sequential patterns, or stages, are driven by a range of factors that we have previously described in our conceptual model. These include economic growth, industrialisation and urbanisation and technological and cultural change.

Latin American countries are mostly going through pattern 4, the pattern of degenerative diseases, mainly since the 1990s and with a rapid progression. This stage of nutrition transition is characterised primarily by increased fat, sugar and processed food intakes and increasing adoption of new technology at work and home, favouring physical inactivity. This pattern results in an increasing prevalence of obesity, higher life expectancy at birth and a rising prevalence of non-communicable diseases (e.g., Popkin, 1994, 2002).

To sum up, nutrition transition is a progressive process accompanying the demographic and epidemiologic transitions that characterise modern demographics worldwide. It complements the conceptual model of the determinants of obesity explained in the previous section, as the determinants of obesity are driving up the nutrition transition.

The change in diet in the LAC region caused by income increase is undeniable. According to Bermudez and Tucker (2003), the main foods contributing to total caloric energy consumption in LA countries for 1995 were cereals; they represented the main contributor to dietary energy, ranging from 59% in Guatemala, 56% in El Salvador and 53% in Nicaragua, to less than 30% in Argentina and Uruguay. Root crops and tubers were essential components of the caloric energy supply in 1995, contributing 16% in Paraguay and 13% in Peru. Meat was also a vital caloric energy contributor, with this kind of source contributing 22% in Uruguay and 17% in Argentina. Meat contributed 2% in El Salvador, 3% in Honduras, 3% in Nicaragua and 4% in Peru. Sugars contributed 10% in Paraguay and 24% of caloric energy consumption in Cuba. Moreover, vegetable oil contributions ranged from 5% in Guatemala and El Salvador and 6% in Bolivia, Peru and Uruguay to 19% in Ecuador.

However, in 2015, the main foods contributing to total caloric energy consumption in LA countries, according to Kovalskys et al. (2018), were the grains, pasta and bread group. This was the primary source of caloric energy for all countries, contributing about 28% of the total caloric energy. Similarly, in Peru, it contributed 35.9% and 32.9% in Chile. Within the grains pasta, and bread group, the primary source in Peru was rice, contributing 21%, while in Costa Rica, it was 16%, Ecuador 15% and Colombia 10%; in Venezuela, bread was the primary caloric energy source with a percentage of about 4.5%, while in Peru it was 4.3%, in Chile 3.6% and in Brazil 3.4%. In Argentina, that percentage was 3%. Refined-grain products were the primary caloric energy sources in Chile, contributing 25%, and 17% of the total caloric energy consumption in Venezuela.

Meat and eggs were the second main source of caloric energy consumption in all countries, with a share of around 18.9%. This value varies from 14% to 16% in different countries: Ecuador—16%, Brazil—16%, Colombia—15% and Peru and Argentina—14%. Within this group, non-processed beef and poultry represented the major sources for all countries, contributing 9.7%. Overall, fish was almost non-existent in terms of contribution to caloric energy consumption; the percentage contribution is circa 0.9%. Exceptions were found in Peru and Ecuador, where the intake of this kind of caloric energy source was 1.57% and 1.55%, respectively. Oils and fats were the third major caloric energy source consumption in all countries, contributing to 9.7%. In this group, vegetable oil represented the major source, which contributed to 6.1%, while butter and margarine contributed 1.6%. Indeed, margarine intake was higher than that of butter in Brazil with 3.26%, Venezuela with 2.73%, Colombia with 1.95% and Costa Rica with 1.32%. The opposite occurred in Argentina with 0.24%, Ecuador with 0.65%, Peru with 0.52% and Chile with 0.86%.

Non-alcoholic beverages were the fourth primary source of caloric energy consumption in all countries in the LAC region, contributing 12.1%, with a higher contribution to total caloric energy consumption in Venezuela of 14.6%. Soft drinks were the primary source of total caloric energy consumption from beverages in all countries, contributing 3.9%, followed by natural fruit juices with added sugar at 2.9%, except for Argentina at 2.3% and Chile at 2.2%, where ready-to-drink juices with sugar followed soft drinks.

In Brazil, there were similar proportions of caloric energy consumption from natural fruit juices with added sugar with a contribution of 1.9% and ready-to-drink juices with sugar with a contribution of 1.5%. However, Venezuela and Argentina were the countries with the highest intake of caloric energy from non-alcoholic beverages, where both countries contributed above 14%. In comparison, the rest of the countries consumed between 10% and 12%.

2.4 Conclusions

This chapter described the determinants of obesity, which has an increasing prevalence in LAC countries. Obesity is explained by a set of comprehensive determinants organised in a conceptual model in three different layers. These layers are placed around the centre with obesity, health and physical activity, and as the layers get further from the centre, the more comprehensive and larger the macro-effect is. These layers of influence are sequentially individual effects, neighbourhood effects and technological effects. The umbrella effect of all these layers is economic growth. In addition to this conceptual model of determinants of obesity, this chapter also presents the nutritional transition phenomenon, a dynamic process on which the determinants of obesity operate, characterising the fast evolution of the obesity epidemic in LAC countries. The determinants described here and the descriptive approach to the nutrition transition will be used in the following chapters to explain the influences and interactions between the different factors and obesity in the LAC region.

References

Ahumada, C., & Andrews, C. W. (1998). The impact of globalisation on Latin American States: The cases of Brazil and Colombia. *Administrative Theory and Praxis, 20*(4), 452–467. URL: http://www.jstor.org/stable/25611309.

Aizenman, J. (2005). *Financial liberalisations in Latin America in the 1990s: A reassessment.* NBER working paper series, 11145:1–30. URL: https://www.nber.org/papers/w11145.

Amarasinghe, A., D'Souza, G., Brown, C., Oh, H., & Borisova, T. (2009). The influence of socioeconomic and environmental determinants on health and obesity: A West Virginia case study. *International Journal of Environmental Research and Public Health, 6*(8), 2271–2287. https://doi.org/10.3390/ijerph6082271

Agência Brasil. (2018). *América Latina tem maior índice de sedentários; Brasil lidera*. URL: https://agenciabrasil.ebc.com.br/internacional/noticia/2018-09/america-latina-tem-maior-indice-de-sedentarios-brasil-lidera.

Arsht, A. (2014). *Urbanization in Latin America. Atlantic council*. URL: https://www.atlanticcouncil.org/commentary/article/urbanization-in-latin-america/#:~:text=Today%2C%20about%2080%20percent%20of,the%20world's%20most%20urbanized%20region.&text=Today%2C%20260%20million%20people%20live,percent%20of%20Latin%20America's%20GDP.

Balakrishnan, R., & Toscani, F. (2018). *How the commodity boom helped tackle poverty and inequality in Latin America*. IMFBlog. URL: https://blogs.imf.org/2018/06/21/how-the-commodity-boom-helped-tackle-poverty-and-inequality-in-latin-america/.

Bassett, M.,T., & Perl, S. (2004). Obesity: The public health challenge of our time. *American Journal Public Health, 94*(9), 1477. https://doi.org/10.2105/ajph.94.9.1477

Bell, A. C., Ge, K., & Popkin, B. M. (2002). The road to obesity or the path to prevention: Motorized transportation and obesity in China. *Obesity Research, 10*, 277–283. https://doi.org/10.1038/oby.2002.38

Bermudez, O. I., & Tucker, K. L. (2003). Trends in dietary patterns of Latin American populations. *Cadernos de Saúde Pública, 19*(1), 1–14. https://doi.org/10.1590/S0102-311X2003000700010

Birch, M. H., & Halton, G. (2008). Foreign direct investment in Latin America in the 1990s: Old patterns, new trends, and emerging issues. *Latin American Business Review, 2*(1–2), 13–31. https://doi.org/10.1300/J140v02n01_03

Böckerman, P., Viinikainen, J., Pulkki-Råback, L., Hakulinen, C., Pitkänen, N., Lehtimäki, T., Pehkonen, J., & Raitakari, O. T. (2017). Does higher education protect against obesity? Evidence using mendelian randomisation. *Preventive Medicine, 101*, 195–198. https://doi.org/10.1016/j.ypmed.2017.06.015

Brahmbhatt, M. (2017). Social and physical determinants of obesity in adults. *Advances in Obesity, Weight Management and Control, 6*(1), 17–23.

Bustelo, M., Frisancho, V., & Viollaz, M. (2020). *Cómo es el mercado laboral para las mujeres en América Latina y el Caribe?* Banco Interamericano de Desarrollo. URL: https://publications.iadb.org/en/what-labor-market-women-latin-america-and-caribbean.

Carneiro, R. M. (2012). *Commodities, choques externos e crescimento: Reflexões sobre a américa latina*, 117. CEPAL. ISSN:1680–8843. URL: http://www.eco.unicamp.br/cecon/images/arquivos/observatorio/Commodities_choques_externos_crescimento.pdf.

Cecchini, S. (2017). *Reducing poverty amidst high levels of inequality: Lessons from Latin America and the Caribbean*. Economic Commission for Latin America and the Caribbean (ECLAC). URL: https://www.un.org/development/desa/dspd/wp-content/uploads/sites/22/2017/04/Simone-Cecchini-Reducing-poverty-amidst-high-levels-of-inequality.pdf.

Chee, V. A., Teran, E., Hernandez, I., Wright, L., Izurieta, R., Reina-Ortiz, M., Flores, M., Bejarano, S., Dào, L. U., Baldwin, J., & Martinez-Tyson, D. (2019). Desculturización, urbanisation, and nutrition transition among urban Kichwas Indigenous communities residing in the Andes highlands of Ecuador. *Public Health, 176*, 21–28. https://doi.org/10.1016/j.puhe.2019.07.015

Dahlgren, G., & Whitehead, M. (1991). *Policies and strategies to promote social equity in health*. Stockholm (Mimeo): Institute for Future Studies. URL: https://www.iffs.se/media/1326/20080109110739filmz8uvqv2wqfshmrf6cut.pdf.

Den-Hoed, M., & Loos, R. J. F. (2014). Genes and the predisposition to obesity. In G. Bray, & C. Bouchard (Eds.), *Handbook of obesity, epidemiology, etiology, and physiopathology* (3rd ed., Vol. 1). Boca Raton: CRC Press, Taylor & Francis Group.

Economic Commission for Latin America and the Caribbean (ECLAC). (2016). *Social Panorama of Latin America 2015. Santiago de Chile: Economic commission for Latin America and the Caribbean (ECLAC)*. URL: https://www.cepal.org/en/publications/39964-social-panorama-latin-america-2015.

Ewing, R., & Meakins, G. (2014). Urban environment, building design, and obesity. In G. Bray, & C. Bouchard (Eds.), *Handbook of obesity, epidemiology, etiology, and physiopathology* (3rd ed., Vol. 1). Boca Raton: CRC Press, Taylor & Francis Group.

Fox, A., Feng, W., & Asal, V. (2019). What is driving global obesity trends? Globalisation or "modernisation"? *Globalization and Health, 15*(32), 1–16. https://doi.org/10.1186/s12992-019-0457-y

Franco, M., Ordunez, P., Caballero, B., Granados, J. A. T., Lazo, M., Bernal, J. L., Guallar, E., & Cooper, R. S. (2007). Impact of energy intake, physical activity, and population-wide weight loss on cardiovascular disease and diabetes mortality in Cuba, 1980–2005. *American Journal of Epidemiology, 166*(12), 1374–1380.

Fuinhas, J. A., Koengkan, M., & Santiago, R. (2021). *Physical capital development and energy transition in Latin America and the Caribbean*. Elsevier. https://doi.org/10.1016/C2020-0-01491-X

Goryakin, Y., & Suhrcke, M. (2014). Economic development, urbanisation, technological change and overweight: What do we learn from 244 demographic and health surveys? *Economics and Human Biology, 14*(1), 109–127. https://doi.org/10.1016/j.ehb.2013.11.003

Grossman, M. (1972). On the concept of health capital and the demand for health. *Journal of Political Economy, 80*, 223–255. URL: https://www.jstor.org/stable/1830580.

Hallal, P. C., Andersen, L. B., Bull, F. C., Guthold, R., Haskell, W., & Ekelund, U. (2012). Global physical activity levels: Surveillance progress, pitfalls, and prospects. *Lancet, 380*(9838), 247–257. https://doi.org/10.1016/S0140-6736(12)60646-1

Hall, J., Moore, S., Harper, S., & Lynch, J. (2009). Global variability in fruit and vegetable consumption. *American Journal of Preventive Medicine, 36*(5), 402–409. https://doi.org/10.1016/j.amepre.2009.01.029

Howson, C., Kinney, M., & Lawn, J. (2012). *Born too soon: The global action report on preterm birth*. Geneva: World Health Organization. URL: https://www.who.int/maternal_child_adolescent/documents/born_too_soon/en/.

Jaitman, L. (2015). Urban infrastructure in Latin America and the Caribbean: Public policy priorities. *Latin American Economic Review, 24*, 13. https://doi.org/10.1007/s40503-015-0027-5

Jiwani, S. S., Carrillo-Larco, R. M., Hernández-Vásquez, A., Barrientos-Gutiérrez, T., Basto-Abreu, A., Gutierrez, L., Irazola, V., Nieto-Martínez, R., Nunes, B. P., Parra, D. C., & Miranda, J. (2019). The shift of obesity burden by socioeconomic status between 1998 and 2017 in LAC countries: A cross-sectional series study. *The Lancet Global Health, 7*(12), 1644–1654. https://doi.org/10.1016/S2214-109X(19)30421-8

Kain, J., Concha, F., Moreno, L., & Leyton, B. (2014). School-based obesity prevention intervention in chilean children: effective in controlling, but not reducing obesity. *Journal of Obesity, 618293*. https://doi.org/10.1155/2014/618293.

Kain, J., Vio, F., & Albala, C. (2003). Obesity trends and determinant factors in Latin America. *Cadernos de Saude Publica, 19*(Suppl. 1), S77–S86. https://doi.org/10.1590/s0102-311x2003000700009

Koengkan, M. (2020). Capital stock development in Latin America and the Caribbean region and their effect on investment expansion in renewable energy. *Journal of Sustainable Finance and Investment*, 1–21. https://doi.org/10.1080/20430795.2020.1796100

Koengkan, M., & Fuinhas, J. A. (2020a). Exploring the effect of the renewable energy transition on CO_2 emissions of Latin American and Caribbean countries. *International Journal of Sustainable Energy*, 1−24. https://doi.org/10.1080/14786451.2020.1731511

Koengkan, M., & Fuinhas, J. A. (2020b). The interactions between renewable energy consumption and economic growth in the Mercosur countries. *International Journal of Sustainable Energy, 39*(6), 594−614. https://doi.org/10.1080/14786451.2020.1732978

Koengkan, M., & Fuinhas, J. A. (2021). Does the overweight epidemic cause energy consumption? A piece of empirical evidence from the European region. *Energy*, 1−19. https://doi.org/10.1016/j.energy.2020.119297

Koengkan, M., Fuinhas, J. A., & Fuinhas, C. (2021). *Does urbanisation process increase the overweight epidemic? The case of Latin America and the Caribbean region*. SSRN. https://doi.org/10.2139/ssrn.3826196

KOF Globalization index. (2021). URL: https://www.kof.ethz.ch/en/forecastsand-indicators/indicators/kof-globalisation-index.html.

Kovalskys, I., Fisberg, M., Gómez, G., Pareja, R., Yépez García, M., Cortés Sanabria, L., & Koletzko, B. (2018). Energy intake and food sources of eight Latin American countries: Results from the Latin American study of nutrition and health (ELANS). *Public Health Nutrition, 21*(14), 2535−2547. https://doi.org/10.1017/S1368980018001222

Lee, A., Cardel, M., & Donahoo, W. T. (2019). Social and environmental factors influencing obesity. *Endotext*. URL: https://www.ncbi.nlm.nih.gov/books/NBK278977/.

Martins, M. L. R. (2002). *Meio ambiente e morada social nos países do Mercosul*. Madrid.

McLaren, L. (2014). Social and economic determinants of obesity. In G. Bray, & C. Bouchard (Eds.), *Handbook of obesity, epidemiology, etiology, and physiopathology* (3rd ed., Vol. 1). Boca Raton: CRC Press, Taylor & Francis Group.

NCD Alliance. (2021). *Physical activity in Latin America: New WHO plan and industry interference*. URL: https://ncdalliance.org/news-events/news/physical-activity-in-latin-america-new-who-plan-and-industry-interference.

Organisation for Economic Co-operation and Development (OECD). (2021a). *Health at a glance: Latin America and the Caribbean 2020: Preterm birth and low birth weight*. URL: https://www.oecd-ilibrary.org/sites/53620b68-en/index.html?itemId=/content/component/53620b68-en.

Organisation for Economic Co-operation and Development (OECD). (2021b). *Health at a glance: Latin America and the Caribbean 2020: Child malnutrition*. URL: https://www.oecd-ilibrary.org/sites/53620b68-en/index.html?itemId=/content/component/53620b68-en.

Our World in Data. (2021). *Obesity*. URL: https://ourworldindata.org/obesity.

The Pan American Health Organization (PAHO). (2021a). *Active living and physical activity*. URL https://www.paho.org/hq/index.php?option=com_content&view=article&id=5198:2011-active-living-physical-activity&Itemid=1969&lang=pt.

Pan American Health Organization (PAHO) URL:https://www3.paho.org/en, 2021b.

Pérusse, L., Rice, T. K., & Bouchard, C. (2014). Genetic component to obesity: Evidence from genetic epidemiology. In G. Bray, & C. Bouchard (Eds.), *Handbook of obesity, epidemiology, etiology, and physiopathology* (3rd ed., Vol. 1). Boca Raton: CRC Press, Taylor & Francis Group.

Popkin, B. M. (1994). The nutrition transition in low-income countries: An emerging crises. *Nutrition Review, 52*, 285−298.

Popkin, B. M. (1998). The nutrition transition and its health implications in low-income countries. *Public Health Nutrition, 1*(1), 5−21. https://doi.org/10.1079/PHN19980004

Popkin, B. M. (1999). Urbanisation, lifestyle changes and the nutrition transition. *World Development, 27*(11), 1905–1916. https://doi.org/10.1016/S0305-750X(99)00094-7

Popkin, B. M. (2002). An overview on the nutrition transition and its health implications: The bellagio meeting. *Public Health Nutrition, 5*, 93–103.

Popkin, B. M. (2006). Technology, transport, globalisation and the nutrition transition food policy. *Food Policy, 31*(6), 554–569. https://doi.org/10.1016/j.foodpol.2006.02.008

Popkin, B. M., Adair, L. S., & Ng, S. W. (2012). Global nutrition transition and the pandemic of obesity in developing countries. *Nutrition Reviews, 70*(1), 3–21. https://doi.org/10.1111/j.1753-4887.2011.00456.x

Popkin, B. M., & Gordon-Larsen, P. (2004). The nutrition transition: Worldwide obesity dynamics. *International Journal of Obesity, 28*, S2–S9.

Popkin, B. M., & Reardon, T. (2018). Obesity and the food system transformation in Latin America. *Obesity Reviews, 19*, 1028–1064. URL: https://pubmed.ncbi.nlm.nih.gov/29691969/.

Powell, L., Slater, S., & Chaloupka, F. (2004). The relationship between community physical activity settings and race, ethnicity and socioeconomic status. *Evidence-Based Preventive Medicine, 1*(2), 135–144. URL: https://impacteen.uic.edu/journal_pub/pub_PDFs/EBPM-1-2-Powell%20et%20al1.pdf.

Ruel, M. T. (2000). Urbanisation in Latin America: Constraints and opportunities for child feeding and care. *Food and Nutrition Bulletin, 21*(1), 1–13. https://doi.org/10.1177/156482650002100103

Santos, B. G. (2015). O ciclo econômico da América Latina dos últimos 12 anos em uma perspectiva de restrição externa. *Revista Do BNDES, 43*, 205–251. URL: https://web.bndes.gov.br/bib/jspui/bitstream/1408/6242/2/RB%2043%20O%20ciclo%20econ%C3%B4mico%20da%20Am%C3%A9rica%20Latina_P%20.pdf.

Siervo, M., Grey, P., Nyan, O. A., & Prentice, A. M. (2006). Urbanisation and obesity in the Gambia: A country in the early stages of the demographic transition. *European Journal of Clinical Nutrition, 60*(4), 455–463. https://doi.org/10.1038/sj.ejcn.1602337

Sobal, J. (2001). Globalisation and the epidemiology of obesity. *International Journal of Epidemiology, 30*(5), 1136–1137. https://doi.org/10.1093/ije/30.5.1136

Terra, M. I. (2003). Trade liberalisation in Latin American countries and the agreement on textiles and clothing in the WTO. *Économie Internationale, 94*(95), 137–154. ISSN 1240–8093. URL: https://www.cairn.inforevue-economie-internationale-2003-2-page-137.htm.

Thaker, V. V. (2017). Genetic and epigenetic causes of obesity. *Adolescent Medicine: State of the Art Reviews, 28*(2), 379–405. URL: https://www.ncbi.nlm.nih.gov/pmc/articles/PMC6226269/.

The Food and Agriculture Organization (FAO). (2021). *Food Balances (old methodology and population)*. URL: http://www.fao.org/faostat/en/#data/FBSH.

Tovar, A., & Must, A. (2014). Influence of culture on obesity. In G. Bray, & C. Bouchard (Eds.), *Handbook of obesity, epidemiology, etiology, and physiopathology* (3rd ed., Vol. 1). Boca Raton: CRC Press, Taylor & Francis Group.

Townsend, M. S. J., Peerson, B., Love, C. A., & Murphy, S. P. (2001). Food insecurity is positively related to overweight in women. *Journal of Nutrition, 131*(6), 1738–1745. URL: https://pubmed.ncbi.nlm.nih.gov/11385061/.

United Nations (UN). (2003). *The challenge of slums: Global report on human settlements 2003*. United Nations Settlements Programme (UN-Habitat). URL: https://unhabitat.org/the-challenge-of-slums-global-report-on-human-settlements-2003+&cd=2&hl=pt-PT&ct=clnk&gl=pt.

United Nations (UN). (2019). *UN spotlights 'explosive' obesity rates, hunger in Latin America and Caribbean*. URL: https://news.un.org/en/story/2019/11/1051211.

Van-Lenthe, F., & Mackenbach, J. (2002). Neighbourhood deprivation and overweight: The globe study. *International Journal of Obesity, 26*(2), 234–240. https://doi.org/10.1038/sj.ijo.0801841

Vásquez, I. (1996). The Brady plan and market-based solutions to debt crises. *Cato Journal, 16*(2), 1–11. URL: https://www.cato.org/sites/cato.org/files/serials/files/catojournal/1996/11/cj16n2-4.pdf.

Ventura-Dias, V., Cabezas, M., & Contador, J. (1999). Trade reforms and trade patterns in Latin America. *CEPAL, 5*, 1–53. ISBN:92-1-121256-1. URL: https://pdfs.semanticscholar.org/56a4/464b039b3e19246894fa2de5fdc8bbf308ef.pdf.

Wang, R., Feng, Z., Liu, Y., & Qiu, Y. (2020). Is lifestyle a bridge between urbanisation and overweight in China? *Cities, 99*, 102616. https://doi.org/10.1016/j.cities.2020.102616

Waterland, R. A. (2014). Epigenetic mechanisms in obesity. In G. Bray, & C. Bouchard (Eds.), *Handbook of obesity, epidemiology, etiology, and physiopathology* (3rd ed., Vol. 1). Boca Raton: CRC Press, Taylor & Francis Group.

World Bank Open Data. (2021). URL: http://www.worldbank.org/.

World Health Organization (WHO). (1998). Obesity: Preventing and managing the global epidemic. In *Working group on obesity*. Geneva: World Health Organization. URL: https://www.who.int/nutrition/publications/obesity/WHO_TRS_894/en/.

Interactions between obesity, economic growth, globalisation, urbanisation and poverty in Latin American and Caribbean countries

3

3.1 Introduction

Latin American (LA) countries are among the most obese low- and middle-income countries. More than half of women in these countries are overweight or obese (Popkin & Reardon, 2018), and one in five children under 20 years old is either overweight or obese (Ng et al., 2014).

In the LA region, the overweight problem reached 49% of the adult population in Argentina, while in Chile, this was 46%, Venezuela 46%, Mexico 45%, Peru 39%, Colombia 39%, Ecuador 36% and Brazil 35% in 1990. However, in 2016, the overweight problem reached the following values: Chile 64%, Mexico 64%, Argentina 63%, Venezuela 63%, Colombia 59%, Brazil 57%, Peru 56% and Ecuador 55%. In all countries of the region, overweight affects at least half the population, with the highest rates registered in Chile (64%), Mexico (64%), Argentina (63%) and Venezuela (RB) (62%) (see Fig. 3.1 below).

Increase rates are well over 1% point per year in many LA countries (e.g., Chile, Mexico, Argentina and Venezuela). In some cases, like Brazil, they appear to be accelerating (Popkin & Reardon, 2018). Over the last 20 years, there has been a rapid increase in overweight problems across the population. Moreover, women are more affected by overweight and obesity than men. About 38% of men and 40% of women were overweight, and in many countries, as in the case of Chile and Mexico, the figures reach two-thirds of the women and over half of men.

However, the obesity problem reached 17% of the adult population in Argentina, while in Chile, this was 17%, Mexico 16%, Venezuela 15%, Colombia 12%, Peru 10%, Brazil 10% and Ecuador 9% in 1990. However, in 2016, the obesity problem reached the following values: Mexico 30%, Chile 28%, Argentina 28%, Venezuela 26%, Colombia 22%, Brazil 22%, Peru 20% and Ecuador 20% (see Fig. 3.2 below).

Moreover, the prevalence of overweight or obesity in children is also remarkable in the LA region. According to World Health Organization (WHO), the share of overweight or obese children and adolescents aged 5–19 rose from 4% in 1975 to around 18% in 2016. Indeed, the upper-middle-income economies such as Argentina, Brazil, Bolivia, Chile, Colombia, Ecuador, Mexico and Peru have likewise registered an increase in the prevalence of overweight or obesity in children (see Fig. 3.3 below).

Obesity Epidemic and the Environment. https://doi.org/10.1016/B978-0-323-99339-5.00005-4

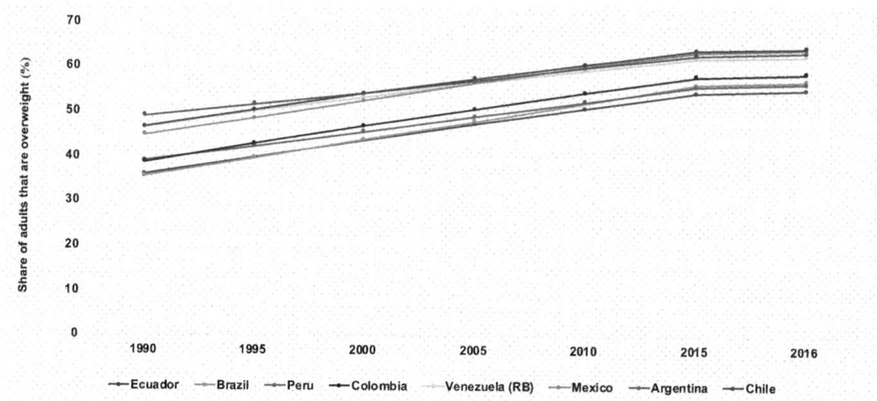

Figure 3.1 Share of adults that are overweight (%) in major Latin American economies, between 1990 and 2016. Being overweight is defined as having a body mass index (BMI) greater than or equal to 25. Obesity is characterised by a BMI greater than or equal to 30. BMI is a person's weight in kilograms divided by their height in meters squared.
The authors created this figure with the Our World in Data (2021). Obesity. https://ourworldindata.org/obesity.

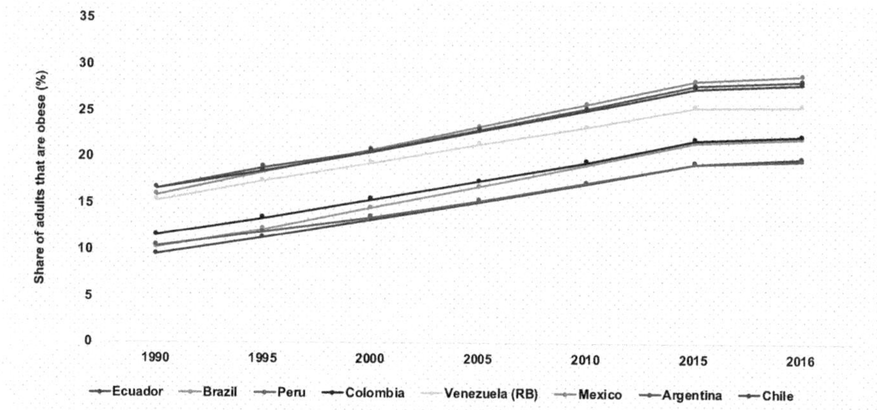

Figure 3.2 Share of adults that are obese (%) in major Latin American economies, between 1990 and 2016. Being overweight is defined as having a body mass index (BMI) greater than or equal to 25. Obesity is characterised by a BMI greater than or equal to 30. BMI is a person's weight in kilograms divided by their height in meters squared.
The authors created this figure with the Our World in Data (2021). Obesity. https://ourworldindata.org/obesity.

The prevalence of overweight or obesity in children in Argentina in 1990 was 10%, and in 2016, this value reached 17%. In Brazil, in 1990, this value was 14% and reached 33% in 2016. In Bolivia, the prevalence of overweight or obesity in children in 1990 was 23%, while in 2016, this value increased to 29%. In Chile, 24% of

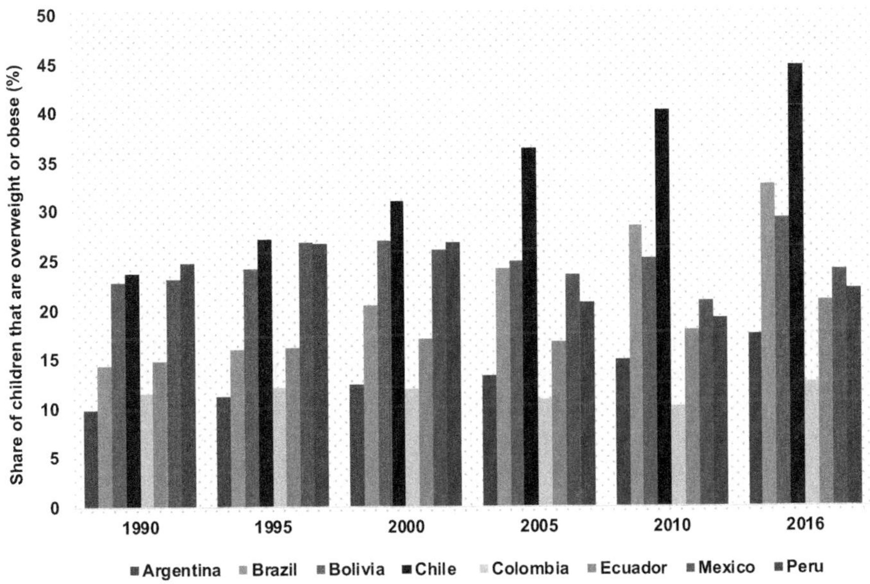

Figure 3.3 Share of children that are overweight or obese in Argentina, Brazil, Bolivia Chile, Colombia, Ecuador, Mexico and Peru between 1990 and 2016; share of children aged 2–4 years old who are defined as overweight or obese; a child is classified as overweight if their weight-for-height is more than two standard deviations from the median of the World Health Organization (WHO) Child Growth Standards.
The authors created this figure with the Our World in Data (2021). Obesity. https://ourworldindata.org/obesity.

children were overweight or obese in 1990, while in 2016, this value reached 45%. Moreover, in Colombia, 12% of children were overweight or obese in 1990, while in 2016, this value reached 13%. In Ecuador in 1990, 15% were overweight or obese, and in 2016, this value reached 21%. In Mexico, 23% of children were overweight or obese in 1990, while in 2016, this value increased to 24%. Finally, in Peru, 25% of children were overweight or obese in 1990, and in 2016, this value dropped to 23%. Exhibit 3.1 discusses the increase of obesity and overweight in the LAC region.

This chapter will focus on the determinants of obesity in Latin American and the Caribbean (LAC) countries mentioned in Chapter 2, introducing a conceptual model for these determinants, as shown in Fig. 3.4 below.

In this model, obesity is placed at the centre around which there are different interrelated layers of determinants. The inner layer concerns individual determinants such as the demographic and socio-economic. The second layer includes the determinants related to the neighbourhood where people are living. This accounts for the social, economic, natural and built surroundings. Urbanisation and poverty are among those determinants. Both are key factors explaining obesity, resulting from the economic development process in the LAC region since the 1980s.

Exhibit 3.1 Obesity and overweight populations in the Latin American and Caribbean region

More than 300 million adults were overweight in the LAC region, and of these, more than 100 million were obese in 2014. Obesity and overweight are defined as abnormal or excessive fat accumulation that may impair health (Garcia-Garcia, 2021). In the world, 39% (2.0 billion) of the adult population (38% of men and 40% of women) were overweight and close to 13% (600 million, 11% of men and 15% of women) were obese in 2014. The global prevalence of obesity more than doubled between 1975 and 2014 (World Health Organization (WHO), 2020; International Food Policy Research Institute, 2016; and NCD-RisC, 2016).

Moreover, obesity has become a significant health challenge in the LAC region. Around 57% (302 million) of the adult population in the LAC region (54% men and 70% of women) are overweight, while 19% (100.8 million) are obese (15% in men and 24% in women) (Garcia-Garcia, 2021).

In other low-middle-income countries, the overweight problem impacts 61% of women and 54% of men and the obesity problem affects 24% of women and 15% of men. The obesity problem is more prevalent in women than in men. In 14 LAC countries the prevalence among females is more significant than 20%. The highest prevalence of obesity problem in the adult population is found in El Salvador (33%) and Paraguay (30%) for women, and in Uruguay (23%) and Chile (22%) for men (Ng et al., 2014).

Moreover, the prevalence of overweight and obesity in children in the LAC region is also high, where it impacts 16% of children. It ranges from more than 12% for girls in Chile, Uruguay and Costa Rica, to less than 5% in Bolivia, Ecuador, Peru, Honduras and Guatemala. Indeed, the highest prevalence of obesity in children is found in Chile (12%) and Mexico (11%) in boys and Uruguay (18%) and Costa Rica (12%) in girls (Ng et al., 2014).

Overweight and obesity are significant risks for non-communicable diseases like cardiovascular disease (heart disease and stroke). Furthermore, the leading cause of death (30% of all causes) in the LAC region is diabetes, hypertension and chronic kidney disease (Garcia-Garcia, 2021).

The movement from rural to urban areas has been ongoing and has fuelled the rapid growth of cities. This has brought about a change in people's lifestyle. They abandoned the physically demanding work of the fields, but they also changed the traditional way of preparing food. The food supply system offers fast food, precooked meals and highly processed food, all with low nutritious quality and contributing to the emergence of and increase in obesity. Unplanned cities are also ill-designed with unfriendly spaces for walking or doing sports, shopping on foot in the neighbourhoods or using fresh food markets nearby. At the same time, people have access to several home appliances and equipment, contributing to less healthy food and less

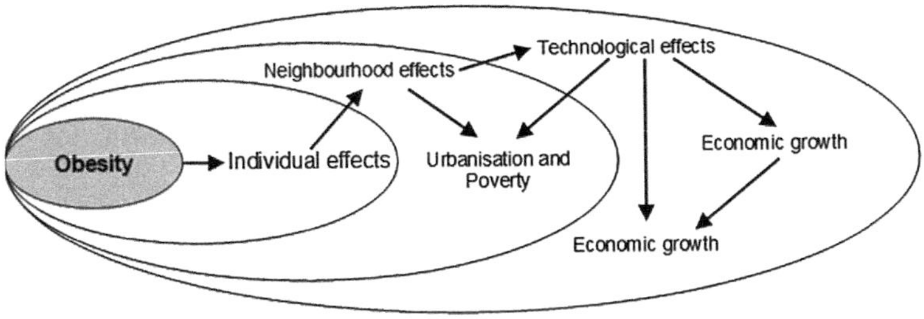

Figure 3.4 A conceptual model for the determinants of obesity.
This figure was adapted from Amarasinghe, A. G., D'Souza, C., Brown, H. O., & Borisova, T., (2009). The influence of socio-economic and environmental determinants on health and obesity: A West Virginia case study. International Journal of Environmental Research and Public Health, 6(8):2271–2287.

exercise. Additionally, unplanned urban growth has generated poor slums and outskirts. The 'urbanisation of poverty,' as referred to in Chapter 2, is strongly associated with the increasing obesity in LA countries.

Poverty is strongly related to obesity (e.g., Popkin, 2006; Popkin and Reardon, 2018). Food insecurity and extremely low levels of food expenditure imply limited and uncertain access to low-quality food. This insecurity and low nutritional value of food leverages body fat accumulation and obesity (e.g., Amarasinghe et al., 2009 and Townsend et al., 2001). The third layer of determinants is focused on technological factors that contribute to energy saving by the human body. Among these are the home appliances and technical devices such as fridges, microwaves, video games and the internet, which have contributed to the change in preferences toward less active life and less physical exercise.

Finally, there is an umbrella factor to all these determinants accounted for in the conceptual model of obesity determinants: economic growth. This factor creates the conditions for increasing urbanisation and poverty, and it interacts directly with the globalisation phenomenon, which characterises contemporary world relations. Globalisation has an effect on society's economic side, which is observable by the increasing levels of financial and trade openness and foreign investment (e.g., Aizenman, 2005, pp. 1–30; Fuinhas et al., 2021, pp. 19–234; Koengkan, 2020; and Koengkan & Fuinhas, 2020a). Nevertheless, globalisation may also have an effect on the cultural side of society, which is reflected in the increasing appropriation of a Western lifestyle in LA countries. Economic growth has been prevalent in LA countries since the 1980s. It generated a change in society that may be conveyed in observing the pattern of people's weight over time.

In sum, obesity is strongly related to economic growth, urbanisation, poverty and globalisation in LA countries since the 1980s, fuelling the nutritional transition. This chapter focuses on the interaction between obesity and these three determinants: urbanisation, poverty and globalisation and the umbrella determinant of economic growth. In the next section, we go through the main results found in previous literature. In Section 3.3, data and methods applied to the econometric analysis performed in this chapter are described. In Sections 3.4 and 3.5, results are presented and discussed.

3.2 Literature review

This section presents a literature review concerning the factors that determine the obesity epidemic in LAC countries, particularly economic growth, urbanisation, poverty and globalisation.

3.2.1 *Obesity and economic growth*

The general relationship between economic growth and obesity is positively correlated, meaning that the number of obese people in the country also increases when per capita income rises. Fig. 3.5 shows the prevalence of obesity among adults (BMI larger than 30) as a percentage of adults and the gross domestic product (GDP) per capita measured in purchasing power parity (PPP) (constant 2011 international dollar) in 2015 for 10 South American countries. A linear trend is also represented in this graph, showing the positive correlation between the two dimensions.

LA countries with lower income per capita, like Honduras, Nicaragua and Bolivia, tend to have smaller numbers of obese people. At the same time, countries with higher levels of income per capita show higher percentages of obese people, such as Argentina and Panama. Moreover, in general, in low- and middle-income countries, the growth of GDP is positively related to an increase in overweight, especially among the poor social groups (Popkin et al., 2012). Economic growth is the natural path for these low- and middle-income countries to move away from poverty. It has undoubtedly resulted in health improvements since the 19th century (e.g., Jack & Lewis, 2009; and Riley, 2001).

Despite the health benefits of economic development, economic development has generated several negative well-being and health returns in contemporary societies,

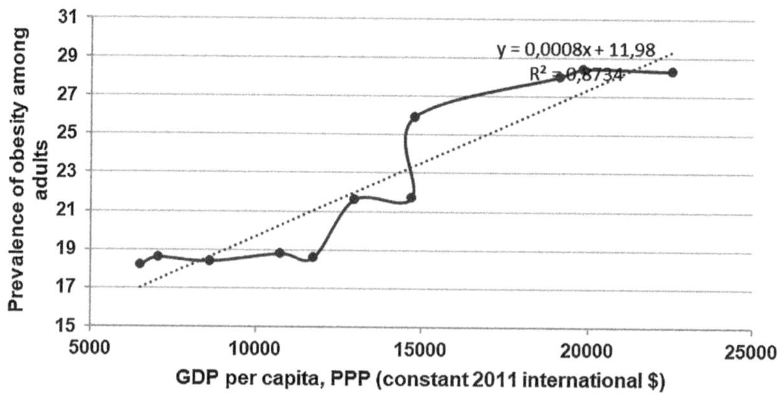

Figure 3.5 Relationship of the prevalence of obesity among adults in the Latin American region (BMI larger than 30) as a percentage of adults and GDP per capita, PPP (constant 2011 international dollar) in 2015.
The authors created this graph with data collected from Our World in Data (2021) Obesity. https://ourworldindata.org/obesity.

including obesity, especially in liberalised market economies. These adverse effects are observable in LA countries. Economic growth promotes changes in diet and lifestyle, which affect peoples' weight. On the one hand, increasing income and a decline in relative food prices overcome the poverty hurdle. Still, on the other hand, the globalised economy and its social influence contribute to savings in the energy intake to energy expenditures promoting obesity (e.g., Aydin, 2019; Garry et al., 2012; Koengkan et al., 2021, pp. 1–7; and Minos et al., 2016). The 'commodities boom' that occurred in the period 2000–15 has also accelerated the process of global openness as well as thc cconomic gains in the region (e.g., Fuinhas et al., 2021, pp. 19–234; Koengkan, 2020; Koengkan et al., 2019a; Koengkan, Fuinhas et al., 2019; and Koengkan & Fuinhas, 2020a). So, concerning the positive association between economic development and obesity in LA countries, empirical evidence by several authors confirmed it in general (e.g., Aydin, 2019; Egger et al., 2012; García, 2019; Goryakin & Suhrcke, 2014; Hoffman, 2001; and Minos et al., 2016).

3.2.2 Obesity and urbanisation

Urbanisation is the effect of industrialisation and economic development, which motivated people to move from rural to urban areas to look for better living conditions. This change has been very prominent in LA countries where the rate of urbanisation is exceptionally large: about 80% of the population lives in cities, while in East Asia and the Pacific, this rate is about 50% (e.g., Arsht, 2014; Jaitman, 2015; and Poveda et al., 2020). Urbanisation supplies infrastructures for a better quality of life, better work opportunities and more services. Urban spaces benefit from economies of scale and network effects, so people may enjoy a more pleasant and complete life (Rauch, 1993). However, the fast, unplanned growth of cities has meant poor living conditions for a significant share of people. Social and building environments are often not appropriate for walking, cycling, or playing outdoors, pushing people to an inactive life and eating choices loaded with excessive calories.

Poverty is another outcome resulting from fast urbanisation. Consequently, it is mainly concentrated in urban areas. In LA countries, 60% of the poor are dwellers in cities, reflecting the well-known phenomenon of 'urbanisation of poverty,' (e.g., Jaitman, 2015; Popkin, 1998; United Nations (UN), 2003, 2015). The socio-economic heterogeneity in urban areas is more significant than in rural areas, and child malnutrition is a worrisome problem in urban areas (Menon et al., 1999).

People living in urban areas need money to buy food, which they no longer produce as they would in rural areas. This expenditure represents about 60%–75% of household income. Mothers are constrained to supply meals to their children and family, not only by money and time but also by food availability close to residential areas. Women work during the day, get little exercise, and prioritise feeding children and husbands, even if with poorly nutritious meals (e.g., Fraser, 2005; and Ruel, 2000). For this reason, women suffer more often from obesity than men, and children end up suffering from malnutrition even if they are not hungry.

The lifestyles and diets in LA countries have changed significantly, and they have converged on what is termed the 'Western diet,' which contributes to an increase in numbers of obese people. The empirical evidence relating to urbanisation and obesity is vast, diversified and covers different countries in the world (e.g., Fox et al., 2019; Popkin, 1999; and Popkin, 2014). The positive correlation between increasing industrialisation and the increasing share of obese people, both children (Pirgon & Aslan, 2015) and adults, are found in several studies. For instance, for country-specific studies: Peru—McCloskey et al. (2017), Taype-Rondan et al. (2017); Brazil—Martins et al. (2018); Argentina—Pou et al. (2020), Tumas et al. (2019); and Chile—Albala et al. (2001).

3.2.3 Obesity and poverty

The economic development in LA countries meant an increase in income and broader income distribution across the population. This movement was translated into a steady decrease in poverty levels since the 1980s (e.g., Balakrishnan & Toscani, 2018; and Cecchini, 2017, pp. 1–8).

Despite the decrease of poverty in LA, poverty is still a social and economic problem in the region, and the chronic poor continues to be poor, despite the economic growth rates from the 2000s on Stampini et al. (2015). Poverty is to be found in rural and urban areas, and the figure has been growing since 2014, with the chronic poor being found in large numbers in urban areas (e.g., Stampini et al., 2015; and Vakis et al., 2015).

Poverty, particularly in urban areas, goes side by side with the new food system that emerged with urbanisation and globalisation. Access to fast food and ultraprocessed food is easier and cheaper. However, this low-nutrition food may contribute to obesity and underfed people (Popkin et al., 2012). This double burden may be found in Bolivia, Guatemala, Honduras, Nicaragua and Peru (Popkin & Reardon, 2018). Additionally, among the poor, often undernourished during pregnancy and childhood, people face obesity in adulthood. Empirical studies found that persistent poverty in LA countries is usually related to low-nutrition food and obesity. The increasing socioeconomic income inequality trends make it easy to see that poor people are more affected by obesity in LA countries (Pedraza, 2009). For specific country studies, see, for instance: Colombia—Álvarez-Castaño et al. (2012); Bolivia—Pérez-Cueto and Kolsteren (2004); Brazil—Schlüssel et al. (2013), and Alves and Magalhães (2019).

3.2.4 Obesity and globalisation

Globalisation is the process that integrates countries in a global network of relations of all kinds, including social, economic, cultural and technological. The 'Westernisation,' life and nutritional transition in developing countries have contributed to the spread of diet-related chronic diseases and obesity. The nutrition transition has changed the consumption of nutrients toward oils, sweeteners and salty ingredients, which characterise fast-food and ultraprocessed food. The statistical information on food intake, in particular concerning the critical nutrients from ultraprocessed food, in seven LA countries

(Argentina, Brazil, Chile, Colombia, Mexico, Peru and Venezuela) in 2014 (PAHO, 2019) shows that 22% of the calories and 50% of free sugars are obtained from carbonated soft drinks, 32% of total fat is eaten in cookies, cakes, desserts and snacks, 33% of saturated fat comes from cookies, cakes, pastries and desserts, and, finally, 50% of salt is ingested in the form of sauces and dressings. These numbers are worrisome as they drive up overweight.

Several factors of influence may link globalisation and dietary patterns (García, 2019), including international food trade, foreign direct investment, global advertising and promotion, restructuring of food retailing, the emergence of agribusiness and transnational food corporations and lastly urbanisation. This transitional diet in LA countries is, in fact, highly associated with the rapid rise in the rates of overweight and obesity.

In the literature, social globalisation is associated with an increase in overweight and obesity because this process influences the adoption of fast food and processed food through a McDonaldisation or Cocalisation processes. These cultural appropriations have resulted in higher consumption of calories and the general increase in people's weight (Fox et al., 2019). Westernisation caused by social globalisation is also influenced by inexpensive transportation means and other activity-sparing mechanisms such as household appliances and technological equipment that minimise physical activity (Sobal, 2001). On the other hand, economic globalisation is related to the entrance of multinational food corporations, fast-food chains and multinational supermarket chains that offer a wide range of processed foods (Fox et al., 2019). Economic globalisation has also brought several modern technologies that minimise physical activity levels and motivate people toward a sedentary and indoor life (Sobal, 2001).

In general, empirical evidence supports the hypothesis that globalisation is driving up the levels of obesity in developing countries (e.g., Aydin, 2019; Egger et al., 2012; Fox et al., 2019; García, 2019; Goryakin & Suhrcke, 2014; and Minos et al., 2016).

3.3 Data and methods

This section will be divided into two parts. The first will approach the group of countries and data/variables used in the chapter, while the second will show the method.

3.3.1 Data

This chapter will use annual data that was collected from 1991 to 2016 on 23 countries from the LAC region, i.e., **Argentina**, **Belize**, **Bolivia**, **Brazil**, **Chile**, **Colombia**, **Costa Rica**, **Dominican Republic**, **Ecuador**, **El Salvador**, **Guatemala**, **Guyana**, **Haiti**, **Honduras**, **Jamaica**, **Mexico**, **Nicaragua**, **Panama**, **Paraguay**, **Peru**, **Trinidad and Tobago**, **Uruguay** and **Venezuela**. Time-series between 1991 and 2016 is due to data availability until 2016 for the variable OBESITY for all countries selected. The variables that were chosen to perform this investigation will be shown in Table 3.1 below.

Table 3.1 Variables' description and summary statistics.

Variables' description	
Variable	**Source**
Share of adults that are obese (percentage). Obesity is defined as having a body mass index (BMI) equal to or greater than 30. BMI is a person's weight in kilograms divided by his or her height in metres squared. In this investigation, we called this variable '**OBESITY**'.	Our World in Data (2021)
Gross domestic product (GDP) per capita based on purchasing power parity (PPP). This variable is converted to international dollars using purchasing power parity rates. An international dollar has the same purchasing power over GDP as the US dollar has in the United States. GDP at purchaser's prices is the sum of gross value added by all resident producers in the country plus any product taxes and minus any subsidies not included in the value of the products. It is calculated without deductions for the depreciation of fabricated assets or depletion and degradation of natural resources. This variable is in current 2017 international dollars. In this investigation, we called this variable '**GDP_PPP**'.	World Bank Open Data (2021)
Urban population refers to people living in urban areas as defined by national statistical offices. It is calculated using World Bank population estimates and urban ratios from the United Nations World Urbanisation Prospects. Aggregation of the urban and rural population may not add up to the total population because of different country coverages. This variable is a proxy for the urbanisation index. In this investigation, we called this variable '**URBA**'.	World Bank Open Data (2021)
De facto social globalisation index. This variable measures the interpersonal contact flows of information and the cultural proximity. Interpersonal contact is measured within the de facto segment with reference to international telephone connections, tourist numbers and migration. Flows of information are determined within the de facto segment with reference to international patent applications, international students and trade in high-technology goods. Cultural proximity is measured in the de facto part via trade in cultural goods, international trademark registrations and the number of McDonald's restaurants and IKEA stores. In this investigation, we called this variable '**KOFSoGI**'.	KOF Globalisation Index (KOF, 2021)

Table 3.1 Variables' description and summary statistics.—cont'd

Variables' description	
Variable	**Source**
De facto economic globalisation index. This variable measures trade and financial globalisation. Trade globalisation is determined based on trade in goods and services and financial globalisation includes foreign investment in various categories. In this investigation, we called this variable '**KOFEcoGI**'.	KOF Globalisation Index (KOF, 2021)
Poverty gap at $5.50 a day (2011 PPP) is the mean shortfall in income or consumption from the poverty line of $5.50 a day (counting the non-poor as having zero shortfalls), expressed as a percentage of the poverty line. This measure reflects the depth of poverty as well as its incidence. In this investigation, we called this variable '**POVERTY**'.	World Bank Open Data (2021)

Summary statistics

Variables	Obs.	Mean	Std.-Dev	Min	Max
DLogOBESITY	572	0.0292	0.0080	0.0125	0.0588
DLogGDP_PPP	567	0.0399	0.0388	−0.1249	0.2118
DLogURBA	572	0.0214	0.0111	−0.0085	0.0623
DLogKOFSoGI	572	0.0193	0.0274	−0.1437	0.2039
DLogKOFEcoGI	572	0.0111	0.0526	−0.2011	0.2741
DLogPOVERTY	293	−0.0433	0.1326	−0.4870	0.4592

Obs. denotes the number of observations; Std. Dev. is the standard deviation; Min. and Max. are the minimum and maximum values, respectively; (Dlog) denotes variables in the first differences of the natural logarithms.

'DLog,' denotes variables in first differences of logarithms, 'Obs.' indicates the number of observations in the model, 'Std.-Dev' denotes the standard deviation and 'Min and Max' denote minimum and maximum. These summary statistics were obtained from the command *sum* of **Stata 16.0**. The board below shows how to get the summary statistics of variables.

How to do:

****The summary statistics****

sum dlogobesity dloggdp_ppp dlogurba dlogkofsogi dlogkofecogi dlogpoverty

In this chapter, we opted to use the variable OBESITY because, as mentioned in Chapter 1, the obesity problem in the LA region has been growing since the end of the 1980s, with this value being 9% of the adult population in 1990. In 2016, this problem reached 19% of the adult population. That is an increase of 109% between 1990 and 2016.

The variable GDP_PPP was used in this chapter because the LA region had registered rapid economic growth from the 1980s to 2016. In this period, LA's GDP per capita growth (annual %) had an average annual growth rate of 3.75%. This increase is related to the structural and stabilisation programs imposed on LA countries by the International Monetary Fund (IMF), as mentioned in Chapter 2.

These programs of adjustment were neoliberal policies that consisted of completely opening LA economies to international trade and capital, deregulation of the economy, privatisation, reduction of public expenditures, creation of proper conditions for foreign investment and the reduction of the role of the state in the economy. Moreover, the 'commodities boom,' that occurred between the beginning of the 2000s and the end of 2014 also accelerated economic growth in the region and the process of liberalisation (e.g., Koengkan et al., 2019b; and Koengkan & Fuinhas, 2020b). Exhibit 3.2 discusses the effect of the 'commodities boom,' on economic growth in the LA region.

This increase in economic growth had effects on dietary habits in LA countries. According to Drenowski and Specter (2004), Gerbens-Leenes et al. (2010), and Roskam et al. (2010), the shift in the lower socio-economic growth group toward energy-dense animal sources and fatty foods added to energy consumption and thus contributed to overweight and obesity problems. Indeed, obesity presents a problem in lower socio-economic classes (e.g., Butzlaff, 2016; and Monteiro et al., 2005). The increase in income allows the consumption of food with higher caloric value, contributing to obesity. In countries with low income but with high economic growth and food insecurity, obesity via food consumption is higher than in high-income countries. Therefore, economic growth is associated with obesity, both with its contemporary and future effects on food consumption. In the literature, this variable was used by Koengkan and Fuinhas (2021), and their investigation approached the impact of overweight on the consumption of energy in the European region.

This rapid economic development caused by economic growth has also influenced the rapid urbanisation process in the region. For this reason, we used the variable URBA in this chapter. As well-known, the LAC region is highly urbanised, with four global cities, São Paulo, Buenos Aires, Rio de Janeiro and Caracas (see Chapter 2). The process of urbanisation in the region made the urban population grow to 30% in the 1940s, 60% in the 1970s, and in the 2000s, 70% of people were living in cities. The most significant urban centres, such as Buenos Aires, São Paulo, Rio de Janeiro and Caracas, increased their urban population by 10% in specific periods (Martins, 2002, pp. 303–313). Today, 80% of the region's population lives in cities, making LA the world's most urbanised region. The European Union is 75% urbanised, and East Asia and the Pacific region are 50% urbanised. In 2050, LA cities will include 90% of the region's population (Arsht, 2014).

This increase in urbanisation in the region caused declines in physical activity, which contributed to the overweight or obesity problem in the region. Moreover,

Exhibit 3.2 Latin America's decade-long prosperity

The 2000−14 was marked as an extended period of prosperity in the LA region caused by the 'commodities boom,' as it known. This phenomenon may be described as the general increase of many physical commodity prices (e.g., metal, oil, chemicals and food) during the early 21st century (2000−14). Indeed, this 'boom', occurred due to the rising demand from emerging markets, particularly China, from 1990 to 2014. This great demand from emerging markets increased concerns about the long-term availability of supply of commodities.

In the LA region, this cycle of commodity prices affected the gross domestic product (GDP) per capita of the region, where according to World Bank Open Data (2021), in 2000, the GDP per capita PPP (constant 2017 international US$) was 12,680.14 USD and in 2014, reached 16,326.26 USD. This increase in GDP per capita between 2000 and 2014 was driven by the 'boom' of exports and foreign direct investment (FDI) inflows. According to World Bank Open Data (2021), in 2000, the exports of goods and services (% of GDP) in the LA region were 18.75%, and in 2014, this value reached 20.21%. Regarding foreign direct investment, net inflows (BoP, current USD), in 2000, this value was 97.255 (billion USD), and in 2014, this value increased to 384.955 (billion USD). Indeed, the sharp rise in commodity prices over the past decade allowed a rise in the wages of the poor in the LA countries and permitted a reduction of poverty due to better income distribution in the region, whereby poverty decreased from about 27%to levels around 12% (World Bank Open Data, 2021).

this process also affected the food systems, where supply and demand changed in line with processing and wholesale, retail and transportation methods. The impact in the food system gave rise to several supermarkets, fast food chains and large food processors fed by modernised procurement systems, and the coevolution among these segments. All these mean that rural and urban areas in the LAC region experienced a rapid, omnipresent transformation. The literature uses variable urban population as a proxy of the urbanisation index used by Koengkan and Fuinhas (2021).

Moreover, the process of economic growth caused by neoliberal policies and the 'commodities boom,' that occurred between the beginning of the 2000s and the end of 2014, as mentioned before, influenced a thorough process of globalisation in the region. This empirical investigation opted to use the social (KOFSoGI) and economic globalisation (KOFEcoGI) subcomponents of the globalisation index. Indeed, these subcomponents of globalisation are associated with increased overweight and obesity in the world and the LAC region (see Chapter 2). In the literature, social globalisation is related to an increase in overweight and obesity because this process influences the adoption of a more robust fast-food/processed foods culture through McDonaldisation or Cocalisation processes leading to more consumption of caloric and so the increase of weight gains (Fox et al., 2019). Indeed, this Westernisation caused by social globalisation also can influence the use of inexpensive transportation, communication and other activity-sparing systems through automobiles and household appliances that minimise physical activity levels (Sobal, 2001). Economic globalisation related to the increase due to this process allows the entrance of multinational food corporations, fast-food chains and multinational supermarket chains that offer a ready supply of processed foods (Fox et al., 2019). Moreover, this process also allows access to modern technologies that minimise physical activity levels (Sobal, 2001). Therefore, for this reason, we included the variables KOFSoGI and KOFEcoGI in our model.

Finally, the LAC region saw a decline in poverty from 44% to 28% (Cecchini, 2017, pp. 1–8). This reduction is related to the increase in economic growth that allowed a rise in poor people's wages and better income distribution in the LAC region (Balakrishnan & Toscani, 2018). Indeed, this reduction of poverty caused by income distribution allows a process of dietary transition, where energy intake and total fat, especially saturated fat, increase (Kain et al., 2003). For this reason, this investigation opted to use the variable POVERTY in the model. Exhibit 3.3 discusses the decline of poverty in the LA region between 2002 and 2012.

This section presents the group of countries from the LAC region that this chapter will focus on the variables and their justifications for use. In the following subsection, we will present the method used to carry out the empirical investigation of this chapter.

Exhibit 3.3 The decline of poverty in the LA region between 2002 and 2012

Latin America is the most unequal region in the world. Despite this, poverty and extreme poverty in the LAC region experienced a substantial reduction between 2002 and 2012. The incidence of poverty in the region declined from 44% to 28% of the population. Indeed, in LA countries, poverty is not evenly distributed among population groups, and it is higher among women, rural dwellers and indigenous and Afro-descendent people (Cecchini, 2017, pp. 1–8).

However, with the 'commodities boom,' many commodity-exporting countries, like Bolivia, Brazil, Peru and Venezuela, saw more substantial declines in their poverty and inequality. In contrast, the commodity importers like Honduras, Nicaragua and Panama experienced a more negligible improvement. Therefore, the boom of commodities allowed commodity exporters to experience a significant boost in trade and economic growth. Moreover, the commodity sector expanded and drew in labour, causing wages and employment to rise. The demand for more workers also spilt over to other industries, such as construction and commerce. At the same time, government revenues increased, which supported higher public investment and spurred job creation. All this allowed an increase in employment gains for the lower-skilled workers, as well as the development of income transfer programs (e.g., *Renta Dignidad* in Bolivia, *Pension 65* in Peru, and the conditional cash transfer programme *Bolsa Familia* in Brazil), and so reduced poverty and inequality (Balakrishnan & Toscani, 2018) more than ever.

3.3.2 Method

The best way to analyse the interactions between obesity, economic growth, globalisation, urbanisation and poverty in LAC countries is using the panel vector autoregression (PVAR) model. As is already known, this method was previously developed by Holtz-Eakin et al. (1988) as an alternative to multivariate simultaneous equation models and introduced by Love and Zicchino (2006). The PVAR model is used in various research fields but is most used by macroeconomists working with data for many countries and a long period (e.g., Koengkan & Fuinhas, 2020a; Kroop & Korobilis, 2016; and Santiago et al., 2020).

Indeed, several authors have pointed out several advantages of using this method. For example, Canova and Ciccarelli (2009) suggested that this model is an excellent way to find how shocks are transmitted across countries. Abrigo and Love (2015) emphasised that it can treat all variables as endogenous. However, restrictions based on statistical procedures may be imposed on disentangling the impact of exogenous shocks on the system. This opinion is shared by Koengkan and Fuinhas (2020b) and Santiago et al. (2020), who used this method in their investigations.

Koengkan and Fuinhas (2020a) point out that this method is adequate for panels with extended periods (macro panels); in our case, cointegration between the variables and endogeneity is presently expected. Moreover, Antonakakis et al. (2017) showed that this model has another advantage. For example, this model is helpful in the presence of little theoretical information about the relationship between the variables to guide the specification of the model; it can get around the endogeneity problem among the variables of the model, as also mentioned by Abrigo and Love (2015); it can account for any delayed effects of the variables under consideration. This model includes country-fixed effects that capture the time-invariant components that may affect the dependent variable and global time effects affecting all countries in the same period. Finally, this model can account for any global shocks that impact all countries simultaneously in the model, as pointed out by Canova and Ciccarelli (2009).

For these reasons, in this chapter, it was opted to use this method to carry out this empirical study. The PVAR model is represented by the following linear Eq. (3.1):

$$Y_t = A_{0i}(t) + A_i(l)Y_{t-j} + u_t \tag{3.1}$$

where Y_t is the vector for the six variables used in this empirical investigation (e.g., DLogOBESITY, DLogGDP_PPP, DLogURBA, DLogKOFSoGI, DLogKOFEcoGI and DLogPOVERTY). The use of variables in the first differences of natural logarithms is due to the PVAR model requiring that variables be of order one. The stationarity of variables can be confirmed by the visual analysis of descriptive statistics and the second-generation unit root test shown in Table 3.4 in this chapter. Moreover, A_{0i} is a fixed vector; $A_i(l)$ represents the polynomial matrix Y and u are the vectors of the dependent variables in a panel of fixed effects and idiosyncratic errors, respectively. The following equations below is the PVAR model:

$$\begin{aligned}DLogOBESITY = {} & A_{01}(t) + A_{11}(l)DLogObesity_{t-j} + A_{12}(l)DLogGDP_{PPPt-j} \\ & + A_{13}(l)DLogURBA_{t-j} + A_{14}(l)DLogKOFSoGI_{t-j} \\ & + A_{15}(l)DLogKOFEcoGI_{t-j} + A_{16}(l)DLogPOVERTY_{t-j} + u_{1t}\end{aligned} \tag{3.2}$$

$$\begin{aligned}DLogGDP_{PPP}P_{PPP} = {} & A_{02}(t) + A_{21}(l)DLogGDP_{PPPt-j} \\ & + A_{22}(l)DLogOBESITY_{t-j} + A_{23}(l)DLogURBA_{t-j} \\ & + A_{24}(l)DLogKOFSoGI_{t-j} + A_{25}(l)DLogKOFEcoGI_{t-j} \\ & + A_{26}(l)DLogPOVERTY_{t-j} + u_{2t}\end{aligned} \tag{3.3}$$

$$\begin{aligned}DLogURBA = {} & A_{03}(t) + A_{31}(l)DLogURBA_{t-j} + A_{32}(l)DLogOBESITY_{t-j} \\ & + A_{33}(l)DLogGDP_{PPPt-j} + A_{34}(l)DLogKOFSoGI_{t-j} \\ & + A_{35}(l)DLogKOFEcoGI_{t-j} + A_{36}(l)DLogPOVERTY_{t-j} + u_{3t}\end{aligned} \tag{3.4}$$

$$\begin{aligned}DLogKOFSoGI = {} & A_{04}(t) + A_{41}(l)DLogKOFSoGI_{t-j} + A_{42}(l)DLogOBESITY_{t-j} \\ & + A_{43}(l)DLogGDP_{PPPt-j} + A_{44}(l)DLogURBA_{t-j} \\ & + A_{45}(l)DLogKOFEcoGI_{t-j} + A_{46}(l)DLogPOVERTY_{t-j} + u_{4t}\end{aligned} \tag{3.5}$$

$$\begin{aligned}DLogKOFEcoGI = {} & A_{05}(t) + A_{51}(l)DLogKOFEcoGI_{t-j} \\ & + A_{52}(l)DLogOBESITY_{t-j} + A_{53}(l)DLogGDP_{PPPt-j} + A_{54}(l)DLogURBA_{t-j} \\ & + A_{55}(l)DLogKOFSoGI_{t-j} + A_{56}(l)DLogPOVERTY_{t-j} + u_{5t}\end{aligned} \tag{3.6}$$

$$\begin{aligned}DLogPOVERTY = {} & A_{06}(t) + A_{61}(l)DLogPOVERTY_{t-j} \\ & + A_{62}(l)DLogOBESITY_{t-j} + A_{63}(l)DLogGDP_{PPPt-j} + A_{64}(l)DLogURBA_{t-j} \\ & + A_{65}(l)DLogKOFSoGI_{t-j} + A_{66}(l)DLogKOFEcoGI_{t-j} + u_{6t}\end{aligned} \tag{3.7}$$

Before carrying out the PVAR model regression, it is necessary to perform preliminary tests and, after the model regression, the postestimation tests. To this end, this investigation will use the conceptual framework that was developed by Santiago et al. (2020) and that highlights the methodological approach (e.g., preliminary and postestimation tests) that needs to be used in PVAR models (see Fig. 3.6 below).

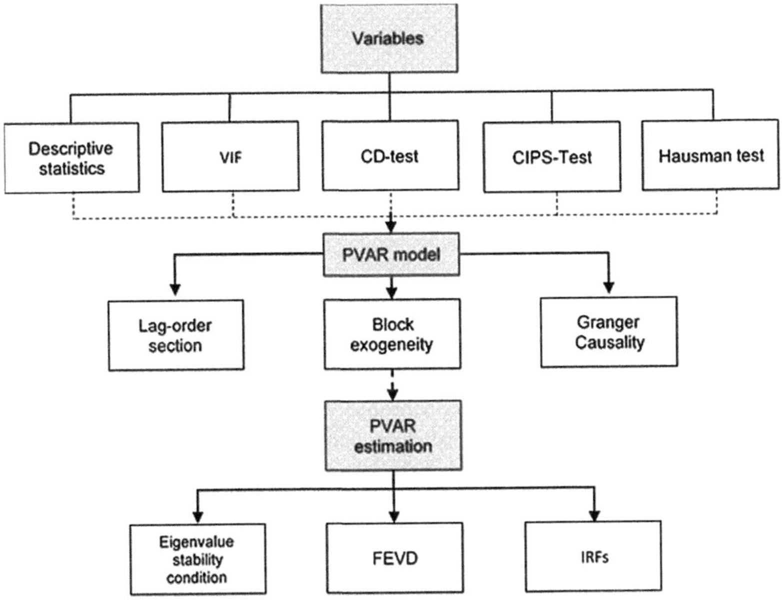

Figure 3.6 Conceptual framework of empirical research.
This figure was based on the conceptual framework developed by Santiago, R., Koengkan, M., Fuinhas, J.A., & Marques, A.C., (2020). The relationship between public capital stock, private capital stock and economic growth in the Latin American and Caribbean countries. International Review of Economics, 67:293–317. https://doi.org/10.1007/s12232-019-00340-x.

Therefore, following the conceptual framework of empirical research developed by Santiago et al. (2020), this investigation will use the following preliminary tests before the PVAR model regression: (1) **variance inflation factor (VIF)** in order to check the existence of multicollinearity between the variables in the panel data; (2) **Pesaran CD-test** in order to check the existence of cross-section dependence in the panel data; (3) **panel unit root test (CIPS-test)** in order to check the presence of unit roots in the variables; (4) **Hausman test** in order to check the presence of heterogeneity, i.e., whether the panel has random effects (RE) or fixed effects (FE); and (5) **panel VAR lag-order selection test** in order to report the overall model coefficients of determination.

Afterward, for the regression, it is necessary to compute some postestimation tests to verify the properties of the model. To this end, some diagnostics tests developed by Abrigo and Love (2015) will be calculated. For example, (6) **the eigenvalue stability condition test** in order to indicate that the panel-VAR model is stable; (7) **the panel Granger causality Wald test,** in order to analyse the causal relationship between the variables of the model; (8) **the forecast-error variance decomposition (FEVD) test** in order to show how a variable responds to shocks in specific variables; and (9) **the impulse-response function (IRF)** in order to indicate the impulse-response function of variables of the model. This section shows the data used, the method, and the preliminary and specification tests. In the next section, the results will be shown.

3.4 Empirical results

As mentioned before, this section will approach the empirical results of our investigation. Therefore, we will show the results from preliminary tests, PVAR model regression and postestimation tests. To find the level of multicollinearity between the variables in our panel data, the VIF tests developed by Belsley et al. (1980, pp. 1–286) were calculated. This test is constructed around the following Eq. (3.8):

$$VIF_i = \frac{1}{1 - R_j^2}, \tag{3.8}$$

where R_j^2 is the coefficient of determination of regression of model in step one. Table 3.2 below shows the outcomes from the VIF test.

Therefore, the results of the VIF-test show that the values are lower than the usually accepted benchmark of 10 in the case of the VIF values and six in the case of the mean VIF values. The results of the VIF-test were obtained from the command *estat vif* in **Stata 16.0.** The board below shows how to carry out and obtain the results from the VIF test.

How to do:

**** The Variance Inflation Factor test****

```
reg dlogobesity dloggdp_ppp dlogurba dlogkofsogi dlogkofecogi dlogpoverty
estat vif
```

Moreover, after performing the VIF test, it is necessary to find cross-sectional dependence (CSD) in the panel data. The Pesaran CD-test developed by Pesaran (2004) was used in this investigation. This test is constructed around the following Eq. (3.9)

Table 3.2 VIF test.

Variables	VIF	1/VIF	Mean VIF
	DLogOBESITY		
DLogGDP_PPP	1.23	0.8141	1.11
DLogURBA	1.04	0.9651	
DLogKOFSoGI	1.06	0.9436	
DLogKOFEcoGI	1.04	0.9600	
DLogPOVERTY	1.20	0.8368	

(DLog) denotes variables in the first differences of the natural logarithms.

$$CD = \sqrt{\frac{2T}{N(N-1)}}\left(\sum_{i=1}^{N-1}\sum_{j=i+1}^{N} pij\right) \tag{3.9}$$

The null hypothesis of this test is the non-presence of cross-section dependence $CD \sim N\ (0,1)$ for $N \rightarrow \infty$ and that T is sufficiently large. Table 3.3 below shows the results from the Pesaran CD test.

Table 3.3 Pesaran CD test.

Variables	CD-test	*P*-value		Corr	Abs (corr)
DLogOBESITY	18.18	0.000	***	0.439	0.456
DLogGDP_PPP	18.62	0.000	***	0.460	0.464
DLogURBA	21.40	0.000	***	0.530	0.618
DLogKOFSoGI	11.20	0.000	***	0.264	0.355
DLogKOFEcoGI	7.05	0.000	***	0.179	0.320
DLogPOVERTY	1.96	0.051	*	0.040	0.257

*** and * denote statistical significance at 1% and 10% level; (DLog) denotes variables in the first differences of the natural logarithms.

Therefore, the results of the CSD-test show the presence of cross-section dependence in all variables of our model. Indeed, the presence of cross-section dependence can signify that the countries selected in our study share the same characteristics and shocks, as mentioned by Fuinhas et al. (2017). The results of the Pesaran CD test were obtained from the command *xtcd* in **Stata 16.0**. The board below shows how to carry out and obtain the results from the Pesaran CD test.

How to do:

****The Pesaran CD-test****

```
xtcd dlogobesity dloggdp_ppp dlogurba dlogkofsogi dlogkofecogi dlogpoverty
```

Nevertheless, in the presence of CSD, it is necessary to verify the order of integration of the variables that will be used in the PVAR regression. To this end, the panel unit root (CIPS test) developed by Pesaran (2007) was calculated. This test is constructed around the following Eq. (3.10):

$$CIPS(N,T) = t{-}^{-}N^{-1}\sum_{i=1}^{N} ti(N,T) \tag{3.10}$$

where $t_i(N,T)$ is the cross-sectionally augmented Dickey-Fuller statistic for the i, the cross-section unit given by the $t-$ ratio of the coefficient of $y_{i,t-1}$ in the CADF regression. Therefore, the null hypothesis of this test is that all series have a unit root. Table 3.4 below shows the results from the CIPS test.

Table 3.4 Panel unit root test (CIPS-test).

	Panel unit root test (CIPS) (Zt-bar)				
	Without trend			With trend	
Variables	**Lags**	**Zt-bar**		**Zt-bar**	
DLogOBESITY	1	−3.618	***	0.074	
DLogGDP_PPP	1	0.279	*	1.676	
DLogURBA	1	−1.166	**	1.044	
DLogKOFSoGI	1	−5.173	***	−3.426	***
DLogKOFEcoGI	1	−3.794	***	−0.496	
DLogPOVERTY	1	−4.838	***	−0.697	

***, ** and * denote statistically significant at 1%, 5% and 10% level, respectively; (DLog) denotes variables in the first differences of the natural logarithms.

The results from the CIPS test obtained indicate that the variable DLogKOFSoGI is I(1), while the variables DLogGDP_PPP, DLogURBA, DLogOBESITY, DLogKOFEcoGI and DLogPOVERITY are on the borderline between I(0) and I(1) of the order of integration. The results of the CIPS test were obtained from the command *multipurt* in **Stata 16.0**. The board below shows how to carry out and obtain the results from the CIPS test.

How to do:

****The CIPS-test****

multipurt dlogobesity dloggdp_ppp dlogurba dlogkofsogi dlogkofecogi dlogpoverty,lags(1)

After finding the presence of the order of integration between the variables of our model, the next step of this investigation is to find the presence of individual effects in the model. To this end, the Hausman test, which compares the random (RE) and fixed effects (FE), was calculated. This test is constructed around the following Eq. (3.11):

$$H = (\beta_{FE} - \beta_{RE})'[\text{Var}\ (\beta_{FE}) - \text{Var}\ (\beta_{RE})]^{-1}(\beta_{FE} - \beta_{RE}) \sim X^2(k) \qquad (3.11)$$

where β_{RE} and β_{FE} are the vectors of coefficient estimates for the random and fixed effects model, respectively. The statistic is $x^2(k)$ distributed under the null hypothesis. The null hypothesis of this test is that the difference in coefficients is not systematic, where the random effects are the most suitable estimator (Fuinhas et al., 2017). The results of this test are presented in Table 3.5 below.

Table 3.5 Hausman test.

Variables	(b) Fixed	(B) random	(b-B) difference	Sqrt(diag(V_b-V-B)) S.E.
DLogGDP_PPP	0.0003	0.0019	−0.0015	0.0008
DLogURBA	0.1348	0.2273	−0.0925	0.0332
DLogKOFSoGI	−0.0070	−0.0035	−0.0035	0.0008
DLogKOFEcoGI	0.0011	0.0009	0.0002	0.0003
DLogPOVERTY	−0.0020	−0.0017	−0.0003	0.0004
Chi2 (5)	37.33***			

*** denotes statistically significant at 1% level; (DLog) denotes variables in the first differences of the natural logarithms.

The results of this test show that the null hypothesis should be rejected (**chi2 (5) = 37.33*****, statistically significant at the 1% level). The results from the Hausman test indicate the presence of fixed effects. Therefore, the results of the Hausman test were obtained from the command *hausman* with option *sigmaless* in **Stata 16.0**. The board below shows how to carry out and obtain the results from the Hausman test.

How to do:

The Hausman test

```
**The Hausman test**

xtreg dlogobesity dloggdp_ppp dlogurba dlogkofsogi dlogkofecogi dlogpoverty,fe
estimates store fixed

xtreg dlogobesity dloggdp_ppp dlogurba dlogkofsogi dlogkofecogi dlogpoverty,re
estimates store random

hausman fixed random, sigmaless
```

After conducting the Hausman test, it is necessary to report the overall model coefficients of determination. To this end, the PVAR lag-order selection test developed by Abrigo and Love (2015) was computed. This test is constructed around the following Eq. (3.12):

$$
\begin{aligned}
&MMSC_{BIC,n}(k,p,q) = J_n\left(k^2p,k^2q\right) - (|q|-|p|)k^2 lnn \\
&MMSC_{AIC,n}(k,p,q) = J_n\left(k^2p,k^2q\right) - 2k^2(|q|-|p|) \\
&MMSC_{HQIC,n}(p,q) = J_n\left(k^2p,k^2q\right) - Rk^2(|q|-|p|) lnlnn, R>2
\end{aligned}
\tag{3.12}
$$

where $J_n(k,p,q)$ is the J statistic of over-identifying restriction for a k-variate PVAR of order p and moment conditions based on q lags of the dependent variables with sample size n. By construction, the above MMSC is available only when $q > p$. The overall coefficient of determination (CD) may be calculated even with just-found GMM models as an alternative criterion. Suppose we denote the $(k\ x\ k)$ unconstrained covariance matrix of the dependent variables by Ψ; CD captures the proportion of variation explained by the PVAR model as (Table 3.6):

Table 3.6 below shows the results from the PVAR lag-order selection test.

Lag	CD	J	J *P*-value	MBIC	MAIC	MQIC
1	0.9959	201.3441	0.0032	−693.9621	−98.6559	−334.615
2	0.9970	164.5619	0.0101	−581.5266	−85.4381	−282.0707
3	0.9970	124.708	0.0477	−472.1628	−75.2920	−232.5981
4	0.9927	77.3320	0.4040	−370.321	−72.6679	−190.6475

The overall coefficient of determination (CD), Hansen's J statistic (J), *P*-value (Jp-value), MMSC-Bayesian information criterion (MBIC), MMSC-Akaike information criterion (MAIC), and MMSC-Hannan and Quinn information criterion (MQIC) were computed.

$$CD = 1 - \frac{det(\sum)}{det(\psi)} \tag{3.13}$$

The PVAR lag-order selection test results point to using one or two lags in the PVAR model. The results of the PVAR lag-order selection test were obtained from the command *pvarsoc* in **Stata 16.0**. The board below shows how to carry out and obtain the results from the PVAR lag-order selection.

How to do:

****The PVAR Lag-order selection test****

pvarsoc dlogobesity dloggdp_ppp dlogurba dlogkofsogi dlogkofecogi dlogpoverty, maxlag (4) pvaropts (instl(1/7))

Before carrying out the PVAR estimation, it is necessary to choose the specification that satisfies Hansen's J statistics, which tests for the overidentification criterion, which means that the specification first needs to 'pass,' this test before choosing between the lag length that minimises the MBIC, MAIC and MQIC. However, the lag length is often selected by focusing on the MAIC criterion supported by Serena and Perron (2001). The PVAR Lag-order selection concentrates on the MAIC criterion using one or two lags in the PVAR model. After carrying out the preliminary tests, the panel-VAR regression is necessary. Table 3.7 shows the results of all equations from the panel-VAR regression model. The lag lengths (1) and (2) were used.

Table 3.7 Panel-VAR model regression.

	Eq. (3.2)					
	1/7 Instrumental lags			1/7 Instrumental lags		
	1 Lag			2 Lags		
Response of DLogOBESITY to:	**Coefficient**	**Heteroskedasticity adjusted t-statistics**		**Coefficient**	**Heteroskedasticity adjusted t-statistics**	
DLogGDP_PPP	0.0041	2.21	**	0.0011	0.26	
DLogURBA	1.0272	38.64	***	0.8269	14.68	***
DLogKOFSoGI	0.0324	13.70	***	0.0407	6.72	***
DLogKOFEcoGI	0.0079	9.65	***	0.0199	7.12	***
DLogPOVERTY	0.0014	3.69	***	0.0017	1.23	
Response of DLogGDP_PPP to:	Eq. (3.3)					
DLogOBESITY	6.2732	14.86	***	7.001	9.51	***
DLogURBA	−7.2643	−15.30	***	−11.3963	−12.95	***
DLogKOFSoGI	−0.3462	−8.08	***	−0.1090	−1.25	
DLogKOFEcoGI	0.2314	12.87	***	0.6161	9.37	***
DLogPOVERTY	−0.0312	−4.46	***	−0.0482	−1.96	**
Response of DLogURBA to:	Eq. (3.4)					
DLogOBESITY	0.2040	15.24	***	0.1080	6.31	***
DLogGDP_PPP	0.0025	4.44	***	0.0115	9.26	***
DLogKOFSoGI	−0.0060	−5.57	***	−0.0118	−4.99	***
DLogKOFEcoGI	−0.0011	−2.74	***	−0.0040	−2.87	***
DLogPOVERTY	0.0011	6.87	***	0.0043	7.07	***
Response of DLogKOFSoGI to:	Eq. (3.5)					
DLogOBESITY	3.0082	14.24	***	3.6201	7.44	***
DLogGDP_PPP	0.1321	11.97	***	0.1420	3.47	***
DLogURBA	−1.0183	−4.45	***	−2.4361	−4.66	***

Continued

Table 3.7 Panel-VAR model regression.—cont'd

	Eq. (3.2)					
	1/7 Instrumental lags			1/7 Instrumental lags		
	1 Lag			2 Lags		
Response of DLogOBESITY to:	**Coefficient**	**Heteroskedasticity adjusted t-statistics**		**Coefficient**	**Heteroskedasticity adjusted t-statistics**	
DLogKOFEcoGI	0.0209	2.66	***	0.0815	2.65	***
DLogPOVERTY	0.0034	1.37		0.0261	2.64	***
Response of DLogKOFEcoGI to:	Eq. (3.6)					
DLogOBESITY	3.7376	10.73	***	4.5443	5.77	***
DLogGDP_PPP	0.0345	1.01		−0.1511	−1.61	
DLogURBA	3.7179	11.04	***	3.3260	3.09	***
DLogKOFSoGI	−0.9583	−16.43	***	−1.1126	−8.16	***
DLogPOVERTY	−0.0388	−4.92	***	−0.1480	−6.79	***
Response of DLogPORVERTY to:	Eq. (3.7)					
DLogOBESITY	11.9232	16.44	***	−1.1458	−0.52	***
DLogGDP_PPP	−0.7564	−7.34	***	−0.2339	−1.02	
DLogURBA	−14.5800	−16.43	***	−5.9776	−2.56	**
DLogKOFSoGI	0.3858	2.92	***	0.2501	0.67	
DLogKOFEcoGI	0.1168	2.24	**	0.4501	2.15	**
	Test of overidentifying restriction: Hansen's J					
	Chi2 (206) = 216.503			**Chi2(180)** = 180.449		

*** and ** denote statistical significance level at 1% and 5%, respectively; Instruments: 1 (1/7) and 2 (1/7) were used; (DLog) denotes variables in the first differences of the natural logarithms.

The results of PVAR regression using 1/7 lags on the instruments and one lag included in the dependent variables list point to endogeneity in the variables. Moreover, the results from Eq. (3.2) show that variables DLogGDP_PPP, DLogURBA, DLogKOFSoGI, DLogKOFEcoGI and DLogPOVERTY, exert a positive effect on variable DLogOBESITY (that is, an increase in the obesity epidemic problem). In Eq. (3.3), the variables DLogOBESITY and DLogKOFEcoGI, exert a positive impact on variable DLogGDP_PPP (that is, an increase in economic growth), while the variables DLogURBA, DLogKOFSoGI and DLogPOVERTY, exert a negative effect (that is, a decrease in economic growth). In Eq. (3.4), the variables DLogOBESITY, DLogGDP_PPP and DLogPOVERTY exert a positive impact on variable DLogURBA (that is, an increase in the process of urbanisation), while the variables DLogKOFSoGI and DLogKOFEcoGI, exert a negative effect (that is, a decrease in the process of urbanisation). In Eq. (3.5), DLogOBESITY, DLogGDP_PPP and DLogKOFEcoGI positively impact variable DLogKOFSoGI (that is, an increase in the process of social globalisation), while the variable DLogURBA exerts a negative impact. In Eq. (3.6), the variables DLogOBESITY and DLogURBA exerts a positive impact on variable DLogKOFEcoGI (that is, an increase in the process of economic globalisation), while the variables DLogKOFSoGI and DLogPOVERTY exert a negative impact (that is, a decrease in the process of economic globalisation). Moreover, in Eq. (3.7), DLogOBESITY, DLogKOFSoGI and DLogKOFEcoGI positively impact DLogPOVERTY and the variables DLogGDP_PPP and DLogURBA exert a negative effect.

The lagged variables in all PVAR equations are at least statistically significant at 1% and 5% levels. Moreover, the use of 1/7 lags on the instruments and one lag produces a Hansen's J statistic connected with a chi2(180) = 180.449, which in this context refers to the case where we cannot reject the null hypothesis, which states that the overidentification restrictions are not valid, thus making the specification valid.

Moreover, the results of PVAR regression using 1/7 lags on the instruments and two lags included in the dependent variables list point to endogeneity in the variables. Indeed, the results from Eq. (3.2) show that the variables DLogURBA, DLogKOFSoGI and DLogKOFEcoGI, exert a positive impact on variable DLogOBESITY. Eq. (3.3) indicates that DLogOBESITY and DLogKOFEcoGI positively impact variable DLogGDP_PPP, while the variables DLogURBA and DLogPOVERTY exert a negative impact. In Eq. (3.4), DLogOBESITY, DLogGDP_PPP and DLogPOVERTY positively affect variable DLogURBA, while the variables DLogKOFSoGI and DLogKOFEcoGI exert a negative impact. In Eq. (3.5), DLogOBESITY, DLogGDP_PPP and DLogKOFEcoGI exert a positive effect on variable DLogKOFSoGI and variable DLogURBA exerts a negative effect. In Eq. (3.6), DLogOBESITY and DLogURBA variables positively affect the variable DLogKOFEcoGI, while the variables DLogKOFSoGI and DLogPOVERTY exert a negative effect. In Eq. (3.7), DLogOBESITY and DLogKOFEcoGI positively impact variable DLogPOVERTY, while the variable DLogURBA exerts a negative impact.

The lagged variables in all PVAR equations are at least statistically significant at 1% and 5% levels. Moreover, the use of 1/7 lags on the instruments and two lags produces a Hansen's J statistic connected with a **chi2(206)** = 216.503, which in this context refers to the case where we can with confidence reject the null hypothesis, which states that the overidentification restrictions are valid, thus making the specification invalid. For this reason, it was opted to use the PVAR regression using 1/7 lags on the instruments and one lag. The results of the PVAR model regression were obtained from the command *pvar* in **Stata 16.0**. The board below shows how to carry out and obtain the results from the PVAR model regression.

How to do:

```
**Panel-VAR model regression**

pvar dlogobesity dloggdp_ppp dlogurba dlogkofsogi dlogkofecogi dlogpoverty,
lags(1) instl (1/7) gmmst overid

pvar dlogobesity dloggdp_ppp dlogurba dlogkofsogi dlogkofsogi dlogpoverty,
lags(2) instl (1/7) gmmst overid
```

As mentioned before, after the model regression, it is necessary to compute some postestimation tests to verify the proprieties of the model. To this end, some diagnostic tests developed by Abrigo and Love (2015) will be computed. The first postestimation test to be computed is the eigenvalue stability condition test. This will be used to find if our PVAR model using 1/7 lags on the instruments and one lag is stable or not. Table 3.8 below shows the results from the eigenvalue stability condition test.

Table 3.8 Eigenvalue stability condition test.

Eigenvalue			Graph
Real	**Imaginary**	**Modulus**	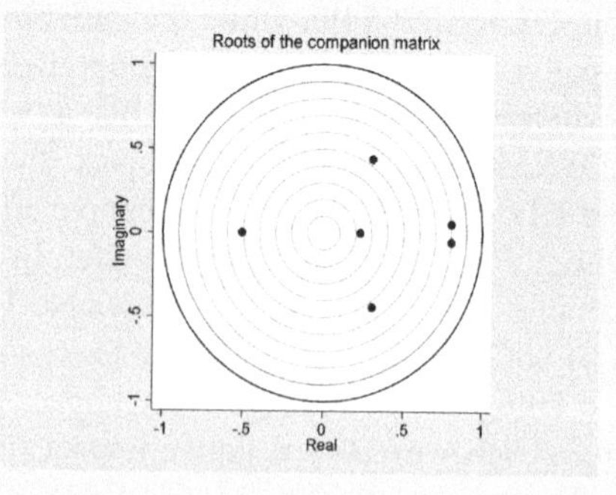
0.7994	−0.0541	0.8013	
0.7994	0.0541	0.8013	
0.3023	0.4407	0.5344	
0.3023	−0.4407	0.5344	
−0.5076	0.0000	0.5076	
0.2265	0.0000	0.2265	

The eigenvalue test shows that the PVAR model is stable because all eigenvalues are inside the unit circle, satisfying the stability condition of the test. The eigenvalue stability condition test results were obtained from the command of *pvarstable* with an option *graph* in **Stata 16.0.** The board below shows how to carry out and obtain the results from the eigenvalue test.

How to do:

****Eigenvalue stability condition test ****

```
pvar dlogobesity dloggdp_ppp dlogurba dlogkofsogi dlogkofecogi dlogpoverty,
lags(1) instl (1/7) gmmst overid

pvarstable, graph
```

After the eigenvalue stability condition test, it is necessary to carry out the panel Granger causality Wald test to analyse the causal relationship between the variables in the PVAR model. Table 3.9 below gives the results from this test.

The results from the panel Granger causality Wald test indicate the presence of a bidirectional relationship between the variables DLogOBESITY and DLogGDP_PPP, DLogOBESITY and DLogURBA, DLogOBESITY and DLogKOFSoGI, DLogOBESITY and DLogKOFEcoGI, DLogOBESITY and DLogPOVERTY, DLogGDP_PPP and DLogURBA, DLogGDP_PPP and DLogKOFSoGI, DLogGDP_PPP and DLogPOVERTY, DLogURBA and DLogKOFSoGI, DLogURBA and DLogKOFEcoGI, DLogURBA and DLogPOVERTY, DLogKOFSoGI and DLogKOFEcoGI, DLogKOFEcoGI and DLogPOVERTY. Moreover, there exists a unidirectional relationship between DLogKOFEcoGI to DLogGDP_PPP and DLogKOFSoGI to DLogPOVERTY. The results from the panel Granger causality Wald test were obtained from the command of *pvargranger* in **Stata 16.0.** The board below shows how to carry out and obtain the results from this test.

How to do:

**** Panel Granger causality Wald test ****

```
pvar dlogobesity dloggdp_ppp dlogurba dlogkofsogi dlogkofecogi dlogpoverty,
lags(1) instl (1/7) gmmst overid

pvargranger
```

Fig. 3.7 summarises the causalities between the variables. This figure was based on results from the panel Granger causality Wald test (see Table 3.9) and the results of PVAR estimation (see Table 3.7).

Table 3.9 Panel Granger causality Wald test.

Equation	Excluded	Chi2	Df	Prob >chi2	
DLogOBESITY	DLogGDP_PPP	4.898	1	0.027	**
	DLogURBA	1493.290	1	0.000	***
	DLogKOFSoGI	187.822	1	0.000	***
	DLogKOFEcoGI	93.057	1	0.000	***
	DLogPOVERTY	13.640	1	0.000	***
	All	1797.730	5	0.000	***
DLogGDP_PPP	DLogOBESITY	220.960	1	0.000	***
	DLogURBA	234.001	1	0.000	***
	DLogKOFSoGI	65.223	1	0.000	***
	DLogKOFEcoGI	165.534	1	0.000	***
	DLogPOVERTY	19.847	1	0.000	***
	All	487.519	5	0.000	***
DLogURBA	DLogOBESITY	232.170	1	0.000	***
	DLogGDP_PPP	19.750	1	0.000	***
	DLogKOFSoGI	31.065	1	0.000	***
	DLogKOFEcoGI	7.493	1	0.000	***
	DLogPOVERTY	47.237	1	0.000	***
	All	317.731	5	0.000	***
DLogKOFSoGI	DLogOBESITY	202.860	1	0.000	***
	DLogGDP_PPP	143.177	1	0.000	***
	DLogUBRA	19.823	1	0.000	***
	DLogKOFEcoGI	7.051	1	0.000	***
	DLogPOVERTY	1.86	1	0.172	
	All	959.127	5	0.000	***
DLogKOFEcoGI	DLogOBESITY	115.227	1	0.000	***
	DLogGDP_PPP	1.019	1	0.313	
	DLogUBRA	121.949	1	0.000	***
	DLogKOFSoGI	269.820	1	0.000	***
	DLogPOVERTY	24.188	1	0.000	***
	All	1364.576	5	0.000	***
DLogPOVERTY	DLogOBESITY	270.270	1	0.000	***
	DLogGDP_PPP	53.892	1	0.000	***
	DLogUBRA	269.864	1	0.000	***
	DLogKOFSoGI	8.515	1	0.000	***
	DLogKOFEcoGI	5.020	1	0.025	**
	All	328.563	5	0.000	***

*** and ** denote statistical significance levels at 1% and 5%, respectively; (DLog) denotes variables in the first differences of the natural logarithms.

After the panel Granger causality Wald test, it is necessary to the compute the forecast-error variance decomposition (FEVD), which shows that one period after the shock, the variables themselves explained all the forecast error variance. Table 3.10 below shows the results from the FEVD test.

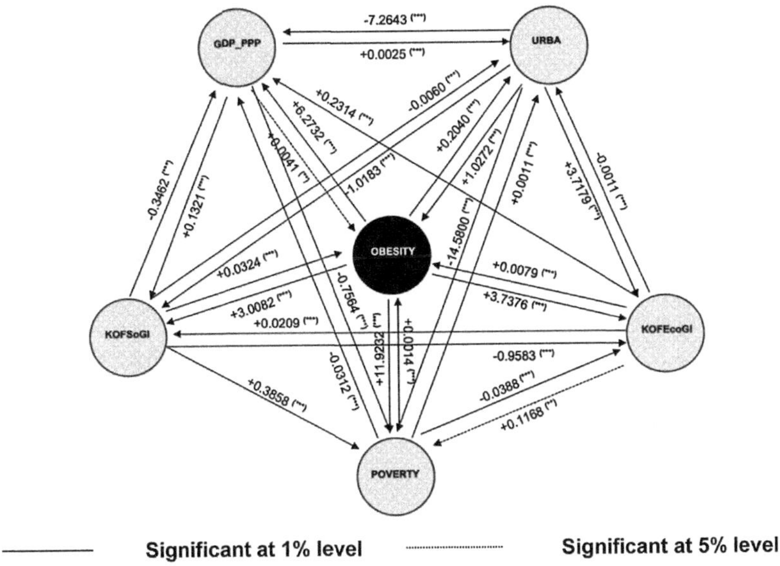

Figure 3.7 Summary of causality of the variables.

Therefore, Eq. (3.2) shows that two periods after a shock in the DLogOBESITY, the variable explains 100% forecast error variance. Two periods after a shock, the variable DLogKOFSoGI, explains 3.6% of forecast error variance, while the variable DLogKOFEcoGI explains 1.42%. Fifteen periods after a shock, the variable DLogGDP_PPP explains 5.41% of forecast error variance, while the variable DLogURBA explains 31%, and the variable DLogPOVERTY 0.71%. Eq. (3.3) shows that one period after a shock in the variable DLogGDP_PPP explains 99% of forecast error variance. Two periods after a shock, the variable DLogOBESIT explains 76.78% of forecast error variance, while the variable DLogKOFSoGI explains 3.61% and DLogKOFEcoGI 1.42% of forecast error variance. Moreover, 15 periods after a shock, the variable DLogURBA explains 31%, while the variable DLogPOVERTY explains 0.71% of forecast error variance.

Eq. (3.4) shows that after a shock in the variable DLogURBA, one period explains 94% of forecast error variance. Two periods after a shock, the variable DLogKOFSoGI explains 0.47% of forecast error variance. Fifteen periods after a shock, the variable DLogOBESITY explains 12% of forecast error variance, while the variable DLogGDP_PPP explains 6%, DLogKOFEcoGI explains 0.4%, and DLogPOVERTY explains 1.68% of forecast error variance. Eq. (3.5) points out that one period after a shock the DLogKOFSoGI explains 87% of forecast error variance. Moreover, 15 periods after a shock, the variable DLogOBESITY explains 13% of forecast error variance, while the variables DLogGDP_PPP explains 21%, DLogURBA explains 6%, DLogKOFEcoGI explains 1.8% and DLogPOVERTY explains 0.1% of forecast error variance.

Table 3.10 Forecast-error variance decomposition (FEVD).

Response variable and forecast horizon	Impulse variables					
	DLogOBESITY	**DLogGDP_PPP**	**DLogURBA**	**DLogKOFSoGI**	**DLogKOFEcoGI**	**DLogPOVERTY**
DLogOBESITY						
1	1	0	0	0	0	0
2	0.7678	0.0432	0.1356	0.0361	0.0142	0.0027
5	0.6888	0.0513	0.2107	0.0305	0.0140	0.0043
10	0.6152	0.0533	0.2851	0.0267	0.0132	0.0063
15	0.5895	0.0541	0.3111	0.0252	0.0127	0.0071
DLogGDP_PPP						
1	0.0060	0.9939	0	0	0	0
2	0.7678	0.0432	0.1356	0.0361	0.0142	0.0027
5	0.6888	0.0513	0.2107	0.0305	0.0140	0.0043
10	0.6152	0.0533	0.2851	0.0267	0.0132	0.0063
15	0.5895	0.0541	0.3111	0.0252	0.0127	0.0071
DLogURBA						
1	0.0085	0.0530	0.9384	0	0	0
2	0.0990	0.0502	0.8371	0.0047	0.0007	0.0081
5	0.1108	0.0572	0.8111	0.0026	0.0031	0.0148
10	0.1211	0.0602	0.7961	0.0020	0.0039	0.0164
15	0.1236	0.0610	0.7924	0.0018	0.0041	0.0168
DLogKOFSoGI						
1	0.0012	0.1282	1.63e-08	0.8704	0	0
2	0.1085	0.1852	0.0012	0.7032	0.0015	0.0002
5	0.1348	0.2143	0.0186	0.6129	0.0189	0.0002
10	0.1349	0.2081	0.0475	0.5898	0.0186	0.0007
15	0.1350	0.2058	0.0590	0.5804	0.0184	0.0010

DLogKOFEcoGI						
1	0.0145	0.0093	0.0295	0.0163	0.9302	0
2	0.0483	0.0126	0.0295	0.1524	0.7500	0.0069
5	0.0702	0.0191	0.0614	0.1634	0.6791	0.0066
10	0.0708	0.0194	0.0681	0.1619	0.6726	0.0070
15	0.0711	0.0196	0.0711	0.1612	0.6697	0.0070
DLogPOVERTY						
1	0.0245	0.0468	0.0070	0.0072	0.0125	0.9016
2	0.0978	0.0978	0.0173	0.0097	0.0130	0.7641
5	0.1227	0.0933	0.0197	0.0272	0.0135	0.7233
10	0.1238	0.0934	0.0199	0.0273	0.0135	0.7219
15	0.1238	0.0934	0.0200	0.0273	0.0135	0.7218

(DLog) denotes variables in the first differences of the natural logarithms.

In Eq. (3.6), one period after a shock, the DLogKOFEcoGI explains 93% of forecast error variance. Moreover, 15 periods after a shock, the variable DLogOBESITY explains 7.1%, while the variables DLogGDP_PPP explains 2%, DLogURBA explains 7.1%, DLogKOFSoGI explains 16% and DLogPOVERTY 0.7% of forecast error variance. Eq. (3.7) points out that one period after a shock, the variable DLogPOVERTY explains 90% of forecast error variance. Fifteen periods after a shock, the variable DLogOBESITY explains 12%, the variable DLogGDP_PPP explains 9%, the variable DLogURBA explains 2%, the variable DLogKOFSoGI explains 3% and the variable DLogKOFEcoGI explains 1.3% of forecast error variance.

The results from the FEVD test were obtained from the command *pvarfevd* with options *mc(1000) st(15)* in Stata 16.0. The board below shows how to carry out and obtain the results from this test.

How to do:

**** Forecast-error variance decomposition (FEVD)****

```
pvar dlogobesity dloggdp_ppp dlogurba dlogkofsogi dlogkofecogi dlogpoverty,
lags(1) instl (1/7) gmmst overid

pvarfevd, mc(1000) st(15)
```

After the FEVD tests, it is necessary to show the impulse-response function of variables of the model. To this end, we computed the impulse-response functions. Fig. 3.8 below shows the impulse-response functions of variables used in this model.

Overall, all variables converge to equilibrium, supporting that the model's variables are I (1). The impulse-response functions are thus in concordance with the FEDV test. The results from the impulse-response functions test were obtained from the command *pvarirf* with options *mc(1000) oirf byopt(yrescale) st(15)* in **Stata 16.0**. The board below shows how to carry out and obtain the results from this test.

How to do:

**** Impulse – response functions****

```
pvar dlogobesity dloggdp_ppp dlogurba dlogkofsogi dlogkofecogi dlogpoverty,
lags(1) instl (1/7) gmmst overid

pvarirf, mc(1000) oirf byopt(yrescale) st(15)
```

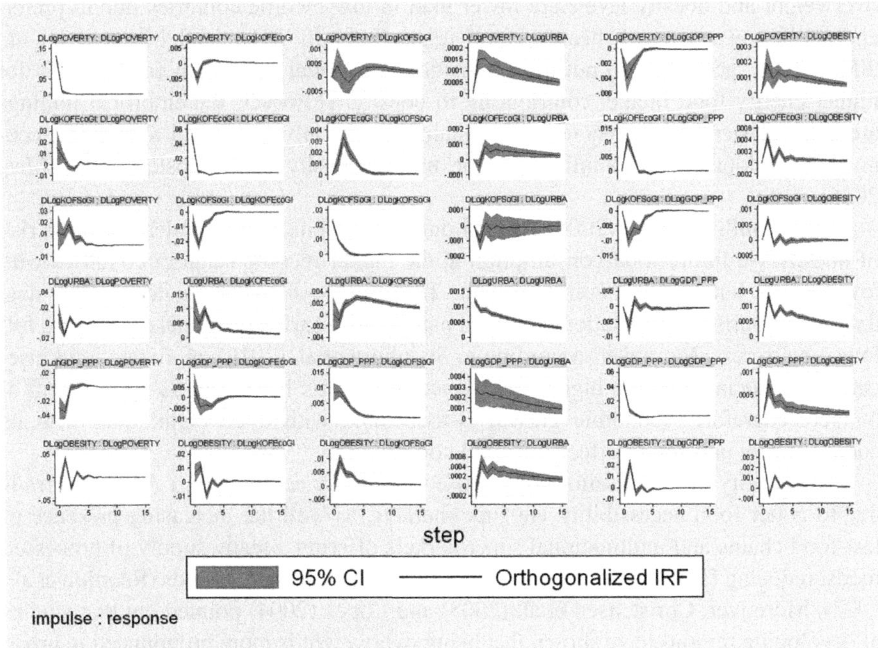

Figure 3.8 Impulse-response functions (IRFs).

In this section, we presented the empirical results of this chapter. The following subsection will present the discussions of the results found.

3.5 Discussion

In this section, we discuss the possible explanations for the results presented in Eq. (3.2). The capacity of economic growth to increase the obesity epidemic in LAC countries is related to the dietary changes caused by economic development (Gerbens-Leenes et al., 2010). According to Drenowski and Specter (2004), Gerben-Leenes et al. (2010), Koengkan and Fuinhas (2021), and Roskam et al. (2010), the nutrition transition approach suggests that initially, lower socio-economic status groups shift toward energy-dense, animal-source, fatty foods adding to the energy consumed and thus contributing to overweight and obesity levels, except for countries where home production of food is prevalent. In developed and upper-middle-income economies, overweight is a problem in lower socio-economic classes, but in these countries, overweight and obesity levels are lower than in low-income countries due to preference shift towards more healthy food (e.g., Butzlaff, 2016; and Monteiro et al., 2005). Thus, according to nutrition transition, the initial income increases allow for higher energy food intake, contributing to obesity. However, the empirical findings are not consistent. Lower economic inequality, commonly associated with higher economic development, also influences the higher obesity rates (Costa-Font & Mas, 2016).

Indeed, Sullivan et al. (2008) pointed out that food insecurity underscores the risk of obesity via future food consumption at the higher income stage in countries with low income but high economic growth. The origins of the mismatch theory state that undernourished populations follow a pattern. The early nutritional deficits are followed by excess. Moreover, according to Swinburn et al. (2019), moderate food insecurity is associated with a higher prevalence of obesity, but not in its severe form of hunger. Therefore, economic growth is associated with overweight, both with its contemporary and future effects on food consumption.

The capacity of urbanisation to increase the obesity epidemic in LAC countries is due to better food accessibility via supermarkets, as well the increasing presence of fast-food chains and multinational supermarkets offering a ready supply of processed foods, reducing farm stands and open markets with more healthy foods (Reardon et al., 2003). Moreover, Christensen et al. (2008) and Lopez (2004) pointed out that studies of developing regions have shown that being overweight is more pronounced in urban areas for both genders. Hawkes (2008) and Toiba et al. (2015) added that urban sprawl contributes to overweight and obesity levels in food desert areas in urban territories, and food marketing can aid overweight. Supermarkets lead to more inequality in food accessibility, aiding overweight. Hawkes (2006) pointed out that food marketing activities in urban areas lead to people's exposure to mass media marketing of food and beverages that can influence traditional diets. Thus, the distribution and marketing of food in urban areas may lead to overweight and obesity.

However, Kjellstrom et al. (2007) and Lindstrom (2008) had a different opinion about the capacity of urbanisation to increase the obesity problem in LAC countries. According to the authors, in urban areas, it is required less food energy expenditure related to commuting and leisure activities. More travelling by car and less walking or biking for transportation or leisure contribute to overweight and obesity. Moreover, Brug et al. (2012) and Pirgon and Aslan (2015) added that in densely populated neighborhoods, there is less recreational space for outdoor activities and more leisure time spent sitting and in screen-viewing leisure activities, also positively affecting overweight.

Additionally, Bell et al. (2002), Fox et al. (2019), and Kjellstrom et al. (2007), pointed that the process of urbanisation is also related to a higher proportion of manufacturing and service sector jobs, which translates into less energy expended in daily work activities for individuals, which consequently reduces the creation of active jobs, such as farming. The urbanisation process reveals a natural process to increase caloric food consumption, cut down the use of calories and increase people's weight.

The capacity of economic globalisation to increase the obesity epidemic in LAC countries is related to its propensity to change the food systems. Therefore, the food value chain extension has made scale-economy produce processed food and a diet richer in energy-dense foods, and high in sugar and salt. This kind of food which is less expensive and thus more accessible to lower-income classes, offering a ready supply of processed foods by multinational food corporations, fast-food chains and multinational supermarket chains (e.g., Fox et al., 2019; and Popkin, 1998). On the demand side of food systems, the socio-economic dimension of globalisation has increased time constraints resulting in less homemade food, contributing to overweight.

Moreover, economic globalisation affects energy expenditure, where the process of globalisation allows the penetration of new technologies that require less physical activity of people via labor-saving innovations in industrial sectors, more accessible home appliances and motorised transportation (e.g., Bell et al., 2002; and Sobal, 2001). The social aspect primarily promotes inexpensive transportation, communication and other activity-sparing systems (Sobal, 2001).

Indeed, the capacity of social globalisation to increase the obesity epidemic is related to the capacity of this process to influence the adoption of a more robust fast-food/processed food culture through McDonaldisation or Cocalisation processes, leading to greater caloric consumption (energy) and consequently an increase in weight gain (Fox et al., 2019). Indeed, this Westernisation caused by social globalisation can also influence the use of inexpensive transportation, communication and other activity-sparing systems through automobiles and household appliances that minimise physical activity levels (Sobal, 2001).

Finally, the capacity of poverty to increase the obesity epidemic in LAC countries is related to the dietary changes caused by economic development as mentioned above, where the reduction of poverty and increase in income allowed the families to consume fast food and ultra-processed foods that are easier and cheaper, caused by urbanisation and globalisation processes. This idea is shared by Costa-Font and Mas (2016), to whom the initial income increases allow for higher energy food intake and contributes to obesity.

3.6 Conclusion

This chapter approached the interactions between the obesity epidemic, economic growth, globalisation, urbanisation and poverty in the 23 countries from the LAC region, i.e., Argentina, Belize, Bolivia, Brazil, Chile, Colombia, Costa Rica, Dominican Republic, Ecuador, El Salvador, Guatemala, Guyana, Haiti, Honduras, Jamaica, Mexico, Nicaragua, Panama, Paraguay, Peru, Trinidad and Tobago, Uruguay and Venezuela over the period between 1990 and 2016. The panel vector autoregression (PVAR) model was the method used.

The results from the preliminary tests indicated the presence of low multicollinearity, where the VIF-test showed that the values are lower than the usually accepted benchmark of 10, in the case of the VIF values and six in the case of the mean VIF values. The results of the CSD-test showed the presence of cross-section dependence in all variables of our model. Indeed, the presence of cross-section dependence can signify that the countries selected in our study share the same characteristics and shocks. The results from the CIPS-test obtained indicated that the variable DLogKOFSoGI is I(1), while the variables DLogGDP_PPP, DLogURBA, DLogOBESITY, DLogKOFEcoGI and DLogPOVERITY are on the borderline between I(0) and I(1) of the order of integration. The results from the Hausman test indicate the presence of fixed effects. Moreover, the PVAR lag-order selection test results pointed out the use of one or two lags in the panel-VAR model.

The results from the PVAR regression that used the 1/7 lags on the instruments and one lag included in the dependent variables list pointed to the existence of endogeneity in the variables. Moreover, the results from Eq. (3.2) show that variables DLogGDP_PPP, DLogURBA, DLogKOFSoGI, DLogKOFEcoGI and DLogPOVERTY exert a positive effect on variable DLogOBESITY (that is, an increase in the obesity epidemic problem). In Eq. (3.3), the variables DLogOBESITY and DLogKOFEcoGI exert a positive impact on variable DLogGDP_PPP (that is, an increase in economic growth), while the variables DLogURBA, DLogKOFSoGI and DLogPOVERTY exert a negative effect (that is, a decrease in economic growth). In Eq. (3.4), the variables DLogOBESITY, DLogGDP_PPP and DLogPOVERTY exert a positive impact on the variable DLogURBA (that is, an increase in the process of urbanisation), while the variables DLogKOFSoGI and DLogKOFEcoGI exert a negative effect (that is, a decrease in the process of urbanisation). In Eq. (3.5), DLogOBESITY, DLogGDP_PPP and DLogKOFEcoGI positively impact the variable DLogKOFSoGI (that is, an increase in the process of social globalisation), while the variable DLogURBA exerts a negative effect. In Eq. (3.6), the variables DLogOBESITY and DLogURB exert a positive impact on the variable DLogKOFEcoGI (that is, an increase in the process of economic globalisation), while the variables DLogKOFSoGI and DLogPOVERTY exert a negative effect (that is, a decrease in the process of economic globalisation). Moreover, in Eq. (3.7), DLogOBESITY, DLogKOFSoGI and DLogKOFEcoGI positively affect the variable DLogPOVERTY, while the variables DLogGDP_PPP and DLogURBA exert a negative effect.

The posts estimation tests indicated that the model is stable, where the eigenvalue test showed that the PVAR model is stable because all eigenvalues are inside the unit circle, satisfying the stability condition of the test. The results from the panel Granger causality Wald test indicated the presence of a bidirectional relationship between the variables DLogOBESITY and DLogGDP_PPP, DLogOBESITY and DLogURBA, DLogOBESITY and DLogKOFSoGI, DLogOBESITY and DLogKOFEcoGI, DLogOBESITY and DLogPOVERTY, DLogGDP_PPP and DLogURBA, DLogGDP_PPP and DLogKOFSoGI, DLogGDP_PPP and DLogPOVERTY, DLogURBA and DLogKOFSoGI, DLogURBA and DLogKOFEcoGI, DLogURBA and DLogPOVERTY, DLogKOFSoGI and DLogKOFEcoGI and between

DLogKOFEcoGI and DLogPOVERTY. Moreover, there is a unidirectional relationship from DLogKOFEcoGI to DLogGDP_PPP and DLogKOFSoGI to DLogPOVERTY. Furthermore, the IRF test indicated that all variables converge to equilibrium, supporting that the model's variables are I (1).

This chapter has shown how obesity emerges from the nutritional transition and how it is interrelated with economic growth, urbanisation, poverty and globalisation, which have been essential features of these countries since the 1980s. Not only did we find causality from economic growth, urbanisation, poverty and globalisation toward obesity, but also there are other implicit and sustained intercausality relations between those determining phenomena which reinforce the effect on obesity. From a public health perspective, this web of causality relations implies that public interventions need to account for secondary or domino effects. So, for instance, it may not be enough to minimise urban poverty to control the increase of obesity in a country because other factors will be pushing up obesity levels. Nevertheless, any intervention on poverty affects other factors such as economic growth and urbanisation, which will impact the growing levels of obesity. From a development perspective, this chapter has shown that economic growth has a negative effect and negative externalities, such as increased obesity, which translates into more health expenditures in the future and lower productivity levels. In this way, development policies need to account for this type of negative effect so that society fully collects the benefits from economic growth.

References

Abrigo, M. R. M., & Love, I. (2015). *Estimation of Panel Vector Autoregression in Stata: A Package of Programs*. The university of Hawai`i at Mānoa Department of Economics Working Paper Series http://www.economics.hawaii.edu/research/working.papers/WP 16-02.pdf

Aizenman, J. (2005). *Financial liberalisations in Latin America in the 1990s: A reassessment*. NBER working paper series, 11145 https://www.nber.org/papers/w11145.

Albala, C., Vio, F., Kain, J., & Uauy, R. (2001). Nutrition transition in Latin America: The case of Chile. *Nutrition Reviews, 59*(6), 170–176. https://doi.org/10.1111/j.1753-4887.2001.tb07008.x

Álvarez-Castaño, L. S., Goez-Rueda, J. D., & Carreño-Aguirre, C. (2012). Factores sociales y económicos asociados a la obesidad: Los efectos de la inequidad y de la pobreza. *Revista Gerencia y Políticas de Salud, 11*(23), 98–110.

Alves, F. V., & Magalhães, R. (2019). Social inequalities, poverty and obesity. In S. G. Taukeni (Ed.), *Psychology of health – biopsychosocial approach*. IntechOpen. https://doi.org/10.5772/intechopen.80987

Amarasinghe, A. G., D'Souza, C., Brown, H. O., & Borisova, T. (2009). The influence of socioeconomic and environmental determinants on health and obesity: A West Virginia case study. *International Journal of Environmental Research and Public Health, 6*(8), 2271–2287.

Antonakakis, N., Chatziantoniou, L., & Filis, G. (2017). Energy consumption, CO_2 emissions, and economic growth: An ethical dilemma. *Renewable and Sustainable Energy Reviews, 68*(1), 808–824. https://doi.org/10.1016/j.rser.2016.09.105

Arsht, A. (2014). *Urbanization in Latin America*. Atlantic Council. https://www.atlanticcouncil.org/commentary/article/urbanization-in-latin-america/#:~:text=Today%2C%20about%2080%20percent%20of,the%20world's%20most%20urbanized%20region.&text=Today%2C%20260%20million%20people%20live,percent%20of%20Latin%20America's%20GDP.

Aydin, M. (2019). The effect of economic growth on obesity for the most obese countries: New evidence from the obesity Kuznets curve. *European Journal Health Economics, 20*, 1349–1358. https://doi.org/10.1007/s10198-019-01099-2

Balakrishnan, R., & Toscani, F. (2018). *How the commodity boom helped tackle poverty and inequality in Latin America*. IMFBlog. https://blogs.imf.org/2018/06/21/how-the-commodity-boom-helped-tackle-poverty-and-inequality-in-latin-america/.

Bell, A. C., Ge, K., & Popkin, B. M. (2002). The road to obesity or the path to prevention: Motorized transportation and obesity in China. *Obesity Research, 10*, 277–283. https://doi.org/10.1038/oby.2002.38

Belsley, D. A., Kuh, E., & Welsch, E. R. (1980). *Regression diagnostics: Identifying influential data and sources of collinearity*. Wiley. https://doi.org/10.1002/0471725153

Brug, J., van Stralen, M. M., Te-Velde, S. J., Chinapaw, M. J. M., De Bourdeaudhuij, I., Lien, N., Bere, E., Maskini, V., Singh, A. S., Maes, L., Moreno, L., Jan, N., Kovacs, E., Lobstein, T., & Manios, Y. (2012). Differences in weight status and energy-balance related behaviors among schoolchildren across Europe: The ENERGY-project. *PLoS ONE, 7*(4). https://doi.org/10.1371/journal.pone.0034742

Butzlaff, I. (2016). BMI growth rates and the nutrition transition: The role of income, inequality and income growth in Russia. *Global Food Discussion Papers, 81*, 1–42. http://hdl.handle.net/10419/129507.

Canova, F., & Ciccarelli, M. (2009). Estimating multi-country VAR models. *International Economic Review, 50*, 929–959. https://doi.org/10.1111/j.1468-2354.2009.00554.x

Cecchini, S. (2017). *Reducing poverty amidst high levels of inequality: Lessons from Latin America and the Caribbean*. Economic Commission for Latin America and the Caribbean (ECLAC). https://www.un.org/development/desa/dspd/wp-content/uploads/sites/22/2017/04/Simone-Cecchini-Reducing-poverty-amidst-high-levels-of-inequality.pdf.

Christensen, D. L., Eis, J., Hansen, A. W., Larsson, M. W., Mwaniki, D. L., & Kilonzo, B. (2008). Obesity and regional fat distribution in Kenyan populations: Impact of ethnicity and urbanisation. *Annals of Human Biology, 35*(2), 232–249. https://doi.org/10.1080/03014460801949870

Costa-Font, J., & Mas, N. (2016). 'Globesity'? The effects of globalisation on obesity and caloric intake. *Food Policy, 64*, 121–132. https://doi.org/10.1016/j.foodpol.2016.10.001

Drenowski, A., & Specter, S. E. (2004). Poverty and obesity: The role of energy density and energy costs. *The American Journal of Clinical Nutrition, 79*(1), 6–16. https://doi.org/10.1093/ajcn/79.1.6

Egger, G., Swinburn, B., & Amirul, I. F. M. (2012). Economic growth and obesity: An interesting relationship with world-wide implications. *Economics and Human Biology, 10*(2), 147–153.

Fox, A., Feng, W., & Asal, V. (2019). What is driving global obesity trends? Globalisation or 'modernisation'? *Globalisation and Health, 15*(32), 1–16. https://doi.org/10.1186/s12992-019-0457-y

Fraser, B. (2005). *Latin America's urbanisation is boosting obesity*. World Report 365.

Fuinhas, J. A., Koengkan, M., & Santiago, R. (2021). *Physical capital development and energy transition in Latin America and the caribbean*. Elsevier. https://doi.org/10.1016/C2020-0-01491-X

Fuinhas, J. A., Marques, A. C., & Koengkan, M. (2017). Are renewable energy policies upsetting carbon dioxide emissions? The case of Latin America countries. *Environmental Science and Pollution Research, 24*(17), 15044–15054. https://doi.org/10.1007/s11356-017-9109-z

García, C. M. (2019). Association of globalisation in its different dimensions with overweight and obesity: An analysis in 10 Latin American and Caribbean countries. *Salud Pública de México, 61*(2), 174–183.

Garcia-Garcia, G. (2021). *Obesity and overweight populations in Latin America*. The Lancet Kidney Campaign. https://www.thelancet.com/campaigns/kidney/updates/obesity-and-overweight-populations-in-latin-america

Gerbens-Leenes, P. W., Nonhebel, S., & Krol, M. S. (2010). Food consumption patterns and economic growth. *Increasing Affluence and the Use of Natural Resources, Appetite, 55*(3), 597–608. https://doi.org/10.1016/j.appet.2010.09.013

Goryakin, Y., & Suhrcke, M. (2014). Economic development, urbanisation, technological change and overweight: What do we learn from 244 Demographic and Health Surveys? *Economics & Human Biology, 14*, 109–127.

Hawkes, C. (2006). Uneven dietary development: Linking the policies and processes of globalisation with the nutrition transition, obesity and diet-related chronic diseases. *Globalisation and Health, 2*(4), 1–18. https://doi.org/10.1186/1744-8603-2-4

Hawkes, C. (2008). Dietary implications of supermarket development: A global perspective. *Development Policy Review, 26*(6), 657–692. https://doi.org/10.1111/j.1467-7679.2008.00428.x

Hoffman, D. (2001). Obesity in developing countries: Causes and implications. *Food Nutrition and Agriculture, 28*, 35–44.

Holtz-Eakin, D., Newey, W., & Rosen, H. S. (1988). Estimating vector autoregressions with panel data. *Econometrica, 56*(6), 1371–1395. https://www.jstor.org/stable/1913103.

International Food Policy Research Institute. (2016). *Global Nutrition Report 2016: From Promise to Impact: Ending Malnutrition by 2030*. Washington, DC.

Jack, W., & Lewis, M. (2009). *Health investments and economic growth, macroeconomic evidence and microeconomic foundations*. World Bank. Policy Research Working Paper 4877.

Jaitman, L. (2015). Urban infrastructure in Latin America and the Caribbean: Public policy priorities. *Latin American Economic Review, 24*, 13. https://doi.org/10.1007/s40503-015-0027-5

Kain, J., Vio, F., & Albala, C. (2003). Obesity trends and determinant factors in Latin America. *Cadernos de Saúde Pública, 19*(1), 1–10. https://doi.org/10.1590/S0102-311X2003000700009

Kjellstrom, T., Hakansta, C., & Hogstedt, C. (2007). Globalisation and public health-overview and a Swedish perspective. *Scandinavian Journal of Public Health, 70*, 2–68. https://doi.org/10.1080/14034950701628494

Koengkan, M. (2020). Capital stock development in Latin America and the Caribbean region and their effect on investment expansion in renewable energy. *Journal of Sustainable Finance & Investment*, 1–21. https://doi.org/10.1080/20430795.2020.1796100

Koengkan, M., & Fuinhas, J. A. (2020a). Exploring the effect of the renewable energy transition on CO_2 emissions of Latin American & Caribbean countries. *International Journal* of Sustainable Energy, 39(3), 515–538. https://doi.org/10.1080/14786451.2020.1731511

Koengkan, M., & Fuinhas, J. A. (2020b). The interactions between renewable energy consumption and economic growth in the Mercosur countries. *International Journal of Sustainable Energy, 39*(6), 594–614. https://doi.org/10.1080/14786451.2020.1732978

Koengkan, M., & Fuinhas, J. A. (2021). Does the overweight epidemic cause energy consumption? A piece of empirical evidence from the European region. *Energy, 236*(1), 119297. https://doi.org/10.1016/j.energy.2020.119297

Koengkan, M., Fuinhas, J. A., & Fuinhas, C. (2021). *Does urbanization process increase the overweight epidemic? The case of Latin America and the Caribbean region*. SSRN. https://doi.org/10.2139/ssrn.3826196

Koengkan, M., Fuinhas, J. A., & Marques, A. C. (2019). The role of financial openness and China's income on fossil fuels consumption: Fresh evidence from Latin American countries. *GeoJournal*, 1–15.

Koengkan, M., Santiago, R., & Fuinhas, J. A. (2019a). The interactions between renewable energy consumption and economic growth in the Mercosur countries. *International Journal of Sustainable Energy, 39*(6), 594–614.

Koengkan, M., Santiago, R., & Fuinhas, J. A. (2019b). The impact of public capital stock on energy consumption: Empirical evidence from Latin America and the Caribbean region. *International Economics, 160*, 43–55. https://doi.org/10.1016/j.inteco.2019.09.001

KOF Globalization index. (2021). URL: https://www.kof.ethz.ch/en/forecastsand-indicators/indicators/kof-globalisation-index.html.

Kroop, G., & Korobilis, D. (2016). Model uncertainty in panel vector autoregressive models. *European Economic Review, 81*, 115–131. https://doi.org/10.1016/j.euroecorev.2015.09.006

Lindstrom, M. (2008). Means of transportation to work and overweight and obesity: A population-based study in southern Sweden. *Preventive Medicine, 46*, 22–28. https://doi.org/10.1016/j.ypmed.2007.07.012

Lopez, R. (2004). Urban sprawl and risk for being overweight or obese. *American Journal of Public Health, 94*(6), 1574–1579. https://doi.org/10.2105/AJPH.94.9.1574

Love, I., & Zicchino, L. (2006). Financial development and dynamic investment behaviour: Evidence from panel VAR. *The Quarterly Review of Economics and Finance, 46*(2), 190–210. https://doi.org/10.1016/j.qref.2005.11.007

Martins, M. L. R. (2002). *Meio ambiente e morada social nos países do Mercosul*. Madrid.

Martins, S. T., Mola, C. L., Vaz, J. D. S., & Tovo-Rodrigues, L. (2018). General and abdominal obesity in adults living in a rural area in Southern Brazil. *Revista de Saúde Pública, 52*(7), 1–23.

McCloskey, M. L., Tarazona-Meza, C. E., & Jones-Smith, J. C. (2017). Disparities in dietary intake and physical activity patterns across the urbanisation divide in the Peruvian Andes. *International Journal of Behavioral Nutrition and Physical Activity, 14*, 90. https://doi.org/10.1186/s12966-017-0545-4

Menon, P., Ruel, M. T., & Morris, S. S. (1999). Socio-economic differentials in child stunting are consistently larger in urban than rural areas: Analysis of 10 DHS data sets. *FASEB Journal, 13*, A453.

Minos, D., Butzlaff, I., Demmler, K. M., & Rischke, R. (2016). Economic growth, climate change, and obesity. *Current Obesity Reports, 5*(4), 441–448. https://doi.org/10.1007/s13679-016-0234-7

Monteiro, C., Conde, W., Moura, E., & Popkin, B. M. (2005). Socio-economic status and obesity in adult populations of developing countries: A review. *Bulletin of the World Health Organization, 82*, 940–946. https://doi.org/10.1590/S0042-96862004001200011

NCD Risk Factor Collaboration (NCD-RisC). (2016). Trends in adult body-mass index in 200 countries from 1975 to 2014: a pooled analysis of 1698 population-based measurement studies with 19·2 million participants. *Lancet, 387*, 377–396.

Ng, M., Fleming, T., Robinson, M., Thomson, B., Graetz, N., & Margono, C. (2014). Global, regional, and national prevalence of overweight and obesity in children and adults during 1980–2013: A systematic analysis for the global burden of disease study 2013. *Lancet, 384*(9945), 766–781.

Our World in Data. (2021). *Obesity*. https://ourworldindata.org/obesity.

PAHO. (2019). *United Nations calls for urgent action to curb the rise in hunger and obesity in Latin America and the Caribbean*. https://www.paho.org/hq/index.php?option=com_content&view=article&id=15587:united-nations-calls-for-urgent-action-to-curb-the-rise-in-hunger-and-obesity-in-latin-america-and-the-caribbean&Itemid=1926&lang=en.

Pedraza, D. F. (2009). Obesidad y pobreza: Marco conceptual para su análisis en latinoamérica. *Saúde e Sociedades, 18*, 103–117.

Pérez-Cueto, F., & Kolsteren, P. (2004). Changes in the nutritional status of Bolivian women 1994–1998: Demographic and social predictors. *European Journal of Clinical Nutrition, 58*, 660–666.

Pesaran, M. H. (2004). *General diagnostic tests for cross-section dependence in panels*. The University of Cambridge, Faculty of Economics. https://doi.org/10.17863/CAM.5113. Cambridge Working Papers in Economics, n. 0435.

Pesaran, M. H. (2007). A simple panel unit root test in the presence of cross-section dependence. *Journal of Applied Econometrics, 22*(2), 256–312. https://doi.org/10.1002/jae.951

Pirgon, Ö., & Aslan, N. (2015). The role of urbanization in childhood obesity. *Journal of Clinical Research in Pediatric Endocrinology, 7*(3), 163–167. https://doi.org/10.4274/jcrpe.1984

Popkin, B. M. (1998). The nutrition transition and its health implications in low-income countries. *Public Health Nutrition, 1*(1), 5–21. https://doi.org/10.1079/PHN19980004

Popkin, B. M. (1999). Urbanisation, lifestyle changes and the nutrition transition. *World Development, 27*(11), 1905–1916.

Popkin, B. M. (2006). Technology, transport, globalisation and the nutrition transition food policy. *Food Policy, 31*(6), 554–569. https://doi.org/10.1016/j.foodpol.2006.02.008

Popkin, B. M. (2014). Synthesis and implications: China's nutrition transition in the context of changes across other low- and middle-income countries. *Obesity Reviews, 15*(1), 60–67. https://doi.org/10.1111/obr.12120

Popkin, B. M., Adair, L. S., & Ng, S. W. (2012). Global nutrition transition and the pandemic of obesity in developing countries. *Nutrition Reviews, 70*(1), 3–21. https://doi.org/10.1111/j.1753-4887.2011.00456.x

Popkin, B. M., & Reardon, T. (2018). Obesity and the food system transformation in Latin America. *Obesity Reviews, 19*, 1028–1064. https://pubmed.ncbi.nlm.nih.gov/29691969/.

Pou, S. A., Tumas, N., & Aballay, L. R. (2020). Nutrition transition and obesity trends in Argentina within the Latin American context. In J. Faintuch, & S. Faintuch (Eds.), *Obesity and diabetes*. Cham: Springer. https://doi.org/10.1007/978-3-030-53370-0_2

Poveda, Y. E., Koengkan, M., & Fuinhas, J. A. (2020). Exploring the interactions between renewable energy, economic growth, agriculture and urbanisation in the Mercosur trade-bloc countries. *Revista Valore, 5*, 5049.

Rauch, J. (1993). Productivity gains from geographic concentration of human capital: Evidence from the cities. *Journal Urban Economics, 34*(3), 380–400.

Reardon, T., Timmer, C. P., Barrett, C. B., & Berdegu, J. (2003). The rise of supermarkets in Africa, Asia, and Latin America. *American Journal of Agricultural Economics, 85*(5), 1140–1146. https://www.jstor.org/stable/1244885.

Riley, J. C. (2001). *Rising life expectancy: A global history*. Cambridge University Press.

Roskam, A. J., Kunst, A. E., Van Oyen, H., Demarest, S., Klumbiene, J., Regidor, E., Helmert, U., Jusot, F., Dzurova, D., & Mackenbach, J. P. (2010). Comparative appraisal of educational inequalities in overweight and obesity among adults in 19 European countries. *International Journal of Epidemiology, 39*(2), 392–404. https://doi.org/10.1093/ije/dyp329

Ruel, M. T. (2000). Urbanisation in Latin America: Constraints and opportunities for child feeding and care. *Food and Nutrition Bulletin, 21*(1). The United Nations University.

Santiago, R., Koengkan, M., Fuinhas, J. A., & Marques, A. C. (2020). The relationship between public capital stock, private capital stock and economic growth in the Latin American and Caribbean countries. *International Review of Economics, 67*, 293–317. https://doi.org/10.1007/s12232-019-00340-x

Schlüssel, M. M., Silva, A., Pérez-Escamilla, R., & Kac, G. (2013). Household food insecurity and excess weight/obesity among Brazilian women and children: A life-course approach. *Cadernos de Saúde Pública, 29*(2), 219–226.

Serena, N., & Perron, P. (2001). Lag length selection and the construction of unit root tests with good size and power. *Econometrica, 69*(6), 1519–1554. http://www.jstor.org/stable/2692266.

Sobal, J. (2001). Commentary: Globalisation and the epidemiology of obesity. *International Journal of Epidemiology, 30*, 1136–1137. https://doi.org/10.1093/ije/30.5.1136

Stampini, M., Robles, M., & Sáenz, M. (2015). Poverty, vulnerability and the middle class in Latin America. In *Inter-American development bank. Social protection and health division. VII. Series. IDB-WP-59.2015.*

Sullivan, M. C., Hawes, K., Winchester, S. B., & Miller, R. (2008). Developmental origins theory from prematurity to adult disease. *Journal of Obstetric, Gynecologic, & Neonatal Nursing, 37*(2), 158–164. https://doi.org/10.1111/j.1552-6909.2008.00216.x

Swinburn, M. D. B. A., Kraak, V. I., Allender, S., Atkins, V. J., Baker, P. I., & Bogard, J. R. (2019). The global syndemic of obesity, undernutrition, and climate change. *The Lancet Commission Report, 393*(10173), 791–846. https://doi.org/10.1016/S0140-6736(18)32822-8

Taype-Rondan, A., Abbs, E. S., & Lazo-Porras, M. (2017). Association between chronic conditions and health-related quality of life: Differences by level of urbanisation in Peru. *Quality Life Research, 26*, 3439–3447.

Toiba, H., Umberger, W. J., & Minot, N. W. (2015). Diet transition and supermarket shopping behaviour: Is there a link? *Bulletin of Indonesian Economic Studies, 51*. https://doi.org/10.1080/00074918.2015.1111997

Townsend, M. S. J., Peerson, B., Love, C. A., & Murphy, S. P. (2001). Food insecurity is positively related to overweight in women. *Journal of Nutrition, 131*(6), 1738–1745. https://pubmed.ncbi.nlm.nih.gov/11385061/.

Tumas, N., Junyent, C. R., Aballay, L. R., Scruzzi, G. F., & Pou, S. A. (2019). Nutrition transition profiles and obesity burden in Argentina. *Public Health Nutrition, 22*(12), 2237–2247.

United Nations (UN). (2003). *The challenge of slums: Global report on human settlements 2003*. United Nations Settlements Programme (UN-Habitat). https://unhabitat.org/the-challenge-of-slums-global-report-on-human-settlements-2003+&cd=2&hl=pt-PT&ct=clnk&gl=pt.

United Nations (UN). (2015). *UN - habitat global activities report 2015*. United Nations Settlements Programme UN-Habitat. URL: https://unhabitat.org/un-habitat-global-country-activities-report-2015-increasing-synergy-for-greater-national-ownership.

Vakis, R., Rigolini, J., & Lucchetti, L. (2015). *Left behind, chronic poverty in Latin America and the Caribbean*. International Bank for Reconstruction and Development/The World Bank.

World Bank Open Data. (2021). http://www.worldbank.org/.

World Health Organization (WHO). (2020). *Obesity and overweight: Key Facts*. http://www.who.int/mediacentre/factsheets/fs311/en/#.

The effect of the urbanisation process on body mass index in Latin American and Caribbean countries

4

4.1 Introduction

The overweight epidemic is defined as abnormal or excessive fat accumulation that may impair health. In 2016, about 39% (2.0 billion) of adults aged 18 years and older were overweight or obese. 38% of men and 40% of women worldwide are overweight (World Bank Open Data, 2021). Indeed, this global problem more than doubled between 1975 and 2016. Fig. 4.1 below shows the evolution of the share of adults that were overweight or obese between 1970 and 2016 in the world.

The Latin American and Caribbean (LAC) region is not different, with the region seeing the overweight or obese epidemic for adults aged 18 years jumping from 27% in 1975 to 57% (302 million) in 2016. Indeed, 57% of the adult population (54% men and 70% of women) are overweight, and 19% (100.8 million) are obese (14.6% in men and 24% in women). Fig. 4.2 below shows the evolution of the share of adults that are overweight or obese between 1970 and 2016 in the LAC region.

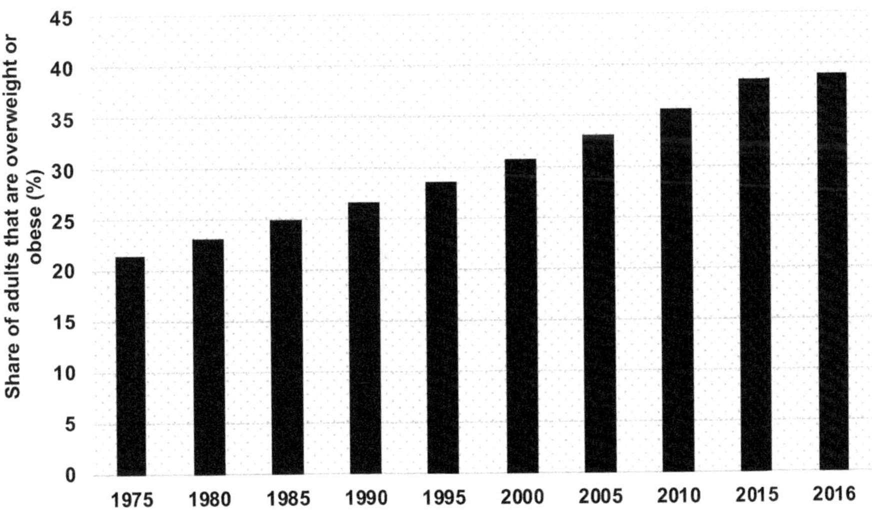

Figure 4.1 Share of adults that are overweight or obese between 1975 and 2016 in the world. The authors created this figure with the World Bank Open Data (2021). http://www.worldbank.org/.

Obesity Epidemic and the Environment. https://doi.org/10.1016/B978-0-323-99339-5.00008-X

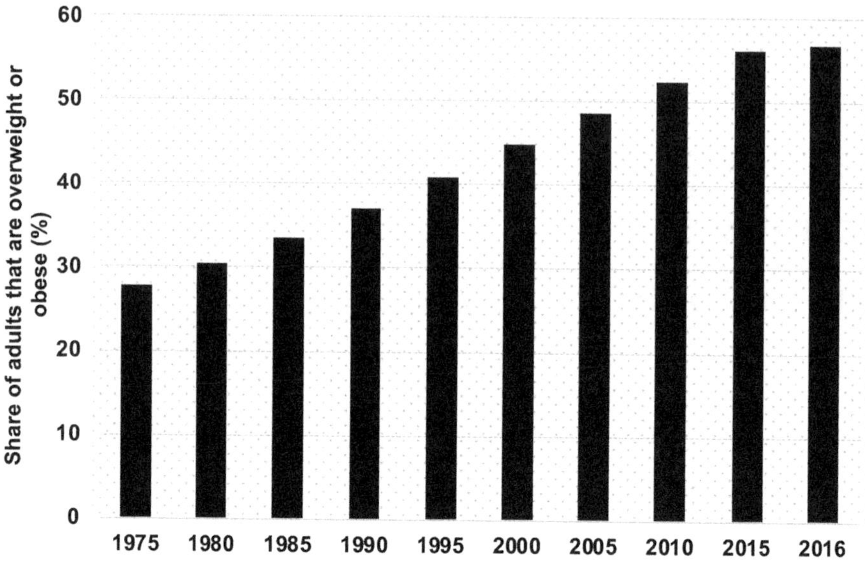

Figure 4.2 Share of adults that are overweight or obese between 1975 and 2016 in the Latin American and Caribbean region.
The authors created this figure with the World Bank Open Data (2021). http://www.worldbank.org/.

The increase in the overweight or obese epidemic has become a major health challenge in the LAC region, where more than 300 million adults in the region are overweight, and more than 100 million are obese. Moreover, 20 Latin American (LA) countries, such as Uruguay (65%), Chile (64%), Mexico (64%), Argentina (63.4%), Venezuela (62.6%), Costa Rica (61.5%), Dominican Republic (59.9%), Colombia (58.6%), Panamá (58.5%), Brazil (56.9%), El Salvador (56.8%), Peru (56.3%), Nicaragua (54.8%), Ecuador (54.9%), Bolivia (53.2%), Honduras (51.9%), Guatemala (51.4%), Haiti (51.1%), Paraguay (50.9%) and Trinidad and Tobago (48.1%) have a high share of overweight or obesity in the adult population (World Bank Open Data, 2021). Exhibit 4.1 discusses the explosion of obesity in the LAC region.

This overweight or obese epidemic in the LAC region is more prevalent in women than in men. The prevalence of this problem is greater than 20% in the female adult population (see Fig. 4.3 below).

In the LAC region, the highest prevalence of obesity in women is found in Jamaica (28.89), Trinidad and Tobago (28.88), Mexico (28.59) and El Salvador (28.50), and the highest prevalence of obesity in men is found in Chile (28.34), Argentina (28.01), Mexico (27.63) and Uruguay (27.59). See Fig. 4.4 below, which shows the mean body mass index (BMI) in men and women in LAC countries.

The increase in the overweight or obese epidemic in the LAC region is related to the rapid process of urbanisation caused by economic development, globalisation and technological revolution. The LA region, in which four global cities are found, such

Exhibit 4.1 Obesity 'explosion' in the Latin America and Caribbean region

According to the Organisation for Economic Co-operation and Development (OECD) and Food and Agriculture Organization (FAO) points out that the LAC region is the victim of an 'obesity epidemic'. At the same time, the number of food-insecure people in the region is also increasing (OECD-FAO, 2019).

Indeed, according to the report 'Agricultural Outlook 2019–28' prepared by OECD and FAO, obesity in the LAC region currently affects almost 25% of the population, and 60% of the inhabitants are overweight. This report points out that the 'triple burden of malnutrition', a mix of malnutrition, obesity and lack of micronutrients is creating 'an increasingly serious public health problem' (OECD-FAO, 2019).

Moreover, according to this report, this phenomenon 'seems to be progressing further', 'especially for poor groups of the population, women, indigenous populations, people of African descent and, in some cases, children'. Indeed, overweight and obesity rates in the LAC region, which are higher than the average world level for more than 40 years, are 'comparable' to those in high-income countries (OECD-FAO, 2019).

The region currently ranks second in the world, behind North America, according to the document. At the same time, despite the surplus in agricultural and food production in Latin America, the number of people in a situation of food insecurity 'has increased for the third consecutive year'. More than food availability, the cost to poor consumers is what explains the worsening situation (OECD-FAO, 2019).

as São Paulo, Buenos Aires, Rio de Janeiro and Caracas, is highly urbanised. The process of urbanisation in the region made the urban population grow 30% in the 1940s, 60% in the 1970s, with 70% of people living in cities in the 2000s. The most significant urban centres, such as Buenos Aires, São Paulo, Rio de Janeiro and Caracas, increased 10% of the urban population in specific periods (Martins, 2002, pp. 303–313). Today, 80% of the region's population lives in cities, making LA the world's most urbanised region. Indeed, in comparison, the European Union is 75% urbanised, and East Asia and Pacific region 50% urbanised (Koengkan, Fuinhas, & Silva, 2021 and Koengkan, Fuinhas, & Fuinhas, 2021, pp. 1–9). In 2050, Latin America's cities will include 90% of the region's population, according to UN-Habitat (Arsht, 2014).

The economic growth registered in the LAC region since the 1980s accelerated the urbanisation process. As can be seen in Fig. 4.5 below, the gross domestic product (GDP) per capita (current USD) in 1970 was 612.40 USD and in 2015 reached a value of 8557.69 USD.

In this period, LAC's annual GDP per capita growth rate had an average value of 3.75%. This increase is related to the structural and stabilisation programmes imposed

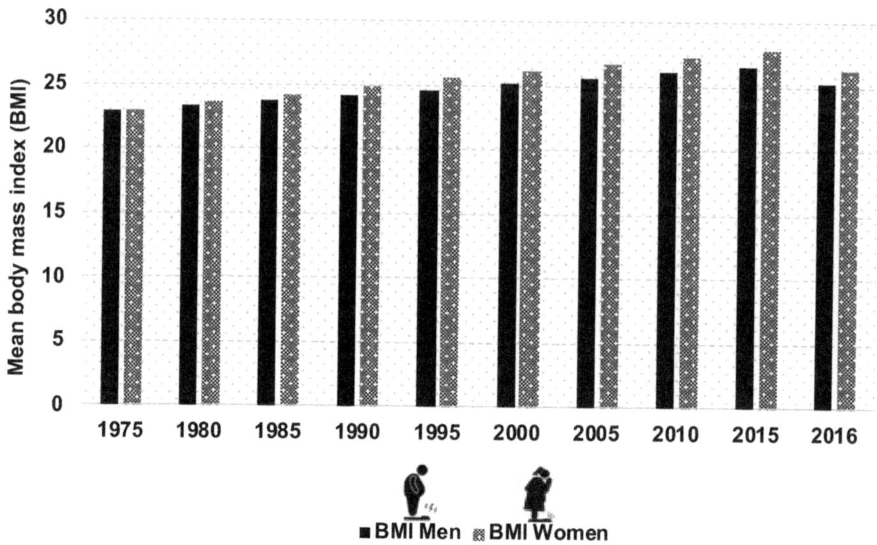

Figure 4.3 Mean body mass index (BMI) in men and women in the Latin America and Caribbean (LAC) region between 1975 and 2016. BMI is measured as a person's weight in kilograms (kg) divided by their height (in metres), squared. The WHO defines a BMI of 30.0 as 'obese'.
The authors created this figure with the World Bank Open Data (2021). http://www.worldbank.org/.

on LAC countries by the International Monetary Fund (IMF). These programmes of adjustment were neoliberal policies that consisted of the complete opening of the economies to international trade and capital, deregulation of the economy, privatisation, reduction of public expenditures, creation of proper conditions for foreign investment and the reduction of the role of the state in the economy. Moreover, the 'commodities boom' that occurred between the 2000s and the end of 2014 also accelerated the process of openness as well as the economic growth in the region (e.g., Fuinhas et al., 2021, pp. 19−234; Poveda et al., 2020 and Koengkan & Fuinhas, 2020a, 2020b).

The process of urbanisation, which began strongly in the 1980s, contributed to the rapid changes in the region's diets. This change was also accompanied by declines in physical activity, which contributed to the overweight or obesity problem in the region. Moreover, this process also affected food systems, where the supply and demand changed through processing and wholesale, retail and transportation methods. The impact on the food system resulted in the emergence of several supermarkets, fast food chains and large food processors that are fed by modernised procurement systems and the coevolution among these segments.

Based on the events described above, we formulate the following research question: Can the process of urbanisation increase the BMI equal to or greater than 25 in LAC countries? A group of 19 countries from the LAC region between 1975 and 2016 was selected to investigate this phenomenon. The panel quantile model with a fixed-effects

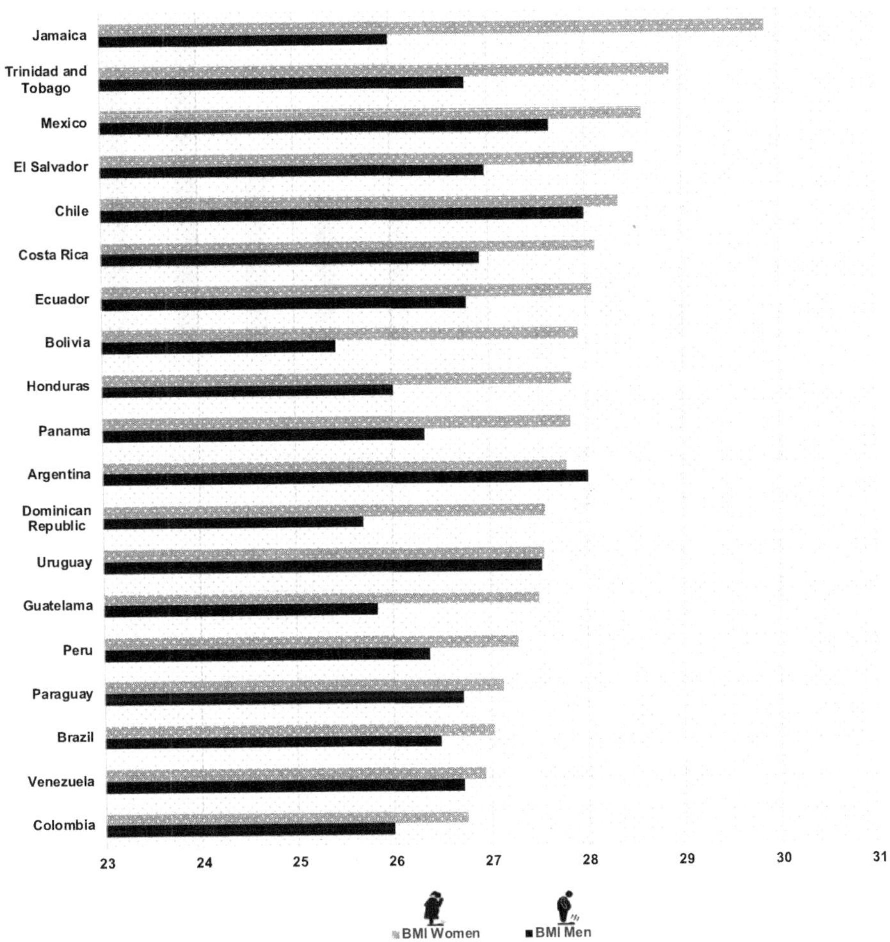

Figure 4.4 Mean body mass index (BMI) in men and women in Latin America and Caribbean (LAC) countries. BMI is measured as a person's weight in kilograms (kg) divided by their height (in meters), squared. The WHO defines a BMI of 30.0 as 'obese'.
The authors created this figure with the World Bank Open Data (2021). http://www.worldbank.org/.

approach developed by Machado and Silva (2019) is used to answer this research question.

There is a broad consensus in the literature that urbanisation leads to a nutritional transition in which a change of eating habits is observed. Therefore, the discussion mostly centres around the mechanisms that operate that transition and how that transition can differ depending on different starting situations.

Market integration and urbanisation contribute to the increase in overweight people among indigenous populations, as Chee et al. (2019) described. A qualitative case study conducted on the Kichwa people of Ecuador concludes that changing children's

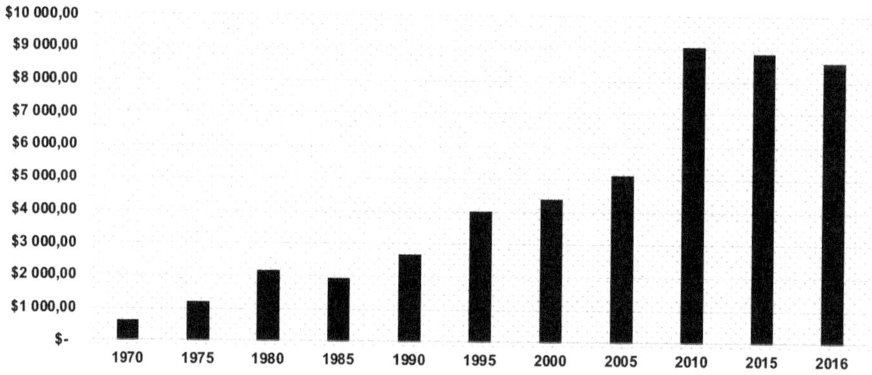

Figure 4.5 Gross domestic product (GDP) per capita (current USD) in the Latin America and Caribbean (LAC) region between 1970 and 2015.
The authors created this figure with the World Bank Open Data (2021). http://www.worldbank.org/.

food preferences, turning away from traditional foods and the poor meal timing associated with urbanisation and market integration pose an increased obesogenic risk. The authors stress the loss of traditional culture as the pivotal issue to the increase in obesity.

The inconsistency of the effects of urbanisation on overweight between high-income countries and low- and middle-income countries is well explained by Popkin (1999). In low- and middle-income countries, urbanisation is linked with increased access to a wide variety of unhealthy food. In contrast, in rich countries, where access to unhealthy foods is widespread both in urban or rural areas, increases in urbanisation tend to imply a greater level of education and access to information that tends to reduce the proportion of overweight people (e.g., Böckerman et al., 2017 and Goryakin & Suhrcke, 2014).

In the context of the difference between developed and developing countries, Wang et al. (2020) found that urbanisation increases the risk of overweight in China since urbanisation shifts peoples' lifestyles in ways that increase calorie consumption and decrease calorie expenditure. In the same vein of research focussed on the relationship between urbanisation and overweight in developing countries. Goryakin and Suhrcke (2014) point out the relationship between urbanisation and overweight is positive across all country wealth groups but stronger in developing countries. Siervo et al. (2006), in a study focused on the Gambia, show a positive relation between urbanisation and obesity.

The difference is striking with developed countries. For example, in a study conducted in Ontario, Ismailov and Leatherdale (2010) show that adolescents living in rural areas are more likely to be overweight or obese than those living in urban or suburban areas. In the same vein, Popkin et al. (2012) point to the growing concern of obesity in low- and middle-income countries. According to the authors, the problems associated with obesity are beginning to surpass the issues associated with malnutrition. The central issue in explaining the increase in the incidence of obesity in developing countries is the mismatch between technological progress and human biology. In other words, the ability to produce more abundant and caloric foodstuff grows faster than the ability of people to adapt to it.

Concerning the relationship between obesity and globalisation, Costa-Font and Mas (2016) found that globalisation has a positive correlation with the proportion of people that are obese. The study approaches globalisation as a multi-dimensional concept and finds that the social dimension of globalisation is the primary driver of the increase in obesity. The change in information flows and personal contact pose the most significant explanation for the expansion of obesity. Hawkes (2006) finds that globalisation contributes to the phenomenon of nutrition transition. According to this author, the primary mechanism that runs this transition is the way that transnational food companies can affect the patterns of food consumption, both by generalising the consumption of some types of foods and at the same time by exploring local niche products. On the relationship between social environment and obesogenic risk, Lake and Townshend (2006) stress that the social environment is a crucial factor in the obesogenic risk that is often overlooked in favour of explanations that only consider calorie intake.

Our research contributes to expanding the literature on the topic. Urbanisation is related to the mean body mass (total, men and women) in LA countries, considering a modelisation that includes several control variables (economic globalisation, social globalisation, gross domestic product per capita, energy consumption per capita, carbon dioxide emissions per capita) that allow the effect of urbanisation on people's weight to be isolated. Furthermore, the econometric technique used, the panel quantile model with a fixed-effects approach, can handle non-linearity between body mass and the explanatory variables. Indeed, it is expected that as we move on to the quantiles, the intensity of the phenomena will evolve in a non-linear path.

The rest of this chapter is organised as follows. In the next section, we describe the method and the data. In Section 4.3, we present the empirical results and the discussion. Finally, Section 4.4 summarises the chapter, presenting the conclusions and policy implications.

4.2 Method and data

This section is divided into two parts. The first one (Section 4.2.1) presents the methodological strategy adopted that will be applied, and the second (Section 4.2.2) describes the data and variables used in this investigation.

4.2.1 Method

As mentioned before, this section will show the methodological strategy that this investigation will use. Therefore, to help answer the central question and answer the specific questions of this investigation, the panel quantile model with a fixed-effects approach was used as a method. This method was introduced by Machado and Silva (2019) as an alternative to quantile regression with non-additive fixed effects developed by Powell (2016). In addition, this method can provide information on how the regressors affect the entire conditional distribution (e.g., Koengkan, 2020; Koengkan, Fuinhas, & Silva, 2021; Koengkan, Fuinhas, & Fuinhas, 2021, pp. 1−9 and Koengkan & Fuinhas, 2020c).

Moreover, the panel quantile model with fixed effects can also be adapted to provide an estimate in the presence of cross-sectional models with endogenous variables (Machado & Silva, 2019). This method is not based on the estimation of conditional means but on moment conditions that find conditional means under exogeneity. This characteristic is closely related to that of the Chernozhukov and Hansen (2008) model. Under suitable conditions, the panel quantile model with fixed effects can find the same structural quantile function. It makes using this method for non-linear models much more straightforward, especially in models with multiple endogenous variables (Machado & Silva, 2019). This method can provide information on how the regressors affect the entire conditional distribution. According to Machado and Silva (2019), the panel quantile model with fixed effects, given data $\{(Y_{it}, X_{it}')'\}$ from a panel of n individuals $i = 1, \ldots,$ n over T periods, is constructed around the following Eq. (4.1):

$$Y_{it} = a_i + X_{it}'\beta + (\delta_i + Z_{it}'\gamma)U_{it}, \tag{4.1}$$

with, $P\{\delta_i + Z_{it}'\gamma > 0\} = 1$. The parameters (α_1, δ_i), $i = 1, \ldots, n$, capture the individual i fixed effects, and Z is a k-vector of known differentiable (with probability 1) transformations of the components of X with element l given by $Z_l = Z_l(X)$, $l = 1, \ldots, k$. In empirical research, we will use a special case of Eq. (4.1), and $Z = X$. The sequence $\{X_{it}\}$ is *i.i.d.* for any fixed i and independent across t. U_{it} are *i.i.d.* (across i and t), statistically independent of X_{it}, and normalised to satisfy the moment condition $E(U) = 0 \wedge E(|U|) = 1$ (Machado & Silva, 2019). Eq. (4.1) implies that the conditional quantile-τ is given by Eq. (4.2):

$$Q_Y(\tau|X_{it}) = (a_i + \delta_i q(\tau)) + X_{it}'\beta + Z_{it}'\gamma q(\tau), \tag{4.2}$$

with $q(\tau) = F_U^{-1}(\tau)$, where F_U is the distribution function of U. The authors propose a recursive estimation method based on a set of moment conditions, which is computationally fast and straightforward and does not require the use of simulations that would render its widespread adoption difficult (Powell, 2016). They also prove that the resulting estimates are consistent and asymptotically normal.

After computing the estimates, the marginal effect of the explanatory variable l on quantile τ of the dependent variable can be retrieved from Eq. (4.3):

$$\beta_l(\tau|X) = \beta_l + \frac{\partial Z'}{\partial X_l}\gamma q(\tau), \tag{4.3}$$

The panel quantile model with fixed effects will be used, and the results will be compared with those from a pooled ordinary least square (OLS) and OLS robust. The pooled OLS and OLS robust regression model will be used as a benchmark. These two regressions were estimated to evaluate the effect of the urbanisation process on male and female overweight and their total overweight, which checks the robustness of results found in the models (male, female and total overweight). Thus, this investigation will estimate three models to answer the central question. Therefore, before the realisation of the panel quantile model with fixed effects and pooled OLS and OLS robust estimations, it is necessary to verify the proprieties of variables that will be used in this investigation, which includes checking the normality, the presence of multicollinearity and fixed or random effects. Consequently, the first tests that need to be applied before the panel quantile model with fixed effects estimation can be seen in Table 4.1 below.

After the panel quantile model with fixed effects and pooled OLS estimations, it is necessary to apply some post-estimation tests (see Table 4.2 below).

These two post-estimation tests need to be applied to verify heteroscedasticity in the models. The estimation and testing procedures are carried out using **Stata 16.0.**

Table 4.1 Preliminary tests.

Test	Finality
Shapiro–Wilk test (Royston, 1983)	This test verifies the normality of the model. The null hypothesis of this test is the presence of normality.
Skewness and Kurtosis test (D'Agostino et al., 1990)	This test checks the normality based on skewness and another based on kurtosis and then combines the two tests into an overall test statistic. The null hypothesis of this test is that the data is normally distributed.
Variance inflation factor (VIF) test (Belsley et al., 1980)	This test verifies the presence of multicollinearity between the variables.
Cross-sectional dependence (CSD) test (Pesaran, 2004)	This test finds the presence of cross-sectional dependence (CSD) in the panel's data. The null hypothesis of this test is the presence of cross-section independence CD $\sim N(0,1)$.
Panel unit root test (CIPS) test (Pesaran, 2007)	This test verifies the presence of unit roots in the variables. The panel unit root test (CIPS) null hypothesis is that all series have a unit root.
Hausman test	This test finds heterogeneity, i.e., whether the panel has random effects (REs) or fixed effects (FEs).

4.2.2 Data

This investigation used annual data collected from 1975 to 2016 on 19 countries from the LAC region, i.e., **Argentina**, **Bolivia**, **Brazil**, **Chile**, **Colombia**, **Costa Rica**, **Dominican Republic**, **Ecuador**, **El Salvador**, **Guatemala**, **Haiti**, **Mexico, Nicaragua**, **Panama**, **Paraguay**, **Peru**, **Trinidad and Tobago**, **Uruguay** and **Venezuela**. The use of time-series between 1975 and 2016 is due to the availability of data until 2016 for the variables of carbon dioxide emissions in kilotons (Kt) per

Table 4.2 Post-estimation tests.

Test	Finality
Breusch-Pagan/Cook-Weisberg test for heteroskedasticity (Breusch & Pagan, 1979; Cook & Weisberg, 1983)	This test finds the presence of heteroscedasticity. Therefore, the null hypothesis of this test is the presence of homoscedasticity.
Wald test (Agresti, 1990)	This test verifies the global significance of the estimated models. The null hypothesis of the Wald test is that all the coefficients are equal to zero.

capita, consumption of energy in (kWh) per capita and the World Bank Data Open urbanisation index (2021) for all countries selected. The variables which were chosen to perform this investigation are:

- **Mean BMI equal to or greater than 25 in adult men and women (aged 18 years and older) (BMI_MAN** and **BMI_WOMAN)** retrieved from World Bank Open Data (2021). The BMI is measured as a person's weight in kilograms (kg) divided by his height (in metres) squared. The WHO defines a BMI $\leq$ 18.5 as 'underweight'; 18.5 to <25 as 'normal/healthy'; 25.0 to <30 as 'overweight'; and >30.0 as 'obese';
- **Total of mean BMI (BMI_TOTAL)** that is the sum of mean BMI equal to or greater than 25 in adult men and women (aged 18 years and older). This variable was created and used to verify the robustness of results found in the models (male and female overweight);
- **Economic globalisation index (ECO_GLOBA)** in the de facto, retrieved from the KOF Globalisation Index (KOF, 2021). This variable measures trade and financial globalisation. Trade globalisation is determined based on trade in goods and services, and financial globalisation includes foreign investment in various categories;
- **Social globalisation index (SO_GLBA)**, in the de facto, was retrieved from the KOF Globalisation Index (KOF, 2021). This variable measures the interpersonal contact flows of information and the cultural proximity. Interpersonal contact is measured within the de facto segment with reference to international telephone connections, tourist numbers and migration. Flows of information are determined within the de facto segment with reference to international patent applications, international students and trade in high-technology goods. Cultural proximity is measured in the de facto segment via trade in cultural goods, international trademark registrations and the number of McDonald's restaurants and IKEA stores;
- **Gross domestic production** (**GDP_PC**) in constant local currency unity (LCU) and expressed per capita, retrieved from World Bank Open Data (2021);
- **Urbanisation index (URBAN)**, which refers to people living in urban areas as defined by national statistical offices, retrieved from World Bank Open Data (2021);
- **Consumption of energy (ENE)**, from fossil and renewable energy sources in (kWh per capita), retrieved from World Bank Open Data (2021);
- **Carbon dioxide emissions (CO_2)**, in kilotons (Kt) per capita, from the burning of fossil fuels and cement manufacturing. These include carbon dioxide produced during the consumption of solid, liquid and gas fuels and gas flaring retrieved from World Bank Open Data (2021).

The variables **CO_2**, **GDP_PC** and **ENE** were transformed into per capita values with the total population of each cross. The per capita value allows disparities to be

Table 4.3 Descriptive statistics.

Variables	Descriptive statistics				
	Obs.	**Mean**	**Std.-dev.**	**Min.**	**Max.**
LogBMI_MAN	798	3.2056	0.0576	3.0672	3.3328
LogBMI_WOMAN	798	3.2365	0.0665	3.0113	3.3967
LogBMI_TOTAL	798	3.9343	0.0586	3.7679	4.0489
LogECO_GLOBA	798	3.8043	0.2684	2.9700	4.3639
LogSO_GLOBA	798	3.8871	0.2643	3.2271	4.3518
LogGDP_PC	798	11.0167	2.8799	7.2408	17.1658
LogURBAN	798	4.1383	0.2472	3.4691	4.5553
LogENE	798	6.8415	0.7941	4.7573	8.8668
LogCO_2	798	3.0918	4.8998	0.3008	36.0916

(Log) denotes variables in the natural logarithms; *Min. and Max.* denote minimum and maximum; *Obs.* denotes the number of observations in the model; respectively; *Std.-Dev.* denotes the standard deviation.

controlled for population growth over time and within countries (Koengkan et al., 2019). Table 4.3 shows the summary statistics of variables.

Moreover, we should stress that they are already in their natural logarithms (see prefix 'Log'). All variables are in natural logarithms to harmonise the interpretation of results and linearise the relationships between variables. These summary statistics were obtained from the command *sum* of **Stata 16.0**. The board below shows how to obtain the summary statistics of variables.

How to do:

****The summary statistics****

```
sum l_logbmi_man l_logbmi_woman l_logbmi_total l_logeco_globa l_logso_globa
l_loggdp_pc l_logurban l_logene l_logco2
```

4.3 Empirical results and discussion

As previously explained in the introduction, this section will present the empirical results of preliminary and post-estimation tests and the outcomes of the panel quantile model with fixed effects and pooled OLS estimations. Therefore, to verify the normality for each model, the Shapiro–Wilk W-test for normal data was used to test the residuals of models from the pooled OLS regression (e.g., Resid_BMI_MAN, Resid_BMI_WOMAN and Resid_BMI_TOTAL). Table 4.4 below shows the results of the Shapiro–Wilk W-test for normal data for three models that will be estimated.

Regarding the Shapiro–Wilk W-test for normal data, we see that the null hypothesis of normality is rejected in all models used. The results from the Shapiro–Wilk W-test for normal data were obtained from the command *sktest* of **Stata 16.0**. The board below shows how to carry out and obtain the results from the Shapiro–Wilk W-test.

Table 4.4 Shapiro–Wilk W-test for normal data.

Models	Obs.	W	V	Z	Prob > z
Resid_BMI_MAN	798	0.9793	10.620	5.797	0.0000
Resid_BMI_WOMAN	798	0.9778	11.367	5.963	0.0000
Resid_BMI_TOTAL	798	0.9768	11.874	6.070	0.0000

How to do:

****Resid****

reg l_logbmi_man l_logeco_globa l_logso_globa l_loggdp_pc l_logurban l_logene l_logco2

predict resid_bmi_man

**** Shapiro-Wilk W-test for normal data ****

swilk resid

After the Shapiro–Wilk W-test, applying the skewness and kurtosis test is necessary to verify the normality in the residuals of models from the pooled OLS regression (e.g., Resid_BMI_MAN; Resid_BMI_WOMAN and Resid_BMI_TOTAL). The null hypothesis that the data is normally distributed is rejected for all models (see Table 4.5 below).

The results from the skewness/kurtosis tests for normality were obtained from the command *sktest* of **Stata 16.0.** The board below shows how to carry out and obtain the results from the skewness/kurtosis tests for normality.

How to do:

****Resid****

reg l_logbmi_man l_logeco_globa l_logso_globa l_loggdp_pc l_logurban l_logene l_lgoco2
predict resid_bmi_man

**** Skewness/kurtosis tests for normality ****

sktest resid

The results of both tests reject the null hypothesis at a significance level of 1%. Moreover, the results of these tests are added support for the adequacy of using the quantile regression (Afonso et al., 2019, pp. 1–25). Subsequen to the tests of normality, the VIF test that informs on the presence of multicollinearity needs to be computed. The results of the VIF test show that the presence of multicollinearity is not a concern in the estimation of each model, given the low VIF and mean VIF values registered, which are lower than the usually accepted benchmark of 10, in the case of the VIF values, and six in the case of the mean VIF values (see Table 4.6 below). This test helps us understand the degree of multicollinearity present in our models, leading to problems in estimation.

The results of the VIF test were obtained from the command *estat vif* in **Stata 16.0.** The board below shows how to carry out and obtain the results from the VIF test.

Table 4.5 Skewness/kurtosis tests for normality.

Variables	Obs.	Pr(Skewness)	Pr(Kurtosis)	adj chi2(2)	Prob > Chi2
Resid_BMI_MAN	798	0.0000	0.0332	24.18	0.0000
Resid_BMI_WOMAN	798	0.0000	0.0010	24.80	0.0000
Resid_BMI_TOTAL	798	0.0000	0.0089	24.65	0.0000

Table 4.6 Variance inflation factor test.

Variables	VIF	1/VIF	Mean VIF
LogBMI_MAN			
LogECO_GLOBA	2.06	0.4852	2.78
LogSO_GLOBA	3.40	0.2942	
LogGDP_PC	1.06	0.9472	
LogURBAN	2.49	0.4014	
LogENE	5.54	0.1803	
LogCO_2	2.15	0.4642	
LogBMI_WOMAN			
LogECO_GLOBA	2.06	0.4852	2.78
LogSO_GLOBA	3.40	0.2942	
LogGDP_PC	1.06	0.9472	
LogURBAN	2.49	0.4014	
LogENE	5.54	0.1803	
LogCO_2	2.15	0.4642	
LogBMI_TOTAL			
LogECO_GLOBA	2.06	0.4852	2.78
LogSO_GLOBA	3.40	0.2942	
LogGDP_PC	1.06	0.9472	
LogURBAN	2.49	0.4014	
LogENE	5.54	0.1803	
LogCO_2	2.15	0.4642	

(Log) denotes variables in the natural logarithms.

How to do:

**** The variance inflation factor test****

reg l_logbmi_man l_logeco_globa l_log_so_globa l_loggdp_pc l_logurban l_logene l_logco2

estat vif

After finding multicollinearity between the variables, it is necessary to find the presence of CSD in the panel data. To this end, the CSD-test was used. The results of the CSD-test show that all variables have the presence of cross-section dependence (see Table 4.7 below).

The results of the Pesaran CD-test were obtained from the command *xtcd* in **Stata 16.0**. The board below shows how to carry out and obtain the results from the Pesaran CD-test.

How to do:

****The Pesaran CD-test****

xtcd l_logbmi_man l_logbmi_woman l_logbmi_total l_logeco_globa l_logso_globa l_loggdp_pc l_logurban l_logene l_logco2

Nevertheless, in the presence of CSD, it is necessary to verify the order of integration of the variables that will be used. To this end, the panel unit root test (CIPS) developed was used.

Table 4.7 Pesaran CD test.

Variables	CD-test	*P*-value	
LogBMI_MAN	84.19	0.000	***
LogBMI_WOMAN	84.25	0.000	***
LogBMI_TOTAL	84.24	0.000	***
LogECO_GLOBA	55.41	0.000	***
LogSO_GLOBA	82.85	0.000	***
LogGDP_PC	62.81	0.000	***
LogURBAN	75.63	0.000	***
LogENE	80.51	0.000	***
LogCO_2	44.13	0.000	***

*** denotes statistical significance at the 1% level; *(Log)* denotes variables in the natural logarithms.

Table 4.8 Panel unit root test (CIPS-test).

	Panel unit root test (CIPS) (Zt-bar)				
	Without trend			With trend	
Variables	Lags	Zt-bar		Zt-bar	
LogBMI_MAN	1	−11.882	***	−8.167	***
LogBMI_WOMAN	1	−11.731	***	−15.361	***
LogBMI_TOTAL	1	−12.704	***	−16.088	***
LogECO_GLOBA	1	−3.824	***	−2.102	**
LogSO_GLOBA	1	−3.548	***	−2.246	**
LogGDP_PC	1	−2.636	***	−1.040	
LogURBAN	1	−1.352	*	0.453	
LogENE	1	−3.088	***	0.032	
LogCO_2	1	−3.029	***	−2.748	***

***, ** and * denote statistically significant at the 1%, 5%, and 10% levels, respectively; *(Log)* denotes variables in the natural logarithms.

The results from the CIPS-test (see Table 4.8 above) show that some variables are borderline between I (0) and I (1). The results of the CIPS-test were obtained from the command *multipurt* in **Stata 16.0**. The board below shows how to carry out and obtain the results from the CIPS test.

How to do:

****The CIPS-test****

multipurt l_logbmi_man l_logbmi_woman l_logbmi_total l_logeco_globa l_logso_globa l_loggdp_pc l_logurban l_logene l_logco2, lags(1)

Due to the most of the variables appearing to be somewhere borderline between I(0) and I(1), this investigation opted not to perform the Westerlund cointegration test to verify the presence of cointegration between the variables, as this test requires that all variables are I(1).

After realising the unit root test, the next step is to assess the presence of individual effects in each model (e.g., BMI_MAN; BMI_WOMAN; BMI_TOTAL). To this end, the Hausman test, comparing random (RE) and fixed effects (FE), was used. The null hypothesis of this test is that the difference in coefficients is not systematic (i.e., random effects are the most suitable estimator). Table 4.9 below gives the results of the Hausman test for four models that will be estimated.

The Hausman test shows that the null hypothesis should be rejected in all models, and that a fixed-effects model is the most proper for this analysis. The results of the Hausman test were obtained from the command *hausman* with option *sigmamore* in **Stata 16.0**. The board below shows how to carry out and obtain the results from the Hausman test.

How to do:

****The Hausman test****

```
xtreg l_logbmi_man l_logeco_globa l_log_so_globa l_loggdp_pc l_logurban l_logene
l_logco2,fe
estimates store fixed

xtreg l_logbmi_man l_logeco_globa l_log_so_globa l_loggdp_pc l_logurban l_logene
l_logco2,re
estimates store random

hausman fixed random, sigmamore
```

After realising preliminary tests, the panel quantile model with fixed effects and pooled OLS estimations can be made. Therefore, the panel quantile model will be used, and the results will be compared with those from a pooled OLS and robust.

Therefore, the pooled OLS and OLS robust were used as a benchmark. Indeed, the OLS robust was added in this estimation due to the possible presence of heteroscedasticity in the models based on visual analysis of descriptive statistics of variables. Moreover, the OLS robust provides corrected standard errors and consequently, the correct coefficient significant level (Afonso et al., 2019, pp. 1–25). Regarding the panel quantile model, the 25th, 50th and 75th quantiles were calculated to assess the impact of the urbanisation process on male and female overweight and their total. The method used does not allow causalities to be concluded; it only allows the effect at the quantiles to be seen. Table 4.10 shows the results of pooled OLS and OLS robust and quantiles of the BMI_MAN model and post-estimation tests (e.g., Breusch-Pagan/Cook-Weisberg test for heteroskedasticity and the Wald test) to confirm the presence of heteroskedasticity in the models.

The pooled OLS and OLS robust estimators show that the economic and social globalisation, urbanisation and energy consumption increase overweight in men, while economic growth decreases overweight. However, the CO_2 emissions are shown to be insignificant statistically and do not cause any impact on overweight in men. Regarding the panel quantile model, the 25th, 50th and 75th quantiles show that the economic and social globalisation, urbanisation, consumption of energy and CO_2 emissions increase overweight in men. Economic growth does not cause any impact on the dependent variable because it is insignificant statistically. Moreover, the post-estimation tests (e.g., Breusch-Pagan/Cook-Weisberg test for heteroskedasticity and the Wald test) show heteroscedasticity in the model and the time fixed-effects are needed. The result from the post-estimation test is an indicator that the estimations that this

Table 4.9 Hausman test.

BMI_MAN	BMI_WOMAN	BMI_TOTAL
chi2(6) = 14.21 **	chi2(6) = 137.85 ***	chi2(6) = 77.22 ***

*** and ** denote statistically significant at the 1% and 5% levels.

Table 4.10 Estimations for BMI_MAN.

Independent variables	Dependent variable (LogBMI_MAN)								
	Pooled			Quantiles					
	OLS		OLS robust	25th		50th		75th	
LogECO_GLOBA	0.0319	***	***	0.0127	***	0.0114	***	0.0101	*
LogSO_GLOBA	0.0923	***	***	0.0694	***	0.0837	***	0.0978	***
LogGDP_PC	−0.0017	***	***	−0.0075		−0.0035		0.0003	
LogURBAN	0.0831	***	***	0.1284	***	0.1270	***	0.1257	***
LogENE	0.0055	*	*	0.0354	***	0.0262	***	0.0173	***
LogCO_2	−0.0000			0.0009	***	0.0010	***	0.0012	***
Constant	2.4022	***	***	NA		NA		NA	
Obs	798		798	798		798		798	
Wald test	F(6791) = 279.22	***	F(5,791) = 301.99 ***	chi2(6) = 5411.88	***	chi2(6) = 7009.89	***	chi2(6) = 3368.97	***
Breusch-Pagan/ Cook-Weisberg test	chi2(1) = 40.60	***	NA	NA		NA		NA	

***, ** and * denote statistically significant at the 1%, 5%, and 10% levels, respectively; *(Log)* denotes variables in the natural logarithms; *NA* denotes not available.

investigation uses are adequate. The pooled OLS and robust models were obtained from the command *reg* and *reg* with option robust, while the results of the panel quantile model with fixed effects were obtained from the command *xtqreg* in **Stata 16.0**. The post-estimation tests, such as Breusch-Pagan/Cook-Weisberg test for heteroskedasticity, were obtained from the command *estat hettest,* while the Wald test was obtained from the command *testparm* in **Stata 16.0**. The board below shows how to carry out and obtain the results from the pooled OLS and robust and the panel quantile models regression and the results from the post-estimation tests.

How to do:

****Pooled OLS****

reg l_logbmi_man l_logeco_globa l_log_so_globa l_loggdp_pc l_logurban l_logene l_logco2

****The Wald test****

testparm l_logbmi_man l_logeco_globa l_log_so_globa l_loggdp_pc l_logurban l_logene l_logco2

****Breusch-Pagan/Cook-Weisberg test****

reg l_logbmi_man l_logeco_globa l_log_so_globa l_loggdp_pc l_logurban l_logene l_logco2

hettest

****Pooled OLS robust****

reg l_logbmi_man l_logeco_globa l_log_so_globa l_loggdp_pc l_logurban l_logene l_logco2, robust

****The Wald test****

testparm l_logbmi_man l_logeco_globa l_log_so_globa l_loggdp_pc l_logurban l_logene l_logco2

****The panel quantile model ****

xtqreg l_logbmi_man l_logeco_globa l_log_so_globa l_loggdp_pc l_logurban l_logene l_logco2, id(country) quantile(.25)

****The Wald test****

testparm l_logbmi_man l_logeco_globa l_log_so_globa l_loggdp_pc l_logurban l_logene l_logco2

**** The panel quantile model****

l_logbmi_man l_logeco_globa l_log_so_globa l_loggdp_pc l_logurban l_logene l_logco2, id(country) quantile(.50)

****The Wald test****

testparm l_logbmi_man l_logeco_globa l_log_so_globa l_loggdp_pc l_logurban l_logene l_logco2

```
** The panel quantile model **

xtqreg l_logbmi_man l_logeco_globa l_log_so_globa l_loggdp_pc l_logurban
l_logene l_logco2, id(country) quantile(.75)

**The Wald test**

testparm l_logbmi_man l_logeco_globa l_log_so_globa l_loggdp_pc l_logurban
l_logene l_logco2
```

After finding the positive impact of the urbanisation process on overweight men, it is necessary to find this same effect on overweight women. Table 4.11 shows the results of the pooled OLS and robust and the quantiles of the BMI_WOMAN model and post-estimation tests (e.g., Breusch-Pagan/Cook-Weisberg test for heteroskedasticity and the Wald test) to confirm the presence of heteroskedasticity in the models.

The pooled OLS and OLS robust results point out that economic and social globalisation and urbanisation increase the number of overweight women, while the economic growth decreases it. However, the consumption of energy and CO_2 emissions are shown to be insignificant statistically and do not cause any impact on the number of overweight women. Regarding the panel quantile model, the 25th, 50th and 70th quantiles show that economic and social globalisation, urbanisation, energy consumption and CO_2 emissions increase overweight in women, while economic growth decreases it. Moreover, the post-estimation tests (e.g., Breusch-Pagan/Cook-Weisberg test for heteroskedasticity and the Wald test) show heteroscedasticity in the model. The result from the post-estimation test is an indicator that the estimations that this investigation uses are adequate. The regression of the pooled OLS and robust models were obtained from the command *reg* and *reg* with option robust, while the results of panel quantile model with fixed-effects regression were obtained from the command *xtqreg* in **Stata 16.0**. Moreover, post-estimation tests, such as Breusch-Pagan/Cook-Weisberg test for heteroskedasticity, were obtained from the command *estat hettest,* while the Wald test was obtained from the command *testparm* in **Stata 16.0**. The board below shows how to carry out and obtain the results from the regression of the pooled OLS and robust and the panel quantile models and the post-estimation tests.

Table 4.11 Estimations for BMI_WOMAN.

Independent variables	Dependent variable (LogBMI_WOMAN)									
	Pooled				Quantiles					
	OLS		OLS robust		25th		50th		75th	
LogECO_GLOBA	0.0750	***	***		0.0398	***	0.0362	***	0.0329	***
LogSO_GLOBA	0.1300	***	***		0.0627	***	0.0901	***	0.1146	***
LogGDP_PC	−0.0017	***	***		−0.0519	***	−0.0460	***	−0.0407	***
LogURBAN	0.0651	***	**		0.2350	***	0.2108	***	0.1891	***
LogENE	−0.0007				0.0572	***	0.0438	***	0.0318	***
CO_2	0.0003				0.0025	***	0.0029	***	0.0032	***
Constant	2.1995	***	***		NA		NA		NA	
Obs	798		798		798		798		798	
Wald test (chi2)	F(6,791) = 376.06	***	F(6,791) = 356.52	***	chi2(6) = 2885.10	***	chi2(6) = 4635.42	***	chi2(6) = 2483.43	***
Breusch-Pagan/ Cook-Weisberg test	chi2(1) = 35.74	***	NA		NA		NA		NA	

*** and ** denote statistically significant at the 1% and 5% level, respectively; *(Log)* denotes variables in natural logarithms; *NA* denotes not available.

How to do:

****Pooled OLS****

reg l_logbmi_woman l_logeco_globa l_log_so_globa l_loggdp_pc l_logurban l_logene l_logco2

****The Wald test****

testparm l_logbmi_woman l_logeco_globa l_log_so_globa l_loggdp_pc l_logurban l_logene l_logco2

****Breusch-Pagan/Cook-Weisberg test****

reg l_logbmi_woman l_logeco_globa l_log_so_globa l_loggdp_pc l_logurban l_logene l_logco2

hettest

****Pooled OLS Robust****

reg l_logbmi_woman l_logeco_globa l_log_so_globa l_loggdp_pc l_logurban l_logene l_logco2, robust

****The Wald test****

testparm l_logbmi_woman l_logeco_globa l_log_so_globa l_loggdp_pc l_logurban l_logene l_logco2

****The panel quantile model ****

xtqreg l_logbmi_woman l_logeco_globa l_log_so_globa l_loggdp_pc l_logurban l_logene l_logco2, id(country) quantile(.25)

****The Wald test****

testparm l_logbmi_woman l_logeco_globa l_log_so_globa l_loggdp_pc l_logurban l_logene l_logco2

**** The panel quantile model ****

xtqreg l_logbmi_woman l_logeco_globa l_log_so_globa l_loggdp_pc l_logurban l_logene l_logco2, id(country) quantile(.50)

****The Wald test****

testparm l_logbmi_woman l_logeco_globa l_log_so_globa l_loggdp_pc l_logurban l_logene l_logco2

```
** The panel quantile model **

xtqreg l_logbmi_woman l_logeco_globa l_log_so_globa l_loggdp_pc l_logurban
l_logene l_logco2, id(country) quantile(.75)

**The Wald test**

testparm l_logbmi_woman l_logeco_globa l_log_so_globa l_loggdp_pc l_logurban
l_logene l_logco2
```

After finding the positive impact of the urbanisation process on the numbers of overweight men and women, it is necessary to identify this same effect on total overweight, that is, the sum of overweight men and women. This estimation works as a robustness check to find the same results from Tables 4.10 and 4.11. Table 4.12 shows the pooled OLS and OLS robust results and the quantiles of the BMI_TOTAL model and the results of post-estimation tests (e.g., Breusch-Pagan/Cook-Weisberg test for heteroskedasticity and the Wald test) to confirm the presence of heteroskedasticity in the models.

The pooled OLS and OLS robust results indicate that economic and social globalisation and urbanisation increase total overweight, while economic growth decreases it. However, the consumption of energy and CO_2 emissions are shown to be insignificant statistically and do not cause any impact on total overweight. Regarding the panel quantile model with fixed effects, the 25th, 50th and 75th quantiles show that economic and social globalisation, urbanisation, energy consumption and CO_2 emissions increase total overweight, while economic growth decreases it. Moreover, the post-estimation tests (e.g., Breusch-Pagan/Cook-Weisberg test for heteroskedasticity and the Wald test) show heteroscedasticity in the model. The result from the post-estimation test is an indicator that the estimations that this investigation uses are adequate. That is, the results that were found in Table 4.12 show that this investigation is robust. The pooled OLS and robust models' regression were obtained from the command *reg* and *reg* with option robust, while the results of the panel quantile model with fixed effects were obtained from the command *xtqreg* in **Stata 16.0**. The results from the post-estimation tests, such as Breusch-Pagan/Cook-Weisberg test for heteroskedasticity, were obtained from the command *estat hettest,* while the Wald test was obtained from the command *testparm* in **Stata 16.0**. The board below shows how to carry out and obtain the results from the pooled OLS and robust and the panel quantile model with fixed effects and the results from the post-estimation tests.

Table 4.12 Estimations for BMI_TOTAL.

	Dependent variable (LogBMI_TOTAL)									
	Pooled				Quantiles					
Independent variables	**OLS**		**Robust**		**25th**		**50th**		**75th**	
LogECO_GLOBA	0.0534	***	***		0.0256	***	0.0235	***	0.0218	***
LogSO_GLOBA	0.1112	***	***		0.0649	***	0.0879	***	0.1054	***
LogGDP_PC	−0.0017	***	***		−0.0287	***	−0.0241	***	−0.0207	***
LogURBAN	0.0739	***	***		0.1814	***	0.1685	***	0.1586	***
LogENE	0.0024				0.0464	***	0.0342	***	0.0249	***
CO_2	0.0001				0.0017	***	0.0020	***	0.0022	***
Constant	2.9950	***	***		NA		NA		NA	
Obs	798		798		798		798		798	
Wald test (chi2)	F(6791) = 410.77	***	F(6791) = 368.10	***	chi2(6) = 3970.79	***	chi2(6) = 6128.82	***	chi2(6) = 3371.21	***
Breusch-Pagan/Cook-Weisberg test	chi2(1) = 49.31	***	NA		NA		NA		NA	

*** denotes statistically significant at the 1% level; *(Log)* denotes variables in natural logarithms; *NA* denotes not available.

How to do:

```
**Pooled OLS**

reg l_logbmi_total l_logeco_globa l_log_so_globa l_loggdp_pc l_logurban l_logene l_logco2

**The Wald test**

testparm l_logbmi_total l_logeco_globa l_log_so_globa l_loggdp_pc l_logurban l_logene l_logco2

**Breusch-Pagan/Cook-Weisberg test**

reg l_logbmi_total l_logeco_globa l_log_so_globa l_loggdp_pc l_logurban l_logene l_logco2
hettest

**Pooled OLS robust**

reg l_logbmi_total l_logeco_globa l_log_so_globa l_loggdp_pc l_logurban l_logene l_logco2, robust


**The Wald test**

testparm l_logbmi_total l_logeco_globa l_log_so_globa l_loggdp_pc l_logurban l_logene l_logco2


**The panel quantile model **

xtqreg l_logbmi_total l_logeco_globa l_log_so_globa l_loggdp_pc l_logurban l_logene l_logco2, id(country) quantile(.25)

**The Wald test**

testparm l_logbmi_total l_logeco_globa l_log_so_globa l_loggdp_pc l_logurban l_logene l_logco2

** The panel quantile model **

xtqreg l_logbmi_total l_logeco_globa l_log_so_globa l_loggdp_pc l_logurban l_logene l_logco2, id(country) quantile(.50)

**The Wald test**

testparm l_logbmi_total l_logeco_globa l_log_so_globa l_loggdp_pc l_logurban l_logene l_logco2
```

**** The panel quantile model****

```
xtqreg l_logbmi_total l_logeco_globa l_log_so_globa l_loggdp_pc l_logurban
l_logene l_logco2, id(country) quantile(.75)
```

****The Wald test****

```
testparm l_logbmi_total l_logeco_globa l_log_so_globa l_loggdp_pc l_logurban
l_logene l_logco2
```

Having found that the urbanisation process exerts a positive effect on the overweight epidemic in LAC countries, the question remains: What are the explanations for the positive impact of the urbanisation process on the BMI index equal to or greater than 25 in LAC countries? The process of urbanisation in the LAC region has been moving side by side with the globalisation process. Indeed, this process of interaction affects food production, access, and consumption. The process of urbanisation allows better food accessibility, caused by the significant presence of fast-food chains, supermarkets, and multinational supermarkets offering a ready supply of processed foods (Reardon et al., 2003). According to the same authors, the presence of these enterprises causes a decline in farm stands and open markets with healthier foods. Hawkes (2006) points out that the population is exposed to mass media marketing of food and beverages that can influence traditional diets in the LAC region in the urban areas.

Moreover, urban areas require less energy caloric expenditure related to commuting and leisure activities. More frequent travel by car and less walking or biking for transportation or leisure contributes to overweight and obesity (e.g., Lindstrom, 2008 and Kjellstrom et al., 2007). This process also encourages a more densely populated neighbourhood that reduces strenuous outdoor activities due to the limited recreational space (Pirgon & Aslan, 2015). Finally, the process of urbanisation influences the creation of more sedentary jobs, such as desk and manufacturing jobs, reducing the creation of active jobs, such as farming (e.g., Fox et al., 2019 and Kjellstrom et al., 2007). All these are influences on the increase of consumption of caloric and the reduction of physical effort and less caloric expenditure and so bring about an increase in weight gain. For this reason, the urbanisation process is viewed as a key underlying driver of overweight trends in LAC countries. Moreover, the capacity of urbanisation to increase obesity in LAC countries was confirmed in Chapter 3, where an impact of 1.0272 was found.

After the possible explanation of the positive impact of the urbanisation process on the overweight epidemic in LAC countries, a second question remains: What are the possible explanations regarding the positive effect of economic and social globalisation, consumption of energy, and CO_2 emissions, as well as the negative effect of economic growth on the BMI index equal to or greater than 25 in LAC countries?

The positive effect of economic globalisation on the BMI index in the region is due to this process of economic globalisation in the region which allows the entrance of multinational food corporations, fast-food chains and multinational supermarket chains that offer a ready supply of processed foods (Fox et al., 2019). Moreover, this process also allows access to modern technologies that minimise physical activity levels (Sobal, 2001). Moreover, the capacity for economic globalisation to increase obesity was found in Chapter 3, where this variable causes an increase of 0.0079.

Indeed, the positive impact of social globalisation is related to the capacity of this process to influence the adoption of a more robust fast-food/processed foods culture through McDonaldisation or Cocalisation processes leading to more consumption of caloric (energy) and so the increase in weight gain (Fox et al., 2019). Indeed, this Westernisation caused by social globalisation can also influence the use of inexpensive transportation, communication, and other activity-sparing systems through automobiles and household appliances that minimise physical activity levels (Sobal, 2001). Moreover, the capacity of social globalisation to increase obesity was found in Chapter 3, where this variable causes an increase of 0.0324.

The positive impact of consumption of energy and CO_2 emissions on the overweight epidemic in the region is related to the intensive use of modern household appliances, as well as the reliance on motor vehicles as a mode of transportation that contributes to reducing physical effort and lower caloric (energy) expenditure and so an increase in weight gain (Fox et al., 2019). On the other hand, the intensive use of modern household appliances and reliance on motor vehicles increases energy consumption and so the emissions of CO_2 in the region. This is influenced by the urbanisation process, economic and social globalisation, and economic development in LAC countries.

Finally, the negative impact of economic growth on the overweight epidemic in LAC countries (that is, economic growth decreases the overweight epidemic) is due to the fact that wealthy societies can pay for healthy food. This effect is expected to be detected among women, who are traditionally associated with health food concerns. Indeed, this effect was statistically significant for women, but the effect of economic growth was revealed to be statistically insignificant as a driver of overweight in men. As the quantiles go up, the effect vanishes in line with that predicted by economic theory reflecting a smaller food weight in the family's income. However, in Chapter 3 of this book, economic growth increases the overweight and obesity problem. This discovery could be related to the use of different countries and methods.

Having found the drivers for the increase in overweight and obesity in LAC countries, it is necessary to ascertain whether this problem can affect the region's economic development. Therefore, Chapter 5 will approach the effect of the obesity problem on economic growth in LAC countries.

4.4 Conclusion

This chapter investigated the effect of the urbanisation process on the overweight epidemic in 19 countries from LAC countries from 1975 to 2016. The panel quantile model with fixed effects was used to answer the initial research question, and the choice of method proved suitable for this kind of study. This approach is possible because of the research aims defined in this study. Furthermore, it was opted to use an unusual structure and approach in this investigation (e.g., availability of Stata commands and didactic figures) to simplify the investigation and understanding for readers, and the replication of this study in another group of countries and a different context, to be more instructive.

The preliminary tests indicated the non-presence of normality in the models, whereby in the Shapiro–Wilk W-test for normal data, the null hypothesis of this test is rejected. Moreover, the skewness and kurtosis test verifying the normality in the models' residuals indicated that the null hypothesis could also be rejected. Therefore, the results from both tests reject the null hypothesis at a significance level of 1% and 5%. The results of these tests are added support for the suitability of using quantile regression. The results of the VIF-test showed that the presence of multicollinearity is not a concern in the estimation of each model, given the low VIF and mean VIF values registered, which are lower than the usually accepted benchmark of 10 in the case of the VIF values, and six in the case of the mean VIF values. The results of the CSD-test showed that all variables have the presence of cross-section dependence.

The results from the CIPS-test indicated that some variables are borderline between I (0) and I (1). Indeed, because most of the variables appear to be borderline between I(0) and I(1), this investigation opted not to perform the Westerlund cointegration test to verify the presence of cointegration between the variables. The Hausman test indicated that the null hypothesis should be rejected in all models and that a fixed-effects model is the most appropriate for this analysis.

The pooled OLS and OLS robust estimators of model I (BMI_MAN) showed that economic and social globalisation, urbanisation, and energy consumption increase the man overweight. In contrast, economic growth decreases the overweight. However, the CO_2 emissions point to be insignificant statistically and do not cause any impact on man overweight. Regarding the Panel quantile model, the 25th, 50th, and 75th quantiles showed that model I, the economic and social globalisation, urbanisation, consumption of energy, and CO_2 emissions increase the man overweight. On the other hand, economic growth does not cause any impact on the dependent variable because it is insignificant statistically. Moreover, the post-estimation tests (e.g., Breusch-Pagan/Cook-Weisberg test for heteroskedasticity and the Wald test) of model I showed the presence of heteroscedasticity in the model. The result from the post-estimation test is an indicator that the estimations that this investigation use is adequate.

The results from the pooled OLS and OLS robust from the model II (BMI_-WOMAN) pointed out that the economic and social globalisation, and urbanisation, increase the number of women with overweight. In contrast, economic growth decreases the overweight. However, the consumption of energy and CO_2 emissions point to be insignificant statistically and does not cause any impact on women overweight. Regarding the Panel quantile model, the 25th, 50th, and 70th quantiles showed that the economic and social globalisation, urbanisation, energy consumption, and CO_2 emissions increase the woman overweight, while economic growth decreases the overweight. Moreover, the post-estimation tests (e.g., Breusch-Pagan/Cook-Weisberg test for heteroskedasticity and the Wald test) of model II showed the presence of heteroscedasticity in the model. The result from the post-estimation test is an indicator that the estimations that this investigation use is adequate.

The results from the pooled OLS and OLS robust from the model III (BMI_TOTAL) pointed out that the economic and social globalisation, and urbanisation, increase the total overweight, while the economic growth decreases the total overweight. However, the consumption of energy and CO_2 emissions point to be insignificant statistically and does not cause any impact on total overweight. Regarding the panel quantile model with fixed effects, the 25th, 50th, and 75th quantiles showed that the economic and social globalisation, urbanisation, energy consumption, and CO_2 emissions increase the total overweight, while economic growth decreases the overweight. Moreover, the post-estimation tests (e.g., Breusch-Pagan/Cook-Weisberg test for heteroskedasticity and the Wald test) from model III showed the presence of heteroscedasticity in the model. The result from the post-estimation test is an indicator that the estimations that this investigation use is adequate.

Therefore, the models' regression showed that the urbanisation process positively correlates with the overweight epidemic in LAC countries, meaning that as urbanisation expands, so does the epidemic. This effect is related to the fact that urbanisation allows for better (unhealthy) food accessibility, caused by the significant presence of fast-food chains, supermarkets, and multinational supermarkets, offering a ready supply of processed foods. This process is also sustained by the increase in travelling by car and less walking or biking for transportation or leisure, which translates to reduced energy caloric expenditure and overweight and obesity. Moreover, this process also promotes sedentary jobs, such as desk and manufacturing jobs, and reduces physically demanding jobs, such as farming or gardening.

This study contributes to discussing this topic in literature and contributes with well-grounded knowledge to policymakers and governments. Furthermore, the findings of this study help to justify and develop more initiatives to reduce the overweight epidemic in LAC countries. Finally, the findings of this study open a new topic of research in the literature concerning the relationship between urbanisation, globalisation, energy consumption, environmental degradation, and obesity.

References

Afonso, T. L., Marques, A. C., & Fuinhas, J. A. (2019). *Energy−growth nexus and economic development: A quantile regression for panel data*. The Extended Energy-Growth Nexus. https://doi.org/10.1016/b978-0-12-815719-0.00001-2

Agresti, A. (1990). *Categorical data analysis*. New York: John Wiley and Sons. ISBN 0-471-36093-7.

Arsht, A. (2014). *Urbanization in Latin America. Atlantic council*. URL: https://www.atlanticcouncil.org/commentary/article/urbanization-in-latin-america/#:∼:text=Today%2C%20about%2080%20percent%20of,the%20world's%20most%20urbanized%20region.&text=Today%2C%20260%20million%20people%20live,percent%20of%20Latin%20America's%20GDP.

Belsley, D. A., Kuh, E., & Welsch, R. E. (1980). *Regression diagnostics: Identifying influential data and sources of collinearity*. New York: Wiley. https://doi.org/10.1002/0471725153

Böckerman, P., Viinikainen, J., Pulkki-Råback, L., Hakulinen, C., Pitkänen, N., Lehtimäki, T., Pehkonen, J., & Raitakari, O. T. (2017). Does higher education protect against obesity? Evidence using mendelian randomisation. *Preventive Medicine, 101*, 195−198. https://doi.org/10.1016/j.ypmed.2017.06.015

Breusch, T. S., & Pagan, A. R. (1979). A simple test for heteroscedasticity and random coefficient variation. *Econometrica, 47*, 1287−1294. https://doi.org/10.2307/1911963

Chee, V. A., Teran, E., Hernandez, I., Wright, L., Izurieta, R., Reina-Ortiz, M., Flores, M., Bejarano, S., Dào, L. U., Baldwin, J., & Martinez-Tyson, D. (2019). Desculturization rbanisation, and nutrition transition among urban Kichwas Indigenous communities residing in the Andes highlands of Ecuador. *Public Health, 176*, 21–28. https://doi.org/10.1016/j.puhe.2019.07.015

Chernozhukov, V., & Hansen, C. (2008). Instrumental variable quantile regression: A robust inference approach. *Journal of Econometrics, 142*, 379–398. https://doi.org/10.1016/j.jeconom.2007.06.005

Cook, R. D., & Weisberg, S. (1983). Diagnostics for heteroscedasticity in regression. *Biometrika, 70*, 1–10. https://doi.org/10.2307/2335938

Costa-Font, J., & Mas, N. (2016). "Globesity"? The effects of globalisation on obesity and caloric intake. *Food Policy, 64*, 121–132. https://doi.org/10.1016/j.foodpol.2016.10.001

D'Agostino, R. B., Belanger, A. J., & D'Agostino, R. B., Jr. (1990). A suggestion for using powerful and informative tests of normality. *American Statistician, 44*, 316–321. https://doi.org/10.2307/2684359

Fox, A., Feng, W., & Asal, V. (2019). What is driving global obesity trends? Globalisation or "modernisation"? *Globalisation and Health, 15*(32), 1–16. https://doi.org/10.1186/s12992-019-0457-y

Fuinhas, J. A., Koengkan, M., & Santiago, R. (2021). *Physical capital development and energy transition in Latin America and the Caribbean*. Elsevier. https://doi.org/10.1016/C2020-0-01491-X

Goryakin, Y., & Suhrcke, M. (2014). Economic development, urbanisation, technological change and overweight: What do we learn from 244 demographic and health surveys? *Economics and Human Biology, 14*(1), 109–127. https://doi.org/10.1016/j.ehb.2013.11.003

Hawkes, C. (2006). Uneven dietary development: Linking the policies and processes of globalisation with the nutrition transition, obesity and diet-related chronic diseases. *Globalisation and Health, 2*, 4. https://doi.org/10.1186/1744-8603-2-4

Ismailov, R. M., & Leatherdale, S. T. (2010). Rural-urban differences in overweight and obesity among a large sample of adolescents in Ontario. *International Journal of Pediatric Obesity, 5*(4), 351–360. https://doi.org/10.3109/17477160903449994

Kjellstrom, T., Hakansta, C., & Hogstedt, C. (2007). Globalisation and public health-overview and a Swedish perspective. *Scandinavian Journal of Public Health, 70*, 2–68. https://doi.org/10.1080/14034950701628494

Koengkan, M. (2020). Capital stock development in Latin America and the Caribbean region and their effect on investment expansion in renewable energy. *Journal of Sustainable Finance and Investment*, 1–18. https://doi.org/10.1080/20430795.2020.1796100

Koengkan, M., & Fuinhas, J. A. (2020a). Exploring the effect of the renewable energy transition on CO_2 emissions of Latin American and Caribbean countries. *International Journal of Sustainable Energy*, 1–24. https://doi.org/10.1080/14786451.2020.1731511

Koengkan, M., & Fuinhas, J. A. (2020b). The interactions between renewable energy consumption and economic growth in the Mercosur countries. *International Journal of Sustainable Energy, 39*(6), 594–614. https://doi.org/10.1080/14786451.2020.1732978

Koengkan, M., & Fuinhas, J. A. (2020c). Does the overweight epidemic cause energy consumption? A piece of empirical evidence from the European region. *Energy, 216*, 119297. https://doi.org/10.1016/j.energy.2020.119297

Koengkan, M., Fuinhas, J. A., & Fuinhas, C. (2021). *Does urbanisation process increase the overweight epidemic? The case of Latin America and the Caribbean region* (pp. 1–9). SSRN. https://doi.org/10.2139/ssrn.3826196

Koengkan, M., Fuinhas, J. A., & Silva, N. (2021). Exploring the capacity of renewable energy consumption to reduce outdoor air pollution death rate in Latin America and the Caribbean region. *Environmental Science and Pollution Research, 28*(2), 1656–1674. https://doi.org/10.1007/s11356-020-10503-x

Koengkan, M., Santiago, R., & Fuinhas, J. A. (2019). The impact of public capital stock on energy consumption: Empirical evidence from Latin America and the Caribbean region. *International Economics*, 1–20. https://doi.org/10.1016/j.inteco.2019.09.001

KOF Globalization index (KOF). (2021). URL: https://www.kof.ethz.ch/en/forecastsand-indicators/indicators/kof-globalisation-index.html.

Lake, A., & Townshend, T. (2006). Obesogenic environments: Exploring the built and food environments. *Journal of The Royal Society for the Promotion of Health, 126*(6), 262–267. https://doi.org/10.1177/1466424006070487

Lindstrom, M. (2008). Means of transportation to work and overweight and obesity: A population-based study in southern Sweden. *Preventive Medicine, 46*, 22–28. https://doi.org/10.1016/j.ypmed.2007.07.012

Machado, J. A. F., & Silva, J. M. C. S. (2019). Quantiles via moments. *Journal of Econometrics*. https://doi.org/10.1016/j.jeconom.2019.04.009Get. forthcoming.

Martins, M. L. R. (2002). *Meio ambiente e morada social nos países do Mercosul*. Madrid.

OECD-FAO. (2019). *OECD-FAO agricultural Outlook 2019-2028: Special focus: Latin America*. URL: http://www.fao.org/3/ca4076en/CA4076EN.pdf.

Our World in Data. (2021). *Obesity*. URL: https://ourworldindata.org/obesity.

Pesaran, M. H. (2004). *General diagnostic tests for cross-section dependence in panels*. The University of Cambridge, Faculty of Economics. https://doi.org/10.17863/CAM.5113. Cambridge Working Papers in Economics, n. 0435.

Pesaran, M. H. (2007). A simple panel unit root test in the presence of cross-section dependence. *Journal of Applied Econometrics, 22*(2), 256–312. https://doi.org/10.1002/jae.951

Pirgon, Ö., & Aslan, N. (2015). The role of urbanization in childhood obesity. *Journal of Clinical Research in Pediatric Endocrinology, 7*(3), 163–167. https://doi.org/10.4274/jcrpe.1984

Popkin, B. M. (1999). Urbanisation, lifestyle changes and the nutrition transition. *World Development, 27*(11), 1905–1916. https://doi.org/10.1016/S0305-750X(99)00094-7

Popkin, B. M., Adair, L. S., & Ng, S. W. (2012). Global nutrition transition and the pandemic of obesity in developing countries. *Nutrition Reviews, 70*(1), 3–21. https://doi.org/10.1111/j.1753-4887.2011.00456.x

Poveda, Y. E. M., Koengkan, M., & Fuinhas, J. A. (2020). Exploring the interactions between renewable energy, economic growth, agriculture and urbanisation in the Mercosur trade-bloc countries. *Revista Valore, 5*, 1–22. https://doi.org/10.22408/reva502020548e-5049

Powell, D. (2016). *Quantile regression with nonadditive fixed effects*. RAND Working Paper. URL: http://works.bepress.com/david_powell/14.

Reardon, T., Timmer, C. P., Barrett, C. B., & Berdegu, J. (2003). The rise of supermarkets in Africa, Asia, and Latin America. *American Journal of Agricultural Economics, 85*(5), 1140–1146. URL: https://www.jstor.org/stable/1244885.

Royston, P. (1983). A simple method for evaluating the Shapiro-Francia W' test for non-normality. *Statistician, 32*, 297–300. https://doi.org/10.2307/2987935

Siervo, M., Grey, P., Nyan, O. A., & Prentice, A. M. (2006). Urbanisation and obesity in the Gambia: A country in the early stages of the demographic transition. *European Journal of Clinical Nutrition, 60*(4), 455–463. https://doi.org/10.1038/sj.ejcn.1602337

Sobal, J. (2001). Commentary: Globalisation and the epidemiology of obesity. *International Journal of Epidemiology, 30*, 1136–1137. https://doi.org/10.1093/ije/30.5.1136

Wang, R., Feng, Z., Liu, Y., & Qiu, Y. (2020). Is lifestyle a bridge between urbanisation and overweight in China? *Cities, 99*, 102616. https://doi.org/10.1016/j.cities.2020.102616

World Bank Open Data. (2021). URL: http://www.worldbank.org/.

Impact of the obesity epidemic on economic growth in Latin American and Caribbean countries

5

5.1 Introduction

Globally, 13% of adults over 18 years were obese in 2016, while in 1990, this value was 6.8%. That means an astonishing increase of 91% between 1990 and 2016. In most high-income countries such as the United States, 36% of adults were obese in 2016, while in the United Arab Emirates, this was 32%, in the United Kingdom, 28%, and in the European region, 23% were obese. However, in upper-middle-income economies, such as China, obesity reached 6.2% of the adult population in 2016, while in the Latin America (LA) region, it reached 19% of the adult population. Moreover, in the lower-middle-income economies, for example, India, the obesity problem reached 4% of the adult population.

Indeed, in LA countries, there has been an increase in the prevalence of obesity since the 1980–90s. The adult obesity prevalence has tripled since 1975 (UN, 2019), and economic growth resulted in a gross domestic product (GDP) per capita 12 times higher in 2016 than in 1970 (see Chapter 4). Therefore, as mentioned before, 19% of the adult population in the LA region was obese in 2016, while this value was 9% in 1990 (see Fig. 5.1 below).

The increase of the share of adults that are obese is related to rapid economic development, as mentioned earlier. Indeed, when we approach leading LA economies (e.g., Argentina, Brazil, Chile, Colombia, Ecuador, Mexico, Peru and Venezuela), we can identify this growth trend in the obesity problem (see Fig. 5.2 below).

Therefore, the obesity problem reached 17% of the adult population in Argentina, while in Chile, it was 17%, Mexico 16%, Venezuela 15%, Colombia 12%, Peru 10% and Brazil 10%, and in Ecuador 9% were obese in 1990. However, in 2016, the obesity problem reached the following values, Mexico 30%, Chile 28%, Argentina 28%, Venezuela 26%, Colombia 22%, Brazil 22%, Peru 20% and Ecuador 20%. Exhibit 5.1 discusses the increase of overweight and obesity problems in the LAC region.

It has been recognised that obesity results from economic growth. The nutritional transition approach explains that as countries experience economic growth and increases in income, there is a decrease in poverty and famine; as the economic structure of the countries changes, people start to adapt to new lifestyles. This new lifestyle is characterised by physical inactivity and a 'Western diet,' particularly in low- and middle-income countries, as explained in Chapter 2. The change toward an unhealthy lifestyle drives up the prevalence of obesity, which rose very fast in LAC countries (Popkin et al., 2012).

Obesity Epidemic and the Environment. https://doi.org/10.1016/B978-0-323-99339-5.00009-1

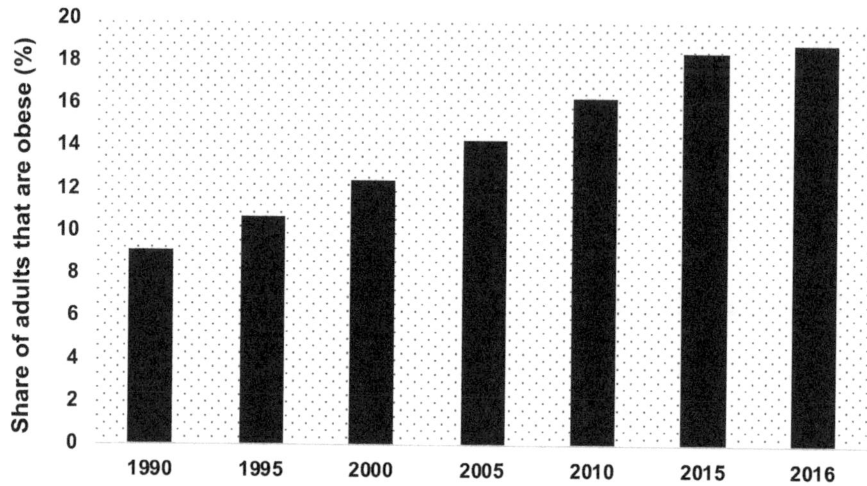

Figure 5.1 Share of adults that are obese (%) in the Latin America region, between 1990 and 2016. Being overweight is defined as having a body mass index greater than or equal to 25. BMI is a person's weight in kilograms divided by their height in meters squared.
The authors created this figure with the Our World in Data database (2021). Obesity. https://ourworldindata.org/obesity.

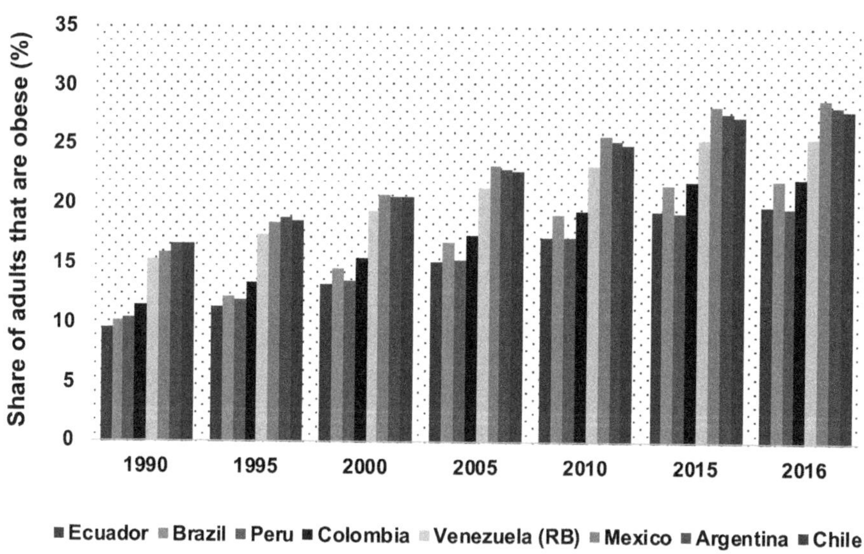

Figure 5.2 Share of adults that are obese (%) in major Latin America economies, between 1990 and 2016. Being overweight is defined as having a body mass index (BMI) greater than or equal to 25. Obesity is characterised by a BMI greater than or equal to 30. BMI is a person's weight in kilograms divided by his or her height in meters squared.
The authors created this figure with the Our World in Data database (2021). Obesity. https://ourworldindata.org/obesity.

Exhibit 5.1 Obesity and overweight problems in the Latin American and Caribbean region

More than 300 million adults were overweight in the LAC region, and of these, more than 100 million were obese in 2014. Indeed, obesity and overweight are defined as abnormal or excessive fat accumulation that may impair health (Garcia-Garcia, 2021). In the world, 39% (2.0 billion) of the adult population (38% of men and 40% of women) were overweight and close to 13% (600 million, 11% of men and 15% of women) were obese in 2014. The global prevalence of obesity more than doubled between 1975 and 2014 (International Food Policy Research Institute, 2016; NCD-RisC, 2016; and World Health Organization (WHO) 2020).

Obesity has become a significant health challenge in the LAC region. Around 57% (302 million) of the adult population in the LAC region (54% men and 70% of women) are overweight, while 19% (100.8 million) are obese (15% in men and 24% in women) (Garcia-Garcia, 2021).

In other low-middle-income countries, the impact of the overweight problem stands a 61% in women and 54% in men and of the obesity problem at 24% in women and 15% in men. Indeed, the obesity problem is more prevalent in women than in men. Fourteen LAC countries have a female prevalence greater than 20%. The highest prevalence of obesity problem in the adult population is found in El Salvador (33%) and Paraguay (30%) for women and in Uruguay (23%) and Chile (22%) for men (Ng et al., 2014).

Moreover, the prevalence of overweight and obesity in children in the LAC region is also high, where it impacts 16% of children. It ranges from more than 12% for girls in Chile, Uruguay and Costa Rica to less than 5% in Bolivia, Ecuador, Peru, Honduras and Guatemala. Indeed, the highest prevalence of obesity in children is found in Chile (12%) and Mexico (11%) in boys and Uruguay (18%) and Costa Rica (12%) in girls (Ng et al., 2014).

Therefore, overweight and obesity are significant risks for non-communicable diseases like cardiovascular disease (heart disease and stroke). Furthermore, the leading cause of death (30% of all causes) in the LAC region is diabetes, hypertension and chronic kidney disease (Garcia-Garcia, 2021).

The question raised in this chapter looks at the relationship between obesity and economic growth from a different perspective. It aims to establish a causal relationship between obesity and economic development. This causality arises because obesity emerges from a particular lifestyle. The 'Western,' lifestyle is based on industrialised and processed food and a low level of physical activity. So, on the one hand, the growth and development of the food industry and its distribution logistics contribute to economic growth.

On the other hand, people look for technology that minimises physical activity like transport and home appliances (Koengkan & Fuinhas, 2021). These transformations were described in Chapter 3 and contributed to the creation of value-added and economic growth. Additionally to this central question, this chapter also explores how globalisation, food production and renewable energy production contribute to economic growth in LAC countries.

This chapter contributes to the literature about the drivers of economic growth. It introduces the role of the obesity pandemic and food production in economic growth, and it also accounts for renewable energy production and globalisation. In the next section, a literature review is presented about these topics. Following the literature review, the methods of analysis and the results are described. The chapter finishes with the conclusions.

5.2 Literature review

5.2.1 *Economic growth and obesity*

The relationship between economic growth and obesity has been discussed in Chapter 3. It is generally accepted that economic growth is associated with obesity, and a causal relationship can also be established between them. Chapter 3 has shown that it is possible to find that obesity positively affects economic growth and economic growth causes obesity. Both these causal relationships are observable in LAC countries.

Economic growth has a crucial role in the improvement of living conditions, in particular in low- and middle-income countries. In these countries, economic growth results in higher incomes, a decrease in poverty, and it guarantees improvements in peoples' health (e.g., Jack & Lewis, 2001; and Riley, 2001).

Nevertheless, economic growth also has a negative side. It causes several losses in different areas of society, like environmental degradation and negative health returns, such as increasing the prevalence of obesity (Costa-Font & Mas, 2016). Thus, obesity is the negative effect created by fast economic growth registered in LAC countries. In general, in low- and middle-income countries, economic development and the increase of GDP per capita correlate with the rise in overweight and obese people, especially among the lower socio-economic groups (Popkin et al., 2012). In these countries, the income transition experienced by people in lower socio-economic groups provides a change and an incentive to consume more fatty food and energy-dense animal-source food as they move away from the food insecurity (e.g., Gerbens-Leenes et al., 2010; Roskam et al., 2010; and Sullivan et al., 2008).

The economic growth in LAC countries was accompanied by a phenomenon of urbanisation where people migrated from rural areas to urban centres. Some of the world's largest cities are found in LAC countries, and 80% of the population lives in cities (e.g., Arsht, 2014; Jaitman, 2015; and Poveda et al., 2020). The effect of urbanisation on the body mass index of these countries is analysed in Chapter 4 of this book.

On the one hand, urbanisation provides suitable living conditions, a better quality of life, more job opportunities, and a larger range of services (Rauch, 1993). On the other hand, the development of urbanisation in LAC happened with no urban planning and very fast. A significant portion of urban dwellers lives in poor conditions and neighborhoods lacking healthy living infrastructures.

Urban migrants arriving from rural areas often ended up living in slums and poor outskirts. Urban poverty is a consequence of the urbanisation process, and about 60% of the poor live in cities (e.g., Jaitman, 2015; Popkin, 1998; and UN, 2003). The decrease in poverty levels in LAC countries in the last 40 years did not mean its extinction. Poverty is still an economic and social problem in the region, and it is mainly concentrated in urban areas (e.g., Stampini et al., 2015; and Vakis et al., 2015).

Urbanisation arising from the economic growth in LAC countries is the reason for the increasing income inequality and prevalence of poverty in cities (Pedraza, 2009). Moreover, urbanisation is linked to increased overweight and obesity (e.g., Fox et al., 2019; Koengkan et al., 2021, pp. 1–7; Popkin et al., 2012; and Popkin, 1999). Chapter 4 establishes a direct relationship between urbanisation and the obesity epidemic in LAC countries. Urbanisation has created space for healthy living among people with a lower socio-economic level (Chee et al., 2019), but it did not provide a system of healthy food supply. People ended up abandoning their traditional diets based on fresh food and have adopted a diet based on precooked meals, fast food, and other sorts of industrialised and processed food (e.g., Popkin et al., 2012; and Popkin, 1999).

Economic growth in LAC countries is also related to globalisation, which explains the exponential increase of obesity in these countries (Costa-Font & Mas, 2016). The relation between economic growth and globalisation is described below. Globalisation plays a vital role in the dynamic process of nutritional transition (Hawkes, 2006). Globalisation is the feature, which explains the changes in people's diet and physical activity. Peoples' living places and environment are crowded with marketing, fast food chains and colourful and tasty highly processed food, an obesogenic risk (Lake & Townshend, 2006) that contributes to the excess and unhealthy calorie intake.

Economic growth is a complex process involving developing other societal processes, such as urbanisation and globalisation. This later process, in turn, contributes to other changes, such as peoples' lifestyles, which end up promoting unhealthy routines and obesity.

5.2.2 Economic growth and globalisation

Globalisation is a multidimensional and dynamic process that most countries have experienced. It is the country's integration into the world market, and this integration happens at different levels of society: capital, finance, consumption, trade, labour and cultural characteristics.

Economic growth and globalisation are strongly related and nearly inseparable economic phenomena. Globalisation impacts economic growth through several channels: international trade, financial integration, international labour flows and technological change. One of the first studies analysing the relationship between globalisation and economic growth was presented by Dreher (2006), who used the well-known globalisation index (KOF globalisation index used in our analysis) on 123 countries over 30 years. Several other studies have confirmed that globalisation contributes to economic growth in developing or developed countries (e.g., Chang & Lee, 2010; Polasek & Sellner, 2011; and Rao et al., 2011). For instance, in the Association of Southeast Asian Nations (ASEAN) countries, Ying et al. (2014) shows that economic globalisation has contributed to economic growth but not social globalisation for nearly 40 years.

LAC countries began to join the global economy and network in the 1980s. Then, after the IMF structural stabilising programs, the countries were under financial liberalisation, with openness to foreign direct investment, an increase in international trade and liberalised economies, which completed the globalisation integration process of these economies. For instance, in the period 1990–2009, the degree of economic openness increased by 50.7% (Carneiro, 2012). Additionally, already in the 21st century, between 2000 and 2014, the 'commodities boom,' accelerated the globalisation of the region even more (e.g., Fuinhas et al., 2021a,b; Koengkan & Fuinhas, 2020a,b; and Poveda et al., 2020). To sum up, economic growth and globalisation are strongly intertwined processes, and they contribute to each other's enlargement.

5.2.3 Economic growth and food production

Economic growth is associated with food production, as discussed in Chapter 8. This chapter showed that economic growth is a driver of food production in LAC countries. Economic growth also promotes the increase of food production. As the population increases, there is a demand for food, which requires an increase in food production (Fukase & Martin, 2020), and this increase in food production contributes to economic growth. On the other hand, as a country that registers economic growth, particularly those low- and middle-income countries, better living conditions and higher income allow people to move away from food insecurity and demand more food.

Economic growth in LAC countries is also related to the urbanisation process. The economic growth registered since the 1980s goes together with the increasing size of the cities. These two processes have changed the food industry and food systems, both production and distribution. The increase in supply has taken place from the production on the farm to the sale in supermarkets or restaurants. National and foreign investment was directed to agriculture—agribusiness—to increase production (Gereffi, 1990); the improved logistic and transportation methods enlarged the possibilities of distribution; the development of supermarket networks and fast food chains are the result of the modernised procurement systems for industrialised and processed food (Fox et al., 2019).

Between 2000 and 2014, the 'commodities boom,' also increased food production and economic growth (e.g., Fuinhas et al., 2021a,b; Koengkan & Fuinhas, 2020a,b; and Poveda et al., 2020). The boom of several physical commodities (see **Exhibit 3.3** in Chapter 3), including food, increased their production to supply the national and foreign markets, increasing exports and foreign direct investment. In this way, food production contributed to the economic growth of LAC countries.

The nutritional transition is another perspective on the relation between economic growth and food production in LAC countries. As countries experience economic growth, there is a change in the lifestyle of people. Physical inactivity becomes a common choice, and nutritional diets lose out to 'Western diets,' mainly based on processed and highly processed food. This change in diet contributes to increased demand for industrialised food (Fukase & Martin, 2020), contributing to economic growth. On the other hand, economic growth increases income per capita and reduces poverty (see **Exhibits 2.2** and **3.3** in Chapters 2 and 3) and food insecurity, contributing to the increased demand for food (Manap & Ismail, 2019). Lastly, there is another path to explain the rise in the consumption of processed foods from multinational food corporations, fast-food chains and multinational supermarket chains, resulting from the prevalence of overweight and obesity itself (Koengkan & Fuinhas, 2021).

5.2.4 Economic growth and renewable energy production

The relationship between economic growth and renewable energy use has long been studied (Kraft & Kraft, 1978). The nexus of this relationship is based on different possible linkages:

(1) The growth hypothesis; (2) the conservation hypothesis; (3) the feedback hypothesis; and (4) the neutrality hypothesis. The growth hypothesis establishes a unidirectional positive association from renewable energy consumption to economic growth; the conservation hypothesis states that economic growth increases the renewable energy consumption; the feedback hypothesis stands for a bidirectional relationship; and finally, the neutrality hypothesis defends the absence of any significant relationship between economic growth and renewable energy consumption. There is evidence for each of these hypotheses, and there may be no clear consensus on the results found for individual countries or groups of countries, as concluded in a survey performed by Payne (2010).

The nexus between economic growth and renewable energy adopted in this chapter is based on the growth hypothesis, that is, renewable energy fosters economic development in LAC countries (e.g., Asafu-Adjaye, 2000; Masih & Masih, 1996; Narayan & Smyth, 2008; Soytas & Sari, 2003; Squalli, 2007; and Yu & Choi, 1985).

Not long ago, the Kyoto Protocol 1997, the Paris Agreement 2015, and the Copenhagen Summit 2009 considered that economies, including emerging and developing, ought to slow down economic growth to reduce carbon emissions to slow down global warming. However, the idea of sacrificing economic growth is not entirely welcome in emerging countries such as Brazil. The aim of maintaining the levels of economic development requires the increasing use of energy, no matter the source of production. Fossil fuels are used to produce energy, but renewable energy is also a credible and efficient energy source to feed economic growth.

In a globalised economy, technologies are fast adopted all around the world. LAC countries have been integrated into the global economy and access to environmental technologies is available at any time. A recent study by Singh et al. (2019) found that renewable energy production contributes positively to economic growth, but these effects were more decisive in developing countries than in developed countries. This result is identical to that found by Bhattacharya et al. (2016). They used the renewable energy country attractiveness index developed by Ernst & Young Global Limited in 38 top renewable energy-consuming countries to explain economic growth in

1991–2012. They confirmed that the long-run elasticities indicate that renewable energy consumption had a significant positive impact on most countries' economic growth. Using a different perspective and only focused on developed countries, (Ntanos et al., 2018) found that among 25 European countries, there is a higher correlation between renewable energy consumption and economic growth in countries with higher GDP than those with lower GDP.

In our work, the growth hypothesis is tested in the set of LAC countries for the period of significant economic growth, followed by a slowdown (1990–2016). Our central hypothesis is that renewable energy production contributes positively to the economic growth in the countries in this region.

5.3 Data and method

This section will be divided into two parts. The first will approach the group of countries and data/variables that the chapter will use, while the second will show the method.

5.3.1 Data

This chapter will use annual data that collected from 1991 to 2016 of 21 countries in the LAC region, i.e., **Argentina**, **Belize**, **Bolivia**, **Brazil**, **Chile**, **Colombia**, **Costa Rica**, **Dominican Republic**, **Ecuador**, **El Salvador**, **Guatemala**, **Guyana**, **Haiti**, **Honduras**, **Jamaica**, **Mexico**, **Nicaragua**, **Panama**, **Paraguay**, **Peru** and **Uruguay**. The use of time-series between 1991 and 2016 is due to data availability until 2016 for the variable OBESITY for all countries selected. The variables, which were chosen to perform this investigation, will be shown in Table 5.1 below.

'Obs.' denotes the number of observations in the model, 'Std.-Dev.' denotes the standard deviation, and 'Min and Max' denote minimum and maximum. All variables are in natural logarithms to harmonise the interpretation of results and linearise the relationships between variables. These summary statistics were obtained from the command *sum* of **Stata 16.0**. The board below shows how to obtain the summary statistics of variables.

How to do:

****The summary statistics****

sum l_loggdp_ppp l_logobesity l_logsogi l_logecogi l_logfood_prod l_logrene,

This section presents countries in the LAC region that this chapter will focus on and the variables used. In the following subsection, we will present the method used to carry out the empirical investigation of this chapter.

Table 5.1 Variables' description and summary statistics.

Variables' description	
Variables	**Source**
Gross domestic product (GDP) per capita based on purchasing power parity (PPP). This variable is converted to international dollars using purchasing power parity rates. An international dollar has the same purchasing power over GDP as the US dollar has in the United States. GDP at purchaser's prices is the sum of gross value added by all resident producers in the country plus any product taxes and minus any subsidies not included in the value of the products. It is calculated without deductions for the depreciation of fabricated assets or depletion and degradation of natural resources. Moreover, this variable is in current 2017 international dollars. In this investigation, we called this variable '**GDP_PPP**'.	World Bank Open Data (2021)
Share of adults that are obese (per cent). Obesity is defined as having a body mass index (BMI) equal to or greater than 30. BMI is a person's weight in kilogrammes divided by his or her height in meters squared. In this investigation, we called this variable '**OBESITY**'.	Our World in Data (2021)
Social globalisation index de facto. This variable measures the interpersonal contact flows of information and cultural proximity. Interpersonal contact is measured within the de facto segment concerning international telephone connections, tourist numbers and migration. Information flows are determined within the de facto segment concerning international patent applications, international students and trade in high-technology goods. Cultural proximity is measured in the de facto segment via trade in cultural goods, international trademark registrations and the number of McDonald's restaurants and IKEA stores. In this investigation, we called this variable '**SoGI**'.	KOF Globalization Index (KOF 2021)
Economic globalisation index de facto. This variable measures trade and financial globalisation. Trade globalisation is determined based on trade in goods and services and financial globalisation includes foreign investment in various categories. In this investigation, we called this variable '**EcoGI**'.	KOF Globalization Index (KOF 2021)
Cereal production (metric tons) per capita. This variable measures the production data on cereals relate to crops harvested for dry grain only. Cereal crops harvested for hay or harvested green for food, feed, or silage and used for grazing are excluded. In this investigation, we called this variable '**FOOD_PROD**'.	World Bank Open Data (2021)

Continued

Table 5.1 Variables' description and summary statistics.—cont'd

Variables' description	
Variables	**Source**
Electricity production from renewable sources, excluding hydroelectric (kWh) per capita. This variable measures electricity production from renewable sources, excluding hydroelectric, including geothermal, solar, tides, wind, biomass and biofuels. In this investigation, we called this variable '**RENE**'.	World Bank Open Data (2021)

Summary statistics

Variables	Obs.	Mean	Std.-Dev.	Min	Max
LogGDP_PPP	546	8.8038	0.5730	7.4326	10.1393
LogOBESITY	546	2.7776	0.2885	1.8870	3.3428
LogSoGI	546	3.9501	0.2512	3.0372	4.3453
LEcoGI	546	3.8888	0.2343	2.9187	4.3154
LogFOOD_PROD	546	−1.9633	1.3754	−7.4333	0.2754
LogRENE	428	3.5302	1.4808	−1.8832	7.0419

Notes: (Log) denotes variables in natural logarithms.

5.3.2 Method

This section will show the method and strategy that this chapter will use. Therefore, the panel quantile regression model approach will be used as a method. As mentioned in Chapter 4, this method was introduced by Machado and Silva (2019) as an alternative for quantile regression with non-additive fixed effects developed by Powell (2016). Additionally, this method can provide information on how the regressors affect the entire conditional distribution.

Moreover, the method can also be adapted to estimate the presence of cross-sectional models with endogenous variables (Machado & Silva, 2019). This method is not based on estimating conditional means but moment conditions under exogeneity into dependent means. This characteristic is closely related to that of Chernozhukov and Hansen (2008) model. In undesirable situations, the panel quantile model with fixed effects can identify the exact structural quantile function. It makes this method used for non-linear models much more straightforward, especially in models with multiple endogenous variables (Machado & Silva, 2019). This method can provide information on how the regressors affect the entire conditional distribution. Moreover, according to Machado and Silva (2019), the panel quantile model with fixed effects, given data $\{(Y_{it}, X'_{it})'\}$ from a panel of n individuals $i = 1,\ldots,$ n over T periods, is constructed around the following Eq. (5.1):

$$Y_{it} = a_i + X'_{it}\beta + \left(\delta_i + Z'_{it}\gamma\right)U_{it}, \tag{5.1}$$

with, $P\{\delta_i + Z'_{it}\gamma > 0\} = 1$. The parameters (α_1, δ_i), $i = 1, \ldots, n$, capture the individual i fixed effects and Z is a k-vector of known differentiable (with probability 1) transformations of the components of X with element l given by $Z_l = Z_l(X)$, $l = 1, \ldots, k$. In empirical research, we will use a special case of Eq. (5.1), a linear heteroskedasticity model, in which $\sigma(\cdot)$ is the identity function, and Z = X. The sequence $\{X_{it}\}$ is *i.i.d.* for any fixed i and independent across t. U_{it} are *i.i.d.* (across i and t), statistically independent of X_{it} and normalised to satisfy the moment condition $E(U) = 0 \wedge E(|U|) = 1$ (Machado & Silva, 2019). Eq. (5.1) implies that the conditional quantile-τ is given by Eq. (5.2):

$$Q_Y(\tau|X_{it}) = (a_i + \delta_i q(\tau)) + X'_{it}\beta + Z'_{it}\gamma q(\tau), \tag{5.2}$$

with $q(\tau) = F_U^{-1}(\tau)$, where F_U is the distribution function of U. The authors propose a recursive estimation method based on a set of moment conditions, which is computationally fast and straightforward and does not require the use of simulations that would render its widespread adoption difficult (Powell, 2016). They also prove that the resulting estimates are consistent and asymptotically normal.

After computing the estimates, the marginal effect of the explanatory variable l on quantile τ of the dependent variable can be retrieved from Eq. (5.3):

$$\beta_l(\tau|X) = \beta_l + \frac{\partial Z'}{\partial X_l}\gamma q(\tau), \tag{5.3}$$

Therefore, before realising the panel quantile regression model, it is necessary to verify the proprieties of variables used in this investigation, including checking the normality and multicollinearity. Consequently, the first tests that need to be applied before the panel quantile model estimation are in Table 5.2 below.

Table 5.2 Preliminary tests.

Test	Finality
Shapiro–Wilk test (Royston, 1983)	This test verifies the normality of the model. The null hypothesis of this test is the presence of normality.
Skewness and Kurtosis test (D'Agostino et al., 1990)	This test checks the normality based on skewness and another based on kurtosis and then combines the two tests into an overall test statistic. The null hypothesis of this test is that the data is normally distributed.
Variance inflation factor (VIF) test (Belsley et al., 1980)	This test verifies the presence of multicollinearity between the variables.
Levin-Lin-Chu unit-root test (LLC-test) (Levin et al., 2002)	This test verifies the presence of unit roots in the variables. The null hypothesis of the LLC-test is that all the panels contain a unit root.
Panel unit root test (CIPS) test (Pesaran, 2007)	This test verifies the presence of unit roots in the variables. The panel unit root test (CIPS) null hypothesis is that all series have a unit root.

After the panel quantile regression model, it is necessary to apply some postestimation tests (see Table 5.3 below).

Table 5.3 Postestimation tests.

Test	Finality
The Wald test (Agresti, 1990)	This test verifies the global significance of the estimated models. The null hypothesis of the Wald test is that all the coefficients are equal to zero.

The estimation and testing procedures are carried out using **Stata 16.0.**

5.4 Empirical results

This section will present the empirical results of preliminary tests, the outcomes of the panel quantile regression model, and the results from the postestimation tests. Therefore, to identify the normality in the model, the Shapiro–Wilk test for normal data was used to test the residuals of pooled ordinary least squares (OLS) model regression. In Table 5.4 below, we can see the results of the Shapiro–Wilk test for normal data.

Regarding the Shapiro–Wilk test for normal data, we see that the null hypothesis of normality is rejected. The results from Shapiro–Wilk test for normal data were obtained from the command *sktest* of **Stata 16.0**. The board below shows how to carry out and obtain the results from the Shapiro–Wilk test.

How to do:

****Resid****

reg l_loggdp_ppp l_logobesity l_logsogi l_logecogi l_logfood_prod l_logrene

predict resid

**** Shapiro-Wilk W-test for normal data ****

swilk resid

Table 5.4 Shapiro−Wilk test for normal data.

Variables	Obs.	W	V	Z	Prob > z
Resid	428	0.9878	2.969	2.597	0.0047

After the Shapiro−Wilk W-test, it is necessary to apply the skewness and kurtosis test to verify the normality in the model's residuals from the pooled OLS regression. The null hypothesis that the data is normally distributed is also rejected (see Table 5.5 below).

The results from skewness/kurtosis tests for normality were obtained from the command *sktest* of **Stata 16.0.** The board below shows how to carry out and obtain the results from the skewness/kurtosis tests for normality.

How to do:

****Resid****

reg l_loggdp_ppp l_logobesity l_logsogi l_logecogi l_logfood_prod l_logrene

predict resid

**** Skewness/kurtosis tests for normality ****

sktest resid

The results of both tests reject the null hypothesis at a significance level of 1% and 5%. The results of these tests are added support for the adequacy of using the quantile regression (Afonso et al., 2019, pp. 1−25). Subsequently, the tests of normality, the VIF test that informs of the presence of multicollinearity, needs to be computed. The results of the VIF test show that the presence of multicollinearity is not a concern in the estimation of each model, given the low VIF and mean VIF values registered, which are lower than the usually accepted benchmark of 10, in the case of the VIF values, and six in the case of the mean VIF values (see Table 5.6 below). This test helps us understand the degree of multicollinearity present in our models, leading to problems in estimation.

Table 5.5 Skewness/kurtosis tests for normality.

Variables	Obs.	Pr(Skewness)	Pr(Kurtosis)	Adj chi2(2)	Prob > Chi2
Resid	428	0.0042	0.4653	8.24	0.0162

Table 5.6 VIF-test.

Variables	VIF	1/VIF	Mean VIF
	LogGDP_PPP		
LogOBESITY	1.64	0.6093	1.64
LogSoGI	2.31	0.4324	
LogEcoGI	1.91	0.5246	
LogFOOD_PROD	1.21	0.8231	
LogRENE	1.15	0.8703	

Notes: (Log) denotes variables in natural logarithms.

The results of the VIF test were obtained from the command *estat vif* in **Stata 16.0.** The board below shows how to carry out and obtain the results from the VIF test.

How to do:

**** The variance inflation factor test****

```
reg l_loggdp_ppp l_logobesity l_logsogi l_logecogi l_logfood_prod l_logrene
estat vif
```

After realising the multicollinearity test, it is necessary to verify the order of integration of the variables. To this end, the panel unit root tests, such as the LLC-test developed by Levin et al. (2002) and the CIPS-test developed by Pesaran (2007), were calculated. Table 5.7 below shows the results from the unit root tests.

Therefore, the results of the LLC test indicate that the variables LogOBESITY, LogEcoGI and LogFOOD_PROD are stationary, while the variables LogGDP_PPP and LogSoGI are on the borderline between I(0) and I(1) of the order of integration. However, variable LogRENE could not be computed by the LLC-test because the test requires strongly balanced data. Moreover, the results from the CIPS-test obtained indicate that the variables LogGDP_PPP, LogSoGI and LogFOOD_PROD, are stationary, while the variable LogOBESITY is on the borderline between I(0) and I(1) of order of integration. However, the variable LogEcogGI is non-stationary. Moreover, the variable LogRENE could not be computed by the CIPS test because the test requires strongly balanced data. The results of the LLC test and CIPS test were obtained

Table 5.7 Unit root tests.

Variables	Levin-Lin-Chu unit-root test (LLC-test)					Panel unit root test (CIPS) (Zt-bar)				
	Without trend			With trend		Without trend			With trend	
	Lags	Adjusted t		Adjusted t		Lags	Zt-bar		Zt-bar	
LogGDP_PPP	1	0.1944		−3.4167	***	1	−1.602	*	−2.108	**
LogOBESITY	1	−19.7918	***	−3.0682	**	1	−1.624	**	−0.760	
LogSoGI	1	−5.7483	***	0.3881		1	−4.517	***	−1.791	**
LogEcoGI	1	−5.4776	***	−2.9241	**	1	−1.226		−0.555	
LogFOOD_PROD	1	−2.7130	**	−1.4554	*	1	−2.549	***	−1.768	**
LogRENE	N.A					NA				

Notes: ***, ** and * denote statistically significant at the 1%, 5% and 10% levels, respectively; (Log), denotes variables in natural logarithms; N.A, denotes not available.

from the commands *xtunitroot* and *multipurt* in **Stata 16.0**. The board below shows how to carry out and ssssa5yuya the LLC and CIPS tests.

How to do:

**** Levin-Lin-Chu unit-root test****

```
xtunitroot llc l_loggdp_ppp
xtunitroot llc l_loggdp_ppp, trend

xtunitroot llc l_logobesity
xtunitroot llc l_logobesity, trend

xtunitroot llc l_logsogi
xtunitroot llc l_logsogi, trend

xtunitroot llc l_logecogi
xtunitroot llc l_logecogi, trend

xtunitroot llc l_logfood_prod
xtunitroot llc l_logfood_prod, trend

xtunitroot llc l_logrene
xtunitroot llc l_logrene, trend
```

****The CIPS-test****

```
multipurt l_loggdp_ppp l_logobesity l_logsogi l_logecogi l_logfood_prod
l_logrene,lags(1)
```

After the realisation of preliminary tests, the panel quantile regression model with fixed effects can be made. The 25th, 50th, 75th and 100th quantiles were calculated to assess the impact of the obesity epidemic on economic growth in the LAC region. The method used does not allow causalities to be performed; it only allows observation of the effect at the quantiles. Table 5.8 shows the results of quantile regressions and the results of the postestimation test (e.g., the Wald test) to confirm the presence of heteroskedasticity in the models.

Therefore, the panel quantile regression model results with fixed effects indicate that in the 25th, 50th and 100th quantiles, the variables obesity epidemic and renewable energy production increase economic growth. Moreover, food production increases the economic growth in the 25th, 50th and 75th quantiles, while the variable social globalisation increases economic growth in 25th and 50th quantiles and economic globalisation in 25th and 100th quantiles. The result from the postestimation test is an indicator that the estimations that this investigation use are adequate. The results of the panel quantile regression model were obtained from the command *xtqreg* in

Table 5.8 Panel quantile estimations.

Independent variables	Dependent variable (LogGDP_PPP)							
	Quantiles							
	25th		50th		75th		100th	
LogOBESITY	0.9923	***	1.0521	***	1.1242	***	0.9444	***
LogSoGI	0.2569	**	0.2666	**	0.2782		0.2492	
LogEcoGI	0.2619	***	0.1426		−0.0013		0.3575	**
LogFOOD_PROD	0.0782	**	0.1105	***	0.1495	**	0.0523	
LogRENE	0.0322	***	0.0340	***	0.0362	**	0.0307	**
Obs	428		428		428		428	
Wald test (chi2)	1591.30	***	1656.67	***	587.12	***	14.55	*

Notes: ***, ** and * denote statistically significant at the 1%, 5% and 10% levels respectively; (Log), denotes variables in natural logarithms.

Stata 16.0. The results from the postestimation test, such as the Wald test, were obtained from the command *testparm* in **Stata 16.0**. The board below shows how to carry out and obtain the results from the panel quantile model regression and the results from the postestimation test.

How to do:

****The panel quantile model regression****

xtqreg l_loggdp_ppp l_logobesity l_logsogi l_logecogi l_logfood_prod l_logrene, id(country) quantile(.25)

****The Wald test****

testparm l_loggdp_ppp l_logobesity l_logsogi l_logecogi l_logfood_prod l_logrene

**** The panel quantile model regression****

xtqreg l_loggdp_ppp l_logobesity l_logsogi l_logecogi l_logfood_prod l_logrene ,id(country) quantile(.50)

****The Wald test****

testparm l_loggdp_ppp l_logobesity l_logsogi l_logecogi l_logfood_prod l_logrene

**** The panel quantile model regression****

xtqreg l_loggdp_ppp l_logobesity l_logsogi l_logecogi l_logfood_prod l_logrene, id(country) quantile(.75)

****The Wald test****

testparm l_loggdp_ppp l_logobesity l_logsogi l_logecogi l_logfood_prod l_logrene

****The panel quantile model regression****

xtqreg l_loggdp_ppp l_logobesity l_logsogi l_logecogi l_logfood_prod l_logrene ,id(country) quantile(.100)

****The Wald test****

testparm l_loggdp_ppp l_logobesity l_logsogi l_logecogi l_logfood_prod l_logrene

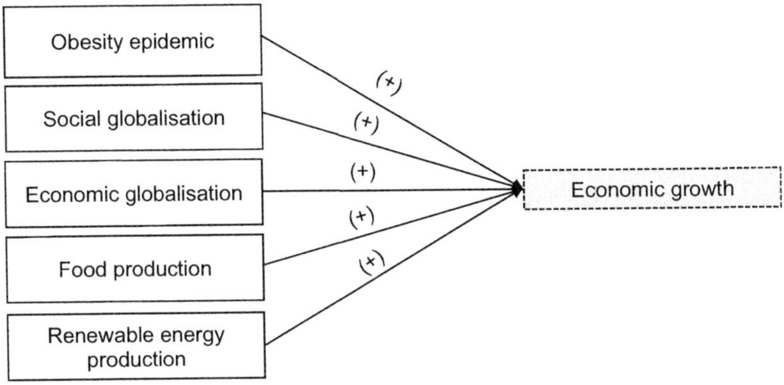

Figure 5.3 Summary of the impact of independent variables on dependent ones. Created by the authors.

Fig. 5.3 below summarises the impact of independent variables on dependent ones. This figure was based on the results from the panel quantile model regression.

In this section, we showed the empirical results of this investigation. The following section will show the discussion of the empirical results.

5.5 Discussion

In this section, we will approach the possible explanations for the results found in the empirical analysis. Therefore, the positive impact of the obesity epidemic on economic growth in LAC countries is related to the capacity of same economic development to affect dietary changes. Indeed, according to Gerbens-Leenes et al. (2010) and Roskam et al. (2010), as the socio-economic groups experience income transition from lower to high income caused by economic development, they tend to consume fatty foods from energy-dense animal sources. Thus, the increase in income contributes to growth in overweight and obesity levels. This phenomenon has occurred in LAC countries where the high economic growth was registered in the last 20 years caused by economic reforms and the commodities boom, as mentioned in Chapters 2, 3 and 4 of this book. Indeed, this idea is shared by Costa-Font and Mas (2016). According to the authors, the reduction of economic inequality, caused by higher economic development, also affects the overweight and obesity rates. Therefore, in countries with high economic growth but low income, food insecurity underscores the risk of overweight and obesity due to food consumption in the following stage of higher income. Moreover, according to Sullivan et al. (2008), nutritional deficits caused by food insecurity are followed by an excess of caloric consumption. Therefore, economic growth is associated with overweight and obesity, both with its contemporary and future effects on food consumption and production.

That is, the obesity epidemic will affect positively food consumption and production and so other agents of the economic system, and subsequently the economic growth. This point of view is shared by Koengkan and Fuinhas (2021). According to the authors, the overweight or obesity epidemic increases processed foods from multinational food corporations, fast-food chains and multinational supermarket chains and food production in farms. This will impact the multinational food corporations and farm production positively to attend to the demand for processed foods. This increase will affect the economic activity. Another possible explanation, according to the authors, is that overweight and obesity reduce physical activities as well as outdoor activities, increasing the intensive use and consumption of home appliances, motorised transportation and screen-viewing leisure activities. All this affects economic activity and energy consumption from fossil fuels positively and the environment negatively by increasing CO_2 emissions.

The positive impact of food production on economic growth in LAC countries as mentioned above is related to obesity, as this problem will increase the consumption of processed foods from multinational food corporations, fast-food chains and multinational supermarket chains, as well the food production in farms and consequently the economic activity and consumption of energy.

The positive impact of globalisation or, more precisely, economic globalisation in LAC countries is related to its capacity to encourage investment by increasing capital stock and reducing financing costs by financial liberalisation and the increases in exports and imports by trade liberalisation. According to Fuinhas et al. (2021a,b), capital inflow in the LAC region resumed after the Brady plan in the early 1990s. The magnitude of the financial liberalisation in the LAC region can be grasped with the index of capital mobility. In the 1980s, the index capital mobility was 40 and in the 1990s rose to about 75, with normalising completely free capital mobility at 100 (Aizenman, 2005, pp. 1–30). Moreover, the financial liberalisation caused by macroeconomic adjustments mentioned in Chapters 2, 3 and 4 promoted flows of foreign direct investments (FDI). As a result, the FDI flows worldwide grew dramatically between 1990 and 1997. Indeed, the increase in FDI flows in the LAC region is visible in the FDI inflows, where in 1985, before the macroeconomic reforms, these inflows were 6.44 billion USD. After the adjustment, these inflows grew to 9.73 billion USD in 2000. Finally, they reached a value of 2.91 trillion USD in 2015 (Fuinhas et al., 2021a,b).

These FDI inflows to the LAC region in the 1990s evolved in three phases (Birch & Halton, 2008). Between 1990 and 1993, investors have seemed to favour acquiring already existing assets. However, between 1994 and 1996, most investment was directed to large-scale projects via restructuring existing foreign firms or modernising recently privatised firms. Finally, in 1997, acquiring existing assets to consolidate the market power became the most common form of foreign investment in the region. In this period, more money was spent on purchasing already existing private assets than on privatisation (Fuinhas et al., 2021a,b). Moreover, during the 1990s, the LAC region had registered an increase in FDI in the industries related to natural resources and the energy sector (Birch & Halton, 2008).

However, economic globalisation began in the 1970s with financial and trade liberalisation and intensified with the 'commodities boom,' between 2004 and 2014, where the region had an average growth rate of 7.40% (Santos, 2015). The cycle of commodity prices in LAC economies impacted the degree of economic openness or, more precisely, the degree of dependence on external demand vis-à-vis domestic demand or markets. Between 1990 and 1993, the degree of economic openness was 28.6, between 1998 and 2001, it was 38.5, and between 2006 and 2009, it was 44.7 on a scale of 0–100, where 100 represents an open economy. In this period between 1990 and 2009, the degree of economic openness had a growth of 50.71% (Carneiro, 2012).

Furthermore, this fact seems to have allowed the region to surpass the problems generated by the 2008–2009 financial crisis. The growth in the degree of economic openness was caused by an increase in the exports and imports in the region, where in 2004 exports of goods and services (BoP current USD) was six billion USD and reached a value of 1.39 trillion USD in 2018, while imports in 2004 were 5.40 billion USD and reached a value of 1.41 trillion USD in 2018. Indeed, the growth in the manufacturing sector caused by the commodities boom led to FDI inflows in the largest economies in the region, with 61% of total FDI inflows in Mexico and 38% in Brazil (Fuinhas et al., 2021a,b). In this period, imports had a vital role in modernising the production process. Thus, modern machines and better industrial inputs contributed to the technological upgrading of the industrial basis in the region.

Moreover, the positive impact of social and economic globalisation on economic growth in the LAC region is linked with obesity. According to Koengkan and Fuinhas (2021), social and economic globalisation also impacts people's energy-caloric expenditure. This impact occurs due to the penetration of new technologies made available by the increase in economic liberalisation. These new technologies reduce general and daily for physical activity. Indeed, innovations in technology have evolved on a path that allows labour-saving behaviours in industrial sectors and home appliances and motorised transportation have become more accessible.

Indeed, the increase of this accessibility will reduce physical activity as well the outdoor activities, consequently increasing the intensive use and consumption of home appliances and motorised transportation. Subsequently it will impact the economic activity and consumption of energy positively and the environment negatively with the increase in CO_2 emissions, as well as increasing the overweight and obesity problem. Moreover, social and economic globalisation also impacts how food is distributed, marketed and consumed. Thus, the social and economic globalisation process caused by trade liberalisation encourages extensive exposure to global eating practices or westernisation of food consumption. A clear example of westernisation of food consumption is McDonaldisation or Cocalisation, which promotes food consumption with higher energy-caloric content, contributing to increased overweight and obesity.

Finally, the positive impact of electricity production from renewable energy sources on economic growth in the LAC region is related to the very economic activity that encourages the consumption of renewable energy sources. Indeed, this increase in renewable energy consumption sources will encourage new investments in this kind of energy to supply the demand (Koengkan, 2017). Therefore, these new investments will positively impact economic activity, thus creating a feedback relationship.

Moreover, the positive impact of renewable energy production on economic growth is linked with the obesity problem. According to Koengkan and Fuinhas (2021), the overweight or obesity epidemic encourages processed foods from multinational food corporations, fast-food chains and multinational supermarket chains and production on farms. This will impact the multinational food corporations and farm production positively to attend to the demand for processed foods. This increase will affect the consumption of energy. Another possible explanation for these phenomena is related to the capacity of overweight to reduce physical activities and outdoor activities, increasing the intensive use of home appliances, motorised transportation and screen-viewing leisure activities. All this affects energy consumption and economic growth positively and the environment negatively with the increase in CO_2 emissions. Fig. 5.4 below, which Koengkan and Fuinhas (2021) developed, shows how overweight and obesity increase energy consumption and environmental degradation.

This section showed the possible explanations for the positive impact of the obesity epidemic, economic globalisation, food production and electricity consumption from renewable energy sources on economic growth. The following section will show the main conclusions of this chapter.

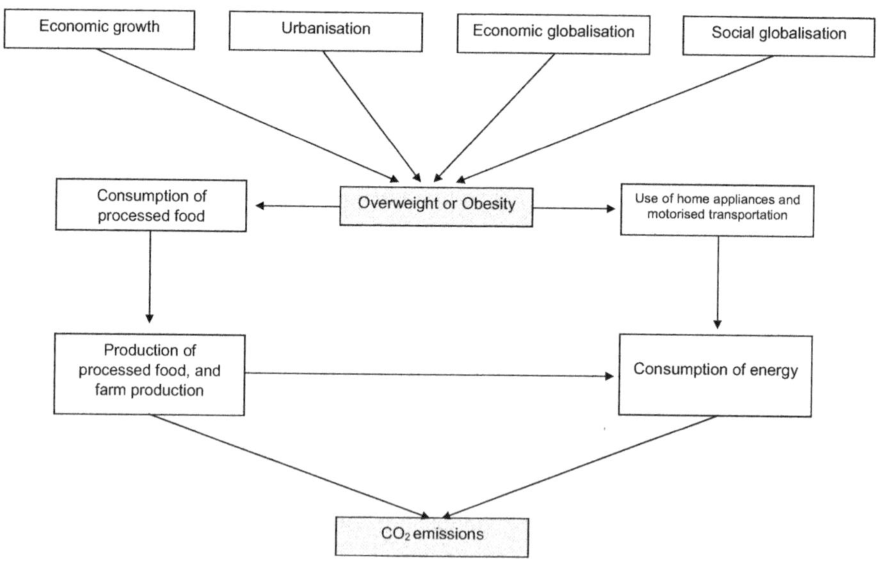

Figure 5.4 Summary of a possible explanation of the positive impact of overweight or obesity on energy consumption and, consequently, CO_2 emissions.
This figure was based on Koengkan and Fuinhas (2021).

5.6 Conclusion

This chapter approached the impact of the prevalence of obesity on the economic growth of LAC countries. This chapter used data for 21 countries in the LAC region for the period between 1991 and 2016. This study used a panel quantile regression model with fixed effects as a method.

The preliminary tests indicated the non-presence of normality between the variables, where in the Shapiro–Wilk W-test for normal data, the null hypothesis of this test is rejected. Moreover, the skewness and kurtosis test that also verifies the normality in the residuals indicated that the null hypothesis also could be rejected. Therefore, the results from both tests reject the null hypothesis at a significance level of 1% and 5%. Moreover, the results of these tests are added support for the adequacy of using the quantile regression. The results of the VIF-test showed that the presence of multicollinearity is not a concern in the estimation of each model, given the low VIF and mean VIF values registered, which are lower than the usually accepted benchmark of 10, in the case of the VIF values and six in the case of the mean VIF values.

The results from the LLC test indicated that the variables LogOBESITY, LogEcoGI and LogFOOD_PROD are stationary, while the variables LogGDP_PPP and LogSoGI are on the borderline between I(0) and I(1) of the order of integration. However, the variable LogRENE could not be computed by the LLC-test because the test requires strongly balanced data. Moreover, the results from the CIPS-test obtained indicated that the variables LogGDP_PPP, LogSoGI and LogFOOD_PROD, are stationary, while the variable LogOBESITY is on the borderline between I(0) and I(1) of the order of integration. However, the variable LogEcogGI is non-stationary. The variable LogRENE could not be computed by the CIPS test because the test requires strongly balanced data.

Therefore, the results from the panel quantile regression model with fixed effects indicated that in the 25th, 50th and 100th quantiles, the obesity epidemic and renewable energy production increase economic growth. Moreover, food production increases economic growth in the 25th, 50th and 75th quantiles, while the variable social globalisation increases economic growth in the 25th and 50th quantiles and economic globalisation in the 25th and 100th quantiles. The result from the postestimation test is an indicator that the estimations used in this investigation are adequate.

Therefore, estimation of a panel quantile model regression allowed us to conclude that obesity does foster economic growth in LAC countries, along with all the GDP quantiles. Income transition and nutritional transition change people's lifestyles; they become less physically active and choose industrialised and processed food. These new lifestyles promote obesity, and obesity is translated into food, technology and equipment production and consumption choices, contributing to countries' economic growth. In the future, policymakers may want to design measures that promote healthy lifestyles, create consumption choices and produce goods and services, contributing to economic growth.

Additionally, this analysis also showed that in parallel with globalisation, two other factors contribute to the economic growth of these countries, namely, food production and renewable energy production. The contribution of food production to economic growth is mainly made through the industrialisation of farming and food production. This production process is producing unhealthy choices. Changes in consumers' choices are already demanding healthier and less industrialised food choices. These new trends in consumers and marketing preferences will contribute differently to economic growth.

Finally, the production of renewable energy in economic growth is already a sign of the new approaches adopted by the government as a source of energy. The result found confirms the likelihood of the growth hypothesis in LAC countries. It also provides evidence of the role that globalisation plays in adopting environmentally friendly technology. This effect from renewable energy on economic growth may also be interpreted as a response to energy increase resulting from the increasing prevalence of obesity. Under the international environment protocols and agreements of CO_2 emissions, governments were pushed to adopt new technologies to produce energy that comply with those international agreements. To conclude, this chapter highlights the effect of obesity on economic growth and the effect of renewable energy production on economic growth no matter the quantile of GDP of LAC countries.

References

Afonso, T. L., Marques, A. C., & Fuinhas, J. A. (2019). *Energy–growth nexus and economic development: A quantile regression for panel data*. The Extended Energy-Growth Nexus. https://doi.org/10.1016/b978-0-12-815719-0.00001-2

Agresti, A. (1990). *Categorical data analysis*. John Wiley and Sons. ISBN 0-471-36093-7.

Aizenman, J. (2005). *Financial liberalizations in Latin America in the 1990s: A reassessment*. NBER working paper series, 11145:1–30. https://www.nber.org/papers/w11145

Arsht, A. (2014). *Urbanization in Latin America*. Atlantic Council. https://www.atlanticcouncil.org/commentary/article/urbanization-in-latin-america/#:~:text=Today%2C%20about%2080%20percent%20of,the%20world's%20most%20urbanized%20region.&text=Today%2C%20260%20million%20people%20live,percent%20of%20Latin%20America's%20GDP.

Asafu-Adjaye, J. (2000). The relationship between energy consumption, energy prices and economic growth: Time series evidence from Asian developing countries. *Energy Economics, 22*, 615–625.

Belsley, D. A., Kuh, E., & Welsch, R. E. (1980). *Regression diagnostics: Identifying influential data and sources of collinearity*. New York: Wiley. https://doi.org/10.1002/0471725153

Bhattacharya, M., Paramati, S. R., & Ozturk, I. (2016). The effect of renewable energy consumption on economic growth: Evidence from top 38 countries. *Applied Energy, 162*, 733–741.

Birch, M. H., & Halton, G. (2008). Foreign direct investment in Latin America in the 1990s: Old patterns, new trends, and emerging issues. *Latin American Business Review, 2*(1–2), 13–31. https://doi.org/10.1300/J140v02n01_03

Carneiro, R. M. (2012). *Commodities, choques externos e crescimento: reflexões sobre a América Latina, 117* pp. 1–47). CEPAL. ISSN:1680-8843 http://www.eco.unicamp.br/cecon/images/arquivos/observatorio/Commodities_choques_externos_crescimento.pdf.

Chang, C. P., & Lee, C. C. (2010). Globalisation and economic growth: A political economy analysis for OECD countries. *Global Economic Review, 39*(2), 151–173.

Chee, V. A., Teran, E., Hernandez, I., Wright, L., Izurieta, R., Reina-Ortiz, M., Flores, M., Bejarano, S., Dào, L. U., Baldwin, J., & Martinez-Tyson, D. (2019). Desculturización, urbanisation, and nutrition transition among urban Kichwas Indigenous communities residing in the Andes highlands of Ecuador. *Public Health, 176*, 21–28. https://doi.org/10.1016/j.puhe.2019.07.015

Chernozhukov, V., & Hansen, C. (2008). Instrumental variable quantile regression: A robust inference approach. *Journal of Econometrics, 142*, 379–398. https://doi.org/10.1016/j.jeconom.2007.06.005

Costa-Font, J., & Mas, N. (2016). 'Globesity'? The effects of globalisation on obesity and caloric intake. *Food Policy, 64*, 121–132. https://doi.org/10.1016/j.foodpol.2016.10.001

D'Agostino, R. B., Belanger, A. J., & D'Agostino, R. B., Jr. (1990). A suggestion for using powerful and informative. *The American Statistician, 44*(4), 316–321. https://doi.org/10.2307/2684359

Dreher, A. (2006). Does globalization affect growth? Empirical evidence from a new index globalization. *Applied Economics, 38*(10), 1091–1110.

Fox, A., Feng, W., & Asal, V. (2019). What is driving global obesity trends? Globalisation or modernisation. *Globalization and Health, 15*(32), 1–16. https://doi.org/10.1186/s12992-019-0457-y

Fuinhas, J. A., Koengkan, M., & Santiago, R. (2021a). Chapter 5 – energy transition in Latin America and the Caribbean region: Initiatives and challenges. In *Physical capital development and energy transition in Latin America and the Caribbean* (1st ed.) ISBN: 9780128244296.

Fuinhas, J. A., Koengkan, M., & Santiago, R. (2021b). *Physical capital development and energy transition in Latin America and the caribbean* (pp. 19–234). Elsevier. https://doi.org/10.1016/C2020-0-01491-X

Fukase, E., & Martin, W. (2020). Economic growth, convergence, and world food demand and supply. *World Development, 132*, 104954. https://doi.org/10.1016/j.worlddev.2020.104954

Garcia-Garcia, G. (2021). *Obesity and overweight populations in Latin America.* https://www.thelancet.com/campaigns/kidney/updates/obesity-and-overweight-populations-in-latin-america.

Gerbens-Leenes, P. W., Nonhebel, S., & Krol, M. S. (2010). Food consumption patterns and economic growth. Increasing affluence and the use of natural resources. *Appetite, 55*(3), 597–608. https://doi.org/10.1016/j.appet.2010.09.013

Gereffi, G. (1990). Chapter 1: Paths of industrialization: An overview. In G. Gereffi, & D. L. Wyman (Eds.), *Manufacturing miracles*. Princeton University Press.

Hawkes, C. (2006). Uneven dietary development: Linking the policies and processes of globalisation with the nutrition transition, obesity and diet-related chronic diseases. *Globalisation and Health, 2*, 4. https://doi.org/10.1186/1744-8603-2-4

International Food Policy Research Institute. (2016). *Global nutrition report 2016: From promise to impact: Ending malnutrition by 2030*. https://www.ifpri.org/publication/global-nutrition-report-2016-promise-impact-ending-malnutrition-2030.

Jack, W., & Lewis, M. (2001). Health investments and economic growth, Macroeconomic evidence and Microeconomic foundations. In *Policy research working paper 4877*. World Bank.

Jaitman, L. (2015). Urban infrastructure in Latin America and the Caribbean: Public policy priorities. *Latin American Economic Review, 24*, 13. https://doi.org/10.1007/s40503-015-0027-5

Koengkan, M. (2017). The nexus between energy consumption, economic growth, and urbanisation in Latin American and Caribbean countries: An approach with PVAR model. *Revista Valore, 6*(2), 202–218. https://doi.org/10.22408/reva22201761%25p

Koengkan, M., & Fuinhas, J. A. (2020a). Exploring the effect of the renewable energy transition on CO_2 emissions of Latin American & Caribbean countries. *International Journal of Sustainable Energy*, 1–24. https://doi.org/10.1080/14786451.2020.1731511

Koengkan, M., & Fuinhas, J. A. (2020b). The interactions between renewable energy consumption and economic growth in the Mercosur countries. *International Journal of Sustainable Energy, 39*(6), 594–614. https://doi.org/10.1080/14786451.2020.1732978

Koengkan, M., & Fuinhas, J. A. (2021). Does the overweight epidemic cause energy consumption? A piece of empirical evidence from the European region. *Energy, 236*(1), 119297. https://doi.org/10.1016/j.energy.2020.119297

Koengkan, M., Fuinhas, J. A., & Fuinhas, C. (2021). *Does urbanisation process increase the overweight epidemic? The case of Latin America and the caribbean region*. SSRN. https://doi.org/10.2139/ssrn.3826196

KOF Globalization index. (2021). URL: https://www.kof.ethz.ch/en/forecastsand-indicators/indicators/kof-globalisation-index.html.

Kraft, J., & Kraft, A. (1978). On the relationship between energy and GNP. *The Journal of Energy and Development, 3*, 401–403. https://www.jstor.org/stable/24806805.

Lake, A., & Townshend, T. (2006). Obesogenic environments: Exploring the built and food environments. *Journal of The Royal Society for the Promotion of Health, 126*(6), 262–267. https://doi.org/10.1177/1466424006070487

Levin, A., Lin, C.-F., & Chu, C.-S. J. (2002). Unit root tests in panel data: Asymptotic and finite-sample properties. *Journal of Econometrics, 108*, 1–24. https://doi.org/10.1016/S0304-4076(01)00098-7

Machado, J. A. F., & Silva, J. M. C. S. (2019). Quantiles via moments. *Journal of Econometrics, 213*(1), 145–173. https://doi.org/10.1016/j.jeconom.2019.04.009

Manap, N. M. A., & Ismail, N. W. (2019). Food security and economic growth. *International Journal of Modern Trends in Social Sciences, 2*(8), 108–118. https://doi.org/10.35631/IJMTSS.280011

Masih, A. M., & Masih, R. (1996). Energy consumption, real income and temporal causality: Results from a multi-country study based on cointegration and error-correction modelling techniques. *Energy Economics, 18*, 165–183.

Narayan, P. K., & Smyth, R. (2008). Energy consumption and real GDP in G7 countries: New evidence from panel cointegration with structural breaks. *Energy Economics, 30*, 2331–2341.

NCD risk factor collaboration (NCD-RisC). (2016). Trends in adult body-mass index in 200 countries from 1975 to 2014: A pooled analysis of 1698 population-based measurement studies with 19·2 million participants. *Lancet, 387*, 377–396. https://www.thelancet.com/journals/lancet/article/PIIS0140-6736(16)30054-X/fulltext.

Ng, M., Fleming, T., Robinson, M., Thomsom, B., & Graetz, N. (2014). Global, regional, and national prevalence of overweight and obesity in children and adults during 1980–2013: A systematic analysis for the global burden of disease study 2013. *Lancet, 384*, 766–781. https://doi.org/10.1016/S0140-6736(14)60460-8

Ntanos, S., Skordoulis, M., & Kyriakopoulos, G. (2018). Renewable energy and economic growth: Evidence from European countries. *Sustainability, 10*, 2626. https://doi.org/10.3390/su10082626

Our World in Data. (2021). *Obesity*. https://ourworldindata.org/obesity.

Payne, J. E. (2010). Survey of the international evidence on the causal relationship between energy consumption and growth. *Journal of Economic Studies, 37*(1), 53–95.

Pedraza, D. F. (2009). Obesidad y pobreza: Marco conceptual para su análisis en latinoamérica. *Saúde e Sociedades, 18*, 103–117.

Pesaran, M. H. (2007). A simple panel unit root test in the presence of cross-section dependence. *Journal of Applied Econometrics, 22*(2), 256–312. https://doi.org/10.1002/jae.951

Polasek, W., & Sellner, R. (2011). *Does globalization affect regional growth? Evidence for NUTS-2 regions in EU-27*. Institute for Advanced Studies, Economics Series 266.

Popkin, B. M. (1998). The nutrition transition and its health implications in low-income countries. *Public Health Nutrition, 1*(1), 5–21. https://doi.org/10.1079/PHN19980004

Popkin, B. M. (1999). Urbanisation, lifestyle changes and the nutrition transition. *World Development, 27*(11), 1905–1916.

Popkin, B. M., Adair, L. S., & Ng, S. W. (2012). Global nutrition transition and the pandemic of obesity in developing countries. *Nutrition Reviews, 70*(1), 3–21. https://doi.org/10.1111/j.1753-4887.2011.00456.x

Poveda, Y. E., Koengkan, M., & Fuinhas, J. A. (2020). Exploring the interactions between renewable energy, economic growth, agriculture and urbanisation in the Mercosur trade-bloc countries. *Revista Valore, 5*, 5049.

Powell, D. (2016). *Quantile regression with nonadditive fixed effects*. RAND Working Paper http://works.bepress.com/david_powell/14.

Rao, B. B., Tamazian, A., & Krishna, V. C. (2011). Growth effects of a comprehensive measure of globalisation with country-specific time series data. *Applied Economics, 43*(5), 551–568.

Rauch, J. (1993). Productivity gains from geographic concentration of human capital: Evidence from the cities. *Journal Urban Economics, 34*(3), 380–400.

Riley, J. C. (2001). *Rising life expectancy: A global history*. Cambridge University Press.

Roskam, A. J., Kunst, A. E., Van Oyen, H., Demarest, S., Klumbiene, J., Regidor, E., Helmert, U., Jusot, F., Dzurova, D., & Mackenbach, J. P. (2010). Comparative appraisal of educational inequalities in overweight and obesity among adults in 19 European countries. *International Journal of Epidemiology, 39*(2), 392–404. https://doi.org/10.1093/ije/dyp329

Royston, P. (1983). A simple method for evaluating the Shapiro-Francia W' test for non-normality. *Statistician, 32*, 297–300. https://doi.org/10.2307/2987935

Santos, B. G. (2015). O ciclo econômico da América Latina dos últimos 12 anos em uma perspectiva de restrição externa. *Revista do BNDES, 43*, 205–251. https://web.bndes.gov.br/bib/jspui/bitstream/1408/6242/2/RB%2043%20O%20ciclo%20econ%C3%B4mico%20da%20Am%C3%A9rica%20Latina_P%20.pdf.

Singh, N., Nyuur, R., & Richmond, B. (2019). Renewable energy development as a driver of economic growth: Evidence from multivariate panel data analysis. *Sustainability, 11*, 2418.

Soytas, U., & Sari, R. (2003). Energy consumption and GDP: Causality relationship in G-7 countries and emerging markets. *Energy Economics, 25*, 33–37.

Squalli, J. (2007). Electricity consumption and economic growth: Bounds and causality analyses of OPEC members. *Energy Economics, 29*, 1192–1205.

Stampini, M., Robles, M., & Sáenz, M. (2015). *Poverty, vulnerability and the middle class in Latin America*. Inter-American Development Bank. Social Protection and Health Division. VII. Series. IDB-WP-59.2015.

Sullivan, M. C., Hawes, K., Winchester, S. B., & Miller, R. (2008). Developmental origins theory from prematurity to adult disease. *Journal of Obstetric, Gynecologic, & Neonatal Nursing, 37*(2), 158–164. https://doi.org/10.1111/j.1552-6909.2008.00216.x

United Nations (UN). (2003). *The challenge of slums: Global report on human settlements 2003*. United Nations Settlements Programme (UN-Habitat). https://unhabitat.org/the-challenge-of-slums-global-report-on-human-settlements-2003+&cd=2&hl=pt-PT&ct=clnk&gl=pt.

United Nations (UN). (2019). *UN spotlights' explosive' obesity rates, hunger in Latin America and Caribbean*. https://news.un.org/en/story/2019/11/1051211.

Vakis, R., Rigolini, J., & Lucchetti, L. (2015). *Left behind, chronic poverty in Latin America and the caribbean*. International Bank for Reconstruction and Development/The World Bank.

World Bank Open Data. (2021). http://www.worldbank.org/.

World Health Organization (WHO). (2020). *Obesity and overweight. Key facts*. http://www.who.int/mediacentre/factsheets/fs311/en/#.

Ying, Y. H., Chang, K., & Lee, C. H. (2014). The impact of globalisation on economic growth. *Romanian Journal of Economic Forecasting –, XVII*(2p), 25–34.

Yu, E. S., & Choi, J. Y. (1985). The causal relationship between energy and GNP: An international comparison. *Journal of Energy and Development, 10*, 249–272.

Environmental degradation in the Latin American and Caribbean region

6

6.1 Introduction

Assessing the environmental degradation of a country or territory is challenging, given the multi-dimensional nature of the concept and the interdependencies between its various components. According to Johnson et al. (1997), environmental degradation is any change or disturbance to the environment perceived as deleterious or undesirable and encompasses five spheres: atmosphere, biosphere, hydrosphere, lithosphere and pedosphere.

Several international institutions conduct periodic assessments of the state and evolution of the environment, both worldwide and regionally. For example, the Global Environment Outlook (UN Environment, 2019) evaluates the environment worldwide, its evolution and the pathways to achieve the Sustainable Development Goals. This analysis encompasses five broad categories (e.g., air, biodiversity, oceans and coasts, land and soil and freshwater). However, it also covers cross-cutting issues that further enlighten the interconnections between them and stress the need for integrated policy responses. In addition, the report 'Environment at a Glance 2020' (OECD, 2020) studies the evolution of a set of key environmental indicators that reflect major environmental trends in areas such as climate change, air quality, biodiversity, water resources and circular economy.

There are also some synthetic indicators of the global impact of human activity on the environment. Rees (1992) was probably the first academic to propose the measure called the ecological footprint. The ecological footprint measures how much biologically productive land and water an individual, population or activity requires to produce all the resources it consumes and absorbs the waste it generates, using prevailing technology and resource management practices. Another indicator that tracks the state and development of the environment in 180 countries worldwide is the Environmental Performance Index (Wendling et al., 2020). This index provides a data-driven summary of the state of sustainability, based on 32 indicators covering 11 issue categories.

The rest of this chapter is organised as follows. In the next section, we assess the overall status of the environment in Latin American (LA) countries using the Environmental Performance Index. Then, in the following ones, we present a disaggregated view of the recent environmental developments in LA in five different areas: climate change, air quality and health, freshwater resources, circular economy and waste materials and biological resources and biodiversity.

Obesity Epidemic and the Environment. https://doi.org/10.1016/B978-0-323-99339-5.00007-8

6.2 The environmental performance index

The Environmental Performance Index (henceforth EPI) ranks 180 countries based on their environmental health and ecosystem vitality. It is a weighted average of 32 indicators covering 11 issue categories: air quality, sanitation and drinking water, heavy metals, waste management, biodiversity and habitat, ecosystem services, fisheries, climate change, pollution emissions, agriculture and water resources.

Table 6.1 exhibits the EPI scores in 2020 for the 10 most populous countries in Latin America, their 10-year change and their ranks. It also shows the average score and its 10-year change for the whole LA region and the world countries. The environmental performance of the LA region is worse than the world average, and its improvement over the last decade also lags behind the world average. Most LA countries' ranks are in the second quartile of the country level, except Chile – the top performer among this group of countries – and Guatemala – the worst one. The evolution of the EPI score was heterogeneous across countries. While some of the largest countries, such as Mexico, Argentina and Brazil, registered marked increases in their scores, others experienced environmental degradation (Venezuela, Bolivia, Peru and Guatemala).

Usually, environmental preservation is not among the main policy priorities in low- and middle-income countries, which need to foster economic growth to lift their people out of poverty. Thus, wealthy European countries occupy the top 10 spots of the EPI ranking. This positive relationship between economic development and environmental performance is also observable in the group of LA countries. The correlation coefficient between the 2020 EPI scores and 2019 per capita gross domestic product (GDP) in purchasing power parity (World Bank Open Data, 2021) equals 69.7%. Fig. 6.1 displays a scatterplot of the relation between EPI scores and GDP per capita,

Table 6.1 Environmental Performance Index in LA countries.

Country/Region	Rank	EPI score	10-year change
Chile	44	55.3	3.7
Colombia	50	52.9	0.9
Mexico	51	52.6	7.4
Argentina	54	52.2	5
Brazil	55	51.2	4.9
Ecuador	56	51	3.9
Venezuela	59	50.3	−0.5
Bolivia	88	44.3	−0.3
Peru	90	44	−0.8
Guatemala	149	31.8	−0.3
Latin America	–	40.67	1.67
World	–	48.26	3.18

The authors created this table with the data collected from Wendling, Z.A., Emerson, J.W., de Sherbinin, A., & Esty, D.C., (2020). 2020 Environmental performance index. Yale Center for Environmental Law & Policy.

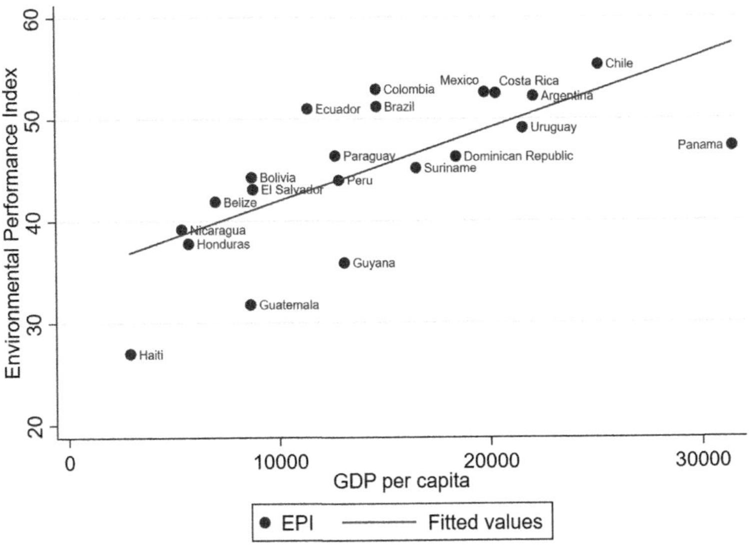

Figure 6.1 Relationship between the 2020 environmental performance index score and the 2019 per capita gross domestic product (GDP) (constant 2011 USD, purchasing power parity adjusted) in Latin America and Caribbean (LAC) countries.
The authors created this figure with the data collected from Wendling, Z.A., Emerson, J.W., de Sherbinin, A., & Esty, D.C., (2020). 2020 Environmental performance index. Yale Center for Environmental Law & Policy and World Bank Open Data (2021). http://www.worldbank.org/.

as well as the linear fit between these variables. Economic development explains a considerable fraction of the EPI scores dispersion, but some countries exhibit a poor performance given their per capita GDP (Haiti, Guatemala, Guyana and Panama), while others outperform (Ecuador, Colombia and Brazil).

6.2.1 Climate change

Humankind faces the difficult challenge of promoting economic development, reducing poverty and increasing living standards without compromising environmental sustainability. The urgency of controlling global warming has been recognised in the United Nations Sustainable Development Goals (UN General Assembly, 2015), which aims to limit the temperature rise this century well below 2°C above pre-industrial levels. A failure to attain this objective may cause severe disruptions in our planet, as climate change increases the frequency and severity of extreme weather events, threatens ecosystems and biodiversity, affects human settlements in coastal areas and compromises our ability to produce food to nourish an increasing human population (United Nations Environment Programme, 2019). Exhibit 6.1 discusses the impact of climate change in the LAC region.

Greenhouse gases (GHGs) are the main anthropogenic source of global warming, as they absorb infrared radiation from the atmosphere. According to the

Exhibit 6.1 Climate change impacts in the LAC region

The LAC region has been going through climate change. Temperatures are rising, precipitation patterns are shifting and some areas are experiencing changes in the frequency and severity of weather extremes, such as heavy rains. Indeed, these impacts range from melting Andean glaciers to devasting floods and droughts (World Wide Fund for Nature, 2020).

Moreover, consistent with the World Wide Fund for Nature (2020), the two great oceans that flank the continent, the Pacific and the Atlantic, are warming and becoming more acidic while the sea level rises. According to the institution, more significant impacts are in store for the region as both the atmosphere and oceans continue to change rapidly. Food and water supplies will be disrupted. Towns and cities and the infrastructure required to sustain them will be increasingly at risk. Human health and welfare will be adversely affected, along with natural ecosystems.

- Extreme weather

  According to the World Wide Fund for Nature (2020), extreme events have severely affected Latin America and the Caribbean region. Indeed, according to institutions, the severe climate and hydro-meteorological events occurred between 2000 and 2013. In hydro-meteorological events, we can include cold spells, heat waves, drought, floods, coastal storm surges, avalanches, heavy snowfall, tornados, blizzards, typhoons, hurricanes, thunderstorms and hailstorms.

  These extreme climates and hydro-meteorological events have resulted in numerous fatalities, displacement of people and significant economic losses. The tropical storms originate in the Atlantic and Pacific and have devastated Central American, the Caribbean and Mexico in the LAC region. Beyond the damage the storms have caused in coastal areas, their torrential rains inland have accounted for much greater devastation.
- Droughts

  Drought conditions in the Amazon, Central America, the Caribbean, Northeastern Brazil and Mexico will increase. Indeed, according to the World Wide Fund for Nature (2020), the prospect of more frequent extreme droughts in the Amazon could push the region to a 'tipping point', increasing the likelihood of large-scale dieback of the Amazon forest.

  Moreover, notable recent droughts afflicted the Amazon in 2005 and 2010 and a drought in Southeastern Brazil extended from 2012 to late 2015. In addition, existing drought conditions in Mexico, Central America and the Caribbean may be intensified by the ongoing strong 2015–2016 El Niño occurring against a backdrop of rising temperatures associated with global warming. For example, after 4 years of below-normal rainfall, São Paulo, Brazil, was experiencing its worst drought in over 80 years by mid-2015. The city's main water system, the Cantareira reservoir, supports the water needs of 5.3 million people. However, by August 2015, it was at record low levels with less than 17% of its normal water capacity, down from the nine million before the drought. This situation illustrates the vulnerability of some LAC cities to drought as climate change alters the frequency and severity of drought in the region (the World Wide Fund for Nature, 2020).

Exhibit 6.1 Climate change impacts in the LAC region—cont'd

- Sea level rise
  The World Wide Fund for Nature (2020) states that oceans expand as they warm and rise further as they receive massive amounts of fresh water from melting glaciers and ice sheets. Moreover, this problem threatens the LAC population, a large proportion of which lives on the coast, by contaminating freshwater aquifers, eroding shorelines, inundating low-lying areas and increasing the risks of storm.

Intergovernmental Panel on Climate Change (Intergovernmental Panel on Climate Change, 2014), carbon dioxide (CO_2) emissions represent 76% of GHGs emissions, while methane (CH_4), nitrous oxide (N_2O) and fluorinated gases account for 16%, 6% and 2% of GHGs emissions measured in tons of CO_2 equivalent. The bulk of these emissions originate in energy use in transport, industry and households. However, deforestation also plays a crucial role as it suppresses carbon sinks that reduce the concentration of carbon dioxide in the atmosphere. This problem is particularly acute in LA, which harbours the largest rainforest in the world, and whose degradation accounts for a non-negligible amount of GHG emissions (Pearson et al., 2017).

Fig. 6.2 presents the evolution of carbon dioxide emissions per capita between 1990 and 2016 in the world, the Organisation for Economic Co-operation and Development

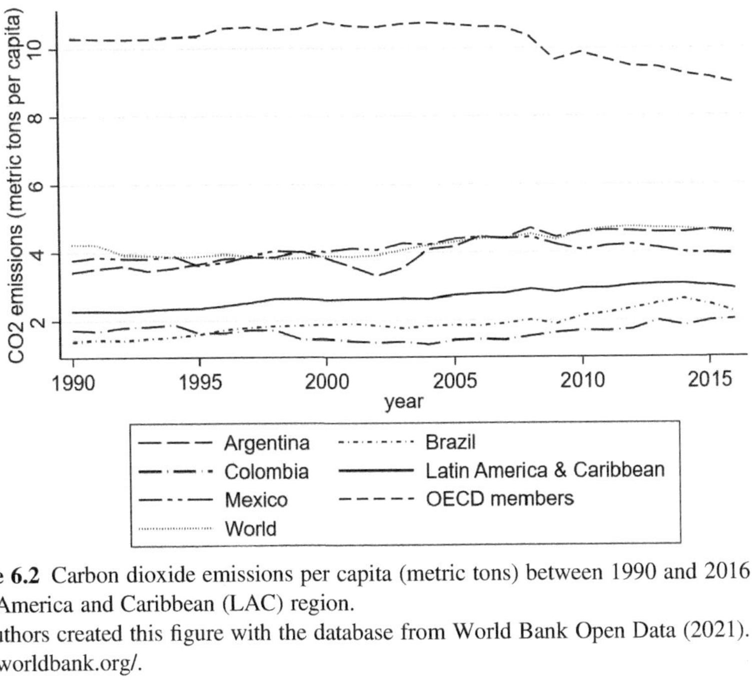

Figure 6.2 Carbon dioxide emissions per capita (metric tons) between 1990 and 2016 in the Latin America and Caribbean (LAC) region.
The authors created this figure with the database from World Bank Open Data (2021). http://www.worldbank.org/.

(OECD) countries, LAC and the four most populous countries in this region. In 2016, CO_2 emissions per capita in LAC (2.92 metric tons) were considerably lower than the world and OECD averages (4.56 and 8.98 metric tons, respectively), but their evolution shows a tendency to close the gap. While emissions per capita showed a cumulative decrease of 12.8% throughout the period in OECD countries, they increased 27.8% in the LAC region, a rate that exceeded the world average of 7.3%. Amongst the largest LAC countries, Argentina and Mexico's 2016 per capita emissions (4.62 and 3.94 metric tons, respectively) were close to the world average, and Brazil and Colombia (2.24 and 2.03 metric tons, respectively) emitted approximately half the world average. However, the pattern of emissions growth is widely different amongst these countries: while Mexico's emissions remained broadly unchanged (3.97% increase), they increased sharply in the other three countries (34.35%, 59.55% and 17.22% in Argentina, Brazil and Colombia, respectively). This evidence is consistent with Román-Collado and Morales-Carrión (2018), Acheampong (2018), Pablo-Romero and De Jesús (2016), Jardón et al. (2017), Van Ruijven et al. (2016), and Koengkan et al. (2021), among others who report a positive relationship between economic growth and CO_2 emissions in the LAC region.

Even though carbon dioxide is the principal GHG generated by human activity, it is essential to study the emissions of other GHGs because they have a higher global warming potential (henceforth GWP). According to the United States Environmental Protection Agency (EPA, 2021), the GWP of methane and nitrous oxide is 28–36 and 265–298 times higher than carbon dioxide, while the GWP from chlorofluorocarbons can be thousands or tens of thousands higher. Thus, if uncontrolled, these gases may cause significant increases in global temperature. These emissions are especially relevant in the LAC region, as they represent approximately 45% of all Kyoto gas emissions, a higher share than the world average (Gütschow et al., 2019).

Fig. 6.3 uses data from the PRIMAP-hist national historical emissions time series to display the evolution of non-CO_2 Kyoto gas emissions in the world, the LAC region and its four largest countries between 1990 and 2018. In the most recent year, the LAC region (2.44 metric tons of CO_2 equivalent per capita) had higher emissions than the world average (1.63 metric tons of CO_2 equivalent per capita). It also exhibited a positive cumulative emissions growth rate of 3.72%, unlike the world, which experienced a 4.18% decrease. Argentina had the highest initial emissions (4.57 metric tons of CO_2 equivalent per capita), but it also managed to reduce them the most (25.71%). Colombia almost closed the gap with the world average, as it managed to decrease its emissions by 17.31%. On the contrary, Mexico, which started in 1990 with an emissions level close to the world average, saw this gap increase to 0.66 metric tons of CO_2 equivalent. Brazil also experienced a surge in emissions, and the gap between this and the world average increased from 15.46% in 1990 to 79.5% in 2018.

Most Kyoto gas emissions stem from energy use, mainly when it is produced from fossil fuels. The positive relationship between energy consumption and greenhouse gas emissions has been established in several studies focussing on the LA region, such as Hanif (2017), Khan et al. (2014), Deng et al. (2020) and Koengkan et al. (2021). Thus, it is essential to reduce the energy intensity of economies to achieve a sustainable growth path.

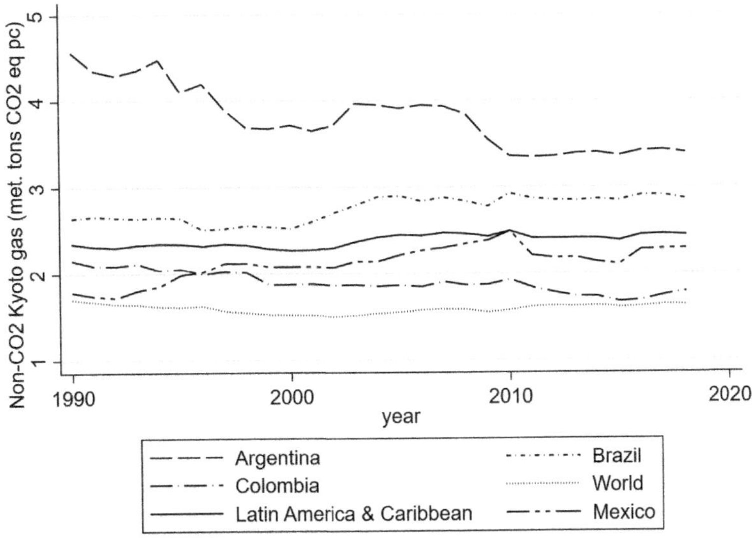

Figure 6.3 Non-CO_2 Kyoto gas emissions per capita metric tons of CO_2 equivalent) between 1990 and 2018 in the Latin America and Caribbean (LAC) region.
The authors created this figure with data collected from Gütschow, J., Jeffery, L., & Gieseke, R., (2019). The PRIMAP-hist national historical emissions time series (1850–2016). V. 2.0. GFZ Data Services. https://doi.org/10.5880/PIK.2019.001.

Fig. 6.4 shows the evolution of the energy intensity in kWh per 2011 constant United States dollars (USD), purchasing power parity adjusted (Our World in Data, 2021). In 2016, the LAC region had an energy intensity that was close to the world average. There was a clear tendency for a decrease in energy intensity throughout the world throughout the period and in the LAC region after 2000. However, the reduction of energy intensity in this region (21.44%) fell short of the world average (35.08%). The four largest countries in the LAC area are more energy efficient than the regional average, with Colombia having the best performance. The cumulative reduction of energy intensity from 1990 to 2016 was similar in Brazil, Colombia and Mexico (between 24.8% and 33.5%), but it was substantially lower in Argentina (7.49%).

Energy efficiency plays an essential yet indecisive role in controlling greenhouse gas emissions, as population growth and economic development cause an ever-growing energy demand. It must be accompanied by a gradual replacement of fossil fuel energy sources by renewable ones to limit global warming. LA countries are privileged in accomplishing this transition because they are endowed with exceptional natural conditions to produce energy from renewable sources (e.g., Aghahosseini et al., 2019; Shahsavari & Akbari, 2018; De Barbosa et al., 2017; and; Griffith-Jones et al., 2017, pp. 1033–1046). However, this potential remains largely underexplored.

Fig. 6.5 displays the evolution of the share of renewable energy in total final energy consumption. In 1990, the renewable percentage in LAC (32.44%) was higher than

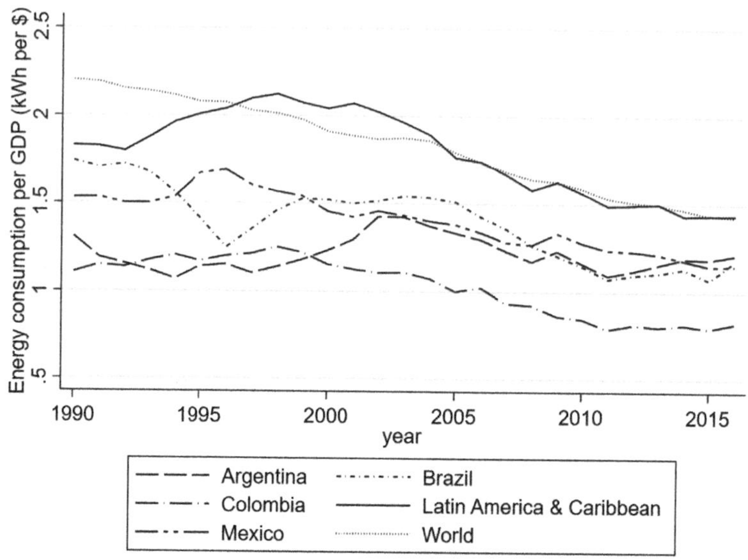

Figure 6.4 Energy intensity (kWh per 2011 constant USD, purchasing power parity adjusted) between 1990 and 2016 in the Latin America and Caribbean (LAC) region.
The authors created this figure with the Our World in Data database (2021). https://ourworldindata.org/.

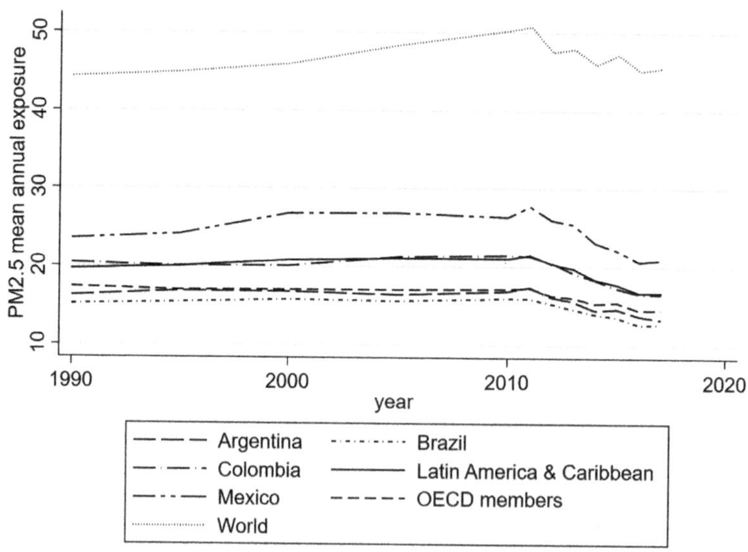

Figure 6.5 Renewable energy consumption in the total final energy consumption percentage between 1990 and 2015 in the Latin America and Caribbean (LAC) region.
The authors created this figure with the Our World in Data database (2021). https://ourworldindata.org/.

that of the world (17.07%) and in OECD countries (7.01%), due to the large hydroelectric installed capacity in the region. The renewable share increased approximately 1% and 5% from 1990 until 2015 in the world and OECD countries. It decreased steadily in LAC, as this region failed to increase renewable capacity enough to respond to the rise in energy demand. Amongst the LAC largest countries, Brazil had the highest renewable share in 1990 (49.86%), followed by Colombia (38.25%), Mexico (14.41%) and Argentina (8.92%). All these countries saw their renewable energy share decrease throughout the period (between 5% and 15%), except for Argentina, whose share rose by 1.08%.

6.3 Air pollution and health

Air pollution is the main factor responsible for environmental degradation in the world and is the leading environmental health risk. Poor air quality increases the incidence of several respiratory and heart diseases (Pope & Dockery, 2013), which reduce the number of healthy life years, increase health costs and raise human mortality (e.g., Pope et al., 2020 and Vodonos et al., 2018). It also causes acid rain that provokes the acidification of freshwater resources, reduces agricultural productivity and damages fragile ecosystems.

The most widely used indicator of air pollution is particulate matter, a mixture of solid particles and liquid droplets found in the air. It originates in a wide range of human activities and sectors, such as transport, industry or agriculture. Its composition, which includes hundreds of different chemicals, depends on the emission sources. With a diameter smaller than 2.5 μm, fine particulates, designated $PM_{2.5}$, are the most damaging to health, as they are inhalable and penetrate deeply into the lungs, causing severe respiratory problems. $PM_{2.5}$ originates from a wide range of human activities and sectors, such as transport, industry or agriculture.

Toxic pollutants, also known as hazardous pollutants, include a wide range of compounds, such as formaldehyde, acrolein, benzene, naphthalene, arsenic, cadmium, mercury, chromium and lead (NACAA, 2021). They cause many adverse health effects such as cancer, reproductive and neurological problems, birth and development defects and heart and respiratory diseases. Furthermore, some of these chemicals are persistent, bioaccumulative toxins, which means that they are transmitted through the food chain and remain indefinitely in the human body, causing long-lasting adverse health effects.

The measurement of the concentration of specific toxic pollutants in the air is not widely available around the world, except for a few developed countries. Furthermore, the reported values of pollutant concentrations are often unreliable, subject to a large margin of error. Thus, we choose to focus our analysis on the evolution of the concentration of $PM_{2.5}$ in LA and compare it to the world average.

Fig. 6.6 presents the evolution of the mean annual concentration of $PM_{2.5}$ in the world, the OECD countries, LAC countries and the four most populous countries in this region. This figure was created using data from World Bank Open Data (2021).

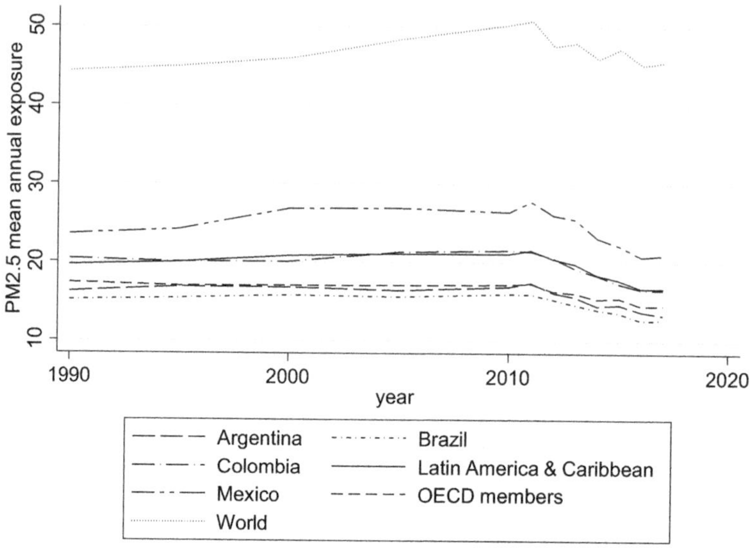

Figure 6.6 $PM_{2.5}$ air pollution means annual exposure (micrograms per cubic metre) between 1990 and 2017 in the Latin America and Caribbean (LAC) region.
The authors created this figure with the database from World Bank Open Data (2021) http://www.worldbank.org/.

Before 2010, $PM_{2.5}$ concentrations are only available every 5 years. Thus, we used linear interpolation to obtain the values in the intermediate years.

In 1990, the $PM_{2.5}$ concentration in the LAC region was less than half the world average (19.61 versus 44.26 μg per cubic metre) and slightly above the average of OECD members (17.35 μg per cubic metre). The $PM_{2.5}$ growth showed a similar pattern in all the regions: it increased until 2011 and then decreased. However, the OECD members and LAC managed to achieve a cumulative reduction of $PM_{2.5}$ relative to 1990 around 15%, while the world experienced a slight increase (2.87%). Amongst the countries considered, Mexico had the highest $PM_{2.5}$ concentration in 1990 (23.52 μg per cubic metre), and Brazil the lowest (15.14 μg per cubic metre). All the countries achieved $PM_{2.5}$ reductions that ranged from 11.04% in Mexico to 18.92% in Colombia.

Even though LA $PM_{2.5}$ concentration is less than half the world average, annual average concentrations provide an incomplete picture of their destructive potential for people's health and mortality booster. Its spatial distribution across countries and the possible presence of peak concentrations during some periods may cause severe health impacts in megacities within the region. Large cities are more prone to suffer from air pollution, as their high road concentration and intensive traffic generate elevated $PM_{2.5}$ levels. Furthermore, air pollution affects many individuals in LAC cities due to their high population density. Several authors reported rises in short-term

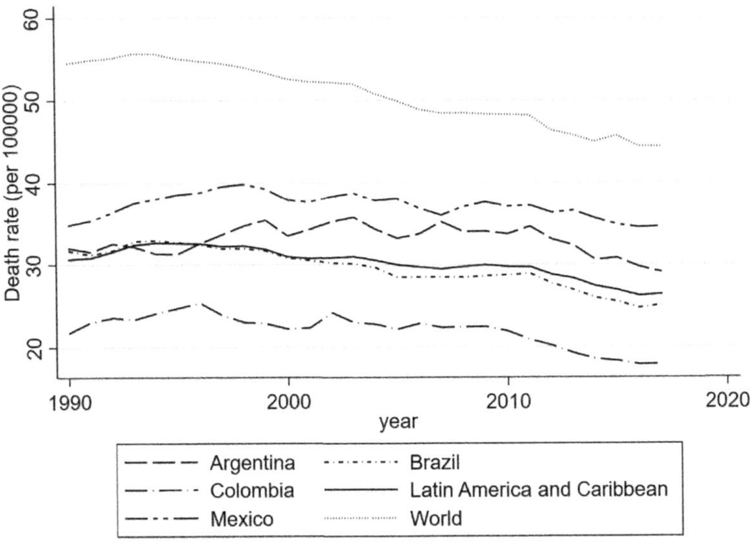

Figure 6.7 Outdoor air pollution death rate (per 100,000 inhabitants) between 1990 and 2017 in the Latin America and Caribbean (LAC) region.
The authors created this figure with the Our World in Data database (2021). https://ourworldindata.org/.

mortality related to particulate matter exposure in São Paulo (Bravo et al., 2016; Costa et al., 2017) and other LA cities (e.g., Liu et al., 2019; and O'Neill et al., 2008).

Fig. 6.7 depicts the annual mean outdoor air pollution death rate (OAPDR) between 1990 and 2017. At the beginning of the sample, OAPDR in the Latin American and the Caribbean region was approximately 56% of the world average (30.67 vs 54.56), which is consistent with the lower $PM_{2.5}$ concentration in LAC relative to the world average (Fig. 6.6). Both the LAC region and the world achieved steady reductions in OAPDR from the mid-nineties until 2017, but this improvement was lower in LAC (13.92% vs 18.66%). Among the countries in the region, Mexico performed the worst: it had the highest OAPDR in 1990, barely improving this rate until 2017. Colombia had the lowest OAPDR, both at the beginning (21.73) and the end of the sample (17.93). Brazil and Argentina had similar OAPDR throughout the period. However, Brazil achieved a substantially higher reduction of this indicator than Argentina (20.82% vs 9.18%).

6.4 Freshwater resources

Freshwater resources are of uttermost importance to human subsistence, environmental preservation and the well-functioning of ecosystems. However, they are unevenly distributed across and within countries, and some have poor quality due to toxic pollutant contamination. Moreover, most human activities require the abstraction of freshwaters, such as agriculture, representing 70% of all worldwide abstractions (OECD, 2020), industrial processes, electric power plants and drinking water supply. Despite the increasing pressure on water demand caused by population growth and economic development, its utilisation should be parsimonious so as not to compromise the renewal of water bodies, ecosystems sustainability, human settlements and agricultural food production.

The LA region is endowed with abundant freshwater resources. However, it faces some problems, such as the lack of integrated management of rivers and water bodies across several countries (Bezerra et al., 2021). Furthermore, access to water is not widespread within countries, and rural areas often lack good quality water and sanitation (San Miguel, 2018).

Table 6.2 shows that LA is well endowed with freshwater resources. Its per capita stock is approximately three times higher and the world average and two and a half times higher than those in OECD countries. Within the region, most countries have resources that are at least four times the world average (Bolivia, Brazil, Chile, Colombia, Ecuador, Peru and Venezuela), while Mexico's are significantly below it. A decreasing tendency in per capita resources over time is observable in all the countries and regions, which may be attributed to population growth.

The sustainability of water bodies depends on the relation between the withdrawals and the available water resources. A high withdrawal rate puts pressure on resources, compromises the ability of nature to renew lakes and rivers and may ultimately lead to desertification. The LAC region compares favourably to the world and the OECD countries in this subject: its freshwater stress is three to four times lower than the

Table 6.2 Renewable internal freshwater resources per capita (cubic metres) in LA countries.

Country/Region	2002	2007	2012	2017
Argentina	7749	7358	6997	6630
Bolivia	34,714	31,752	29,245	27,115
Brazil	31,531	29,774	28,406	27,238
Chile	56,425	53,538	50,861	47,914
Colombia	52,477	49,043	46,554	43,856
Ecuador	33,659	30,945	28,590	26,356
Guatemala	8989	8103	7387	6788
Mexico	4022	3746	3488	3278
Peru	60,551	57,918	55,614	52,188
Venezuela	32,071	29,544	27,416	27,390
LAC	25,920	24,348	23,015	21,874
OECD	10,409	10,033	9711	9402
World	7003	6576	6072	5732

The authors created this table with the World Bank Open Data (2021). http://www.worldbank.org/.

OECD and world averages. All the countries within the region have low water stress levels,[1] except for Argentina, and Mexico whose water stress level is moderate.

The previous analysis depicts a rosy picture of water resources in LA, but it does not cover the full extent of the problem. Besides the quantity of water available, its quality is of uttermost relevance. The lack or inefficiency of sewage treatment facilities, factory disposals and pharma-contamination leads to the accumulation of pollutants in water resources that may compromise their ability to sustain ecosystems and endanger species. Toxic pollutants are especially concerning because they move up through the food chain, causing a wide range of health problems, such as cancer and congenital disabilities. In LA, studies about the prevalence of contaminants in drinking water focussed on a few large countries, mainly Brazil and Mexico (Reichert et al., 2019). However, despite the scarcity of data, there is compelling evidence of the presence of several contaminants, such as pesticides, pharmaceuticals, illicit drugs, hydrocarbons and endocrine disruptors (Peña-Guzmán et al., 2019). Table 6.3 below shows the annual total freshwater withdraws in LA countries.

The widespread[2] access to sanitation and wastewater treatment services is indispensable to avoid contamination of freshwater bodies. Fig. 6.8 depicts the percentage of the population using safely managed sanitation services in the world, the OECD countries, LAC countries and the three most populous countries in this region.[3]

Table 6.3 Annual total freshwater withdraws (% of internal resources) in Latin America countries.

Country/Region	2002	2007	2012	2017
Argentina	10.86	12.00	12.91	12.91
Bolivia	0.66	0.68	0.69	0.69
Brazil	1.00	1.10	1.23	1.16
Chile	3.51	4.00	4.00	4.00
Colombia	0.41	0.53	0.59	0.63
Ecuador	2.17	2.24	2.24	2.24
Guatemala	2.39	3.04	3.04	3.04
Mexico	17.68	19.23	20.23	21.48
Peru	–	–	0.9	0.98
Venezuela	1.62	2.81	2.81	2.81
LAC	–	–	–	2.4
OECD	–	–	–	7.99
World	–	–	–	9.08

The authors created this table with the database from World Bank Open Data (2021). http://www.worldbank.org/.

[1] The OECD (2019) distinguishes four different water stress levels according to the relation between water withdraws and water resources: low (less than 10%), moderate (between 10% and 20%), medium-high (between 20% and 40$) and high (above 40%).

[2] Data for LAC, OECD and the world were unavailable for 2002, 2007 and 2012.

[3] Argentina was excluded from the figure due to the absence of data.

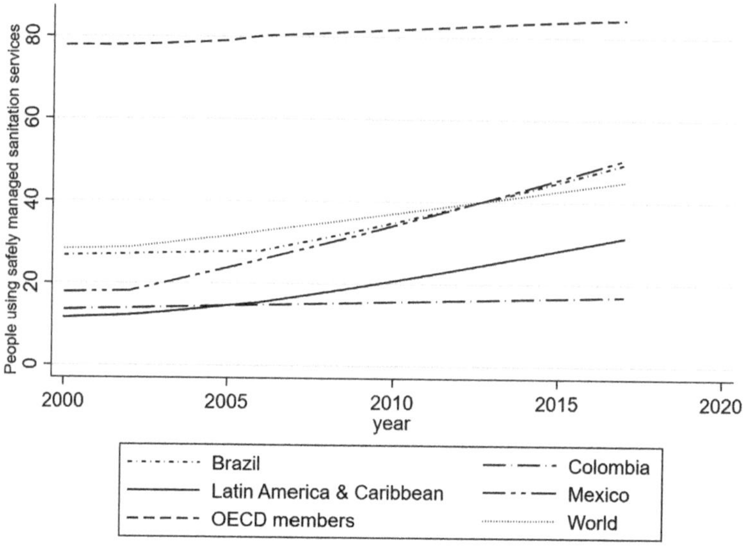

Figure 6.8 Percentage of people using safely managed sanitation systems between 2000 and 2017 in the Latin America and Caribbean (LAC) region.
The authors created this figure with the World Bank Open Data (2021). http://www.worldbank.org/.

At the beginning of this millennium, the LAC region lagged significantly behind OECD countries and the world in its populations' access to sanitation services. A mere 11.53% of its population had access to these services, in sharp contrast to the OECD countries (77.74%) and the world (28.20%). Its economic backwardness may explain this fact relative to the OECD, and its growing population spread across an extensive area. Recently, LA showed an impressive performance in fostering its population's access to sanitation and nearly tripled this percentage from 2000 to 2017, partially closing the gap with the world average. Among the LAC countries, Mexico and Brazil increased access to sanitation to half their populations by 2017, while Colombia showed a dismal performance (rise from 13.47% in 2000 to 16.99% in 2017).

6.5 Circular economy, waste and materials

Materials originating from natural resources form the physical backbone of economies and are indispensable for their well-functioning. They can be classified into four major categories: metals (ferrous, non-ferrous), non-metallic minerals (construction minerals, industrial minerals), biomass (wood, food) and fossil energy carriers (OECD, 2021b). However, the distribution of material resources around the world is uneven. Therefore, their environmental, economic and social impact depends on the processes adopted in their extraction and may spread across borders. Furthermore, population growth raises the pressure for resource extraction, as does economic development because people aim to consume more goods to improve their living standards. Thus, humanity faces the challenge to gradually extract a higher value from natural resources to enhance the well-being of individuals without exhausting them and compromising environmental sustainability.

A circular economy seeks to maximise the value extracted from resources and minimise their consumption. It also aims to properly dispose of the residuals, especially hazardous substances that cause water, soil and air contamination. Recovering materials through recycling and product reuse is another cornerstone of a circular economy, as it reduces the need for resource extraction and the damaging effects of waste disposal.

Table 6.4 shows the total[4] domestic material consumption and the GDP per kilogram of materials consumed for the OECD, the BRIICS countries and a selection of countries from the LA region. Most LA countries have a lower DMC than the OECD and BRIICS averages, except for Peru and Chile. This last country is a clear outlier with a DMC more than two times higher than any other region. Colombia and Guatemala have the lowest DMCs, with values that are less than half the OECD and BRIICS averages. Regarding material energy efficiency, all the countries in the region underperform the OECD, but most are more efficient than BRIICS (Chile and Bolivia being the exceptions). Colombia and Mexico are the most efficient countries, with a GDP per kilogram of DMC around 80% of the OECD average.

Table 6.4 Domestic material consumption in tons per capita (DMC) and gross domestic product (GDP) per domestic material consumption (constant 2015 USD per kg, PPP adjusted) in 2017 in Latin America (LA) countries.

Country	DMC	GPD per DMC
Argentina	11.29	1.75
Bolivia	11.34	0.66
Brazil	12.89	1.10
Chile	40.4	0.56
Colombia	5.89	2.26
Ecuador	8.01	1.35
Guatemala	5.02	1.59
Mexico	8.29	2.27
Peru	15.05	1.59
Venezuela	6.81	1.8
OECD	14.6	2.86
BRIICS	14.60	0.77

The authors created this table with the database from OECD (2021b). Material consumption (indicator). https://doi.org/10.1787/84971620-en.

[4] Brazil, Russia, India, Indonesia, China and South Africa.

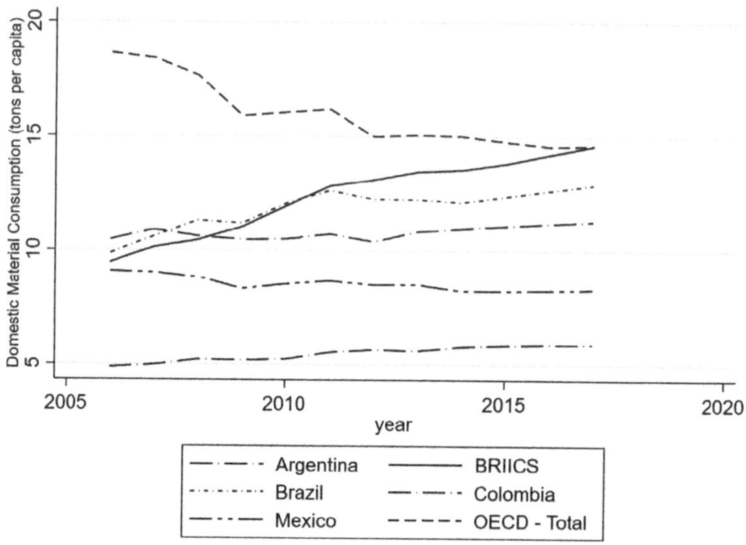

Figure 6.9 Domestic material consumption tons per capita between 2006 and 2017 in Latin American countries.
The authors created this figure with the OECD database (2021b). Material consumption (indicator). https://doi.org/10.1787/84971620-en.

Fig. 6.9 depicts the evolution of domestic material consumption per capita between 2006 and 2017. OECD countries reduced their per capita material consumption (21.57%) as they tilted their economies to high-value services reducing goods production. Most LA countries increased their domestic material consumption per capita (30.72%, 21.1% and 7.98% in Brazil, Colombia and Argentina), except for Mexico, which managed to reduce it by 21.57%. This pattern confirms the conjecture that economies, in their early to middle stages of development, require increasing material resources to grow. BRIICS experienced a sizable rise in DMC per capita (54.13%) because their rapid economic growth, heavily dependent on the manufacturing sector, puts increasing pressure on resources.

Limiting raw materials consumption and increasing the economic value-added from resources should be a top priority for all countries worldwide. However, many materials still end up as waste that needs to be appropriately managed. LA is expected to face a steady increase in waste generation due to population growth, urbanisation and economic development (UNEP, 2018). Nearly 50% of waste generated in the region is not disposed of properly (Arce et al., 2009). Recycling rates are still low (between 1% and 20%), implying that approximately 90% of municipal waste ends up in landfills. Despite the growing public awareness in LAC of the need to implement sustainable solid waste management systems and the acknowledgment by the central and regional governments of the necessity to create long-term plans in this sector (Arce et al., 2009), there is still a long road ahead to attain universal solid waste management in the region. The lack of reliable and up-to-date statistical information, coupled with the high investment needed in countries with scarce economic resources, makes this a challenging but not insurmountable task.[5]

Table 6.5 displays the municipal solid waste (MSW) generation per capita (kg/capita/day) in 2012 and the 2025 projections for the world, the LAC region, a selected set of countries from this region, and four country groups classified according to their income level. This table reveals there is a clear tendency for MSW per capita to increase with the income level. The LA region, whose most countries belong to the lower and upper-middle-income group, generates municipal solid waste per capita compatible with its development level (1.09 kg per capita per day) and is slightly below the world average. However, the projections anticipate a sizable increase until 2025, putting it above the world average. Bolivia and Colombia (0.3 and 0.95 kg per capita per day, respectively) generate the least MSW among the LAC countries, while Guatemala is the worst-performing country. Most countries are expected to increase MSW to a level close to or higher than the world average, except for Bolivia and Venezuela.

Table 6.5 Municipal solid waste (MSW) generation in kilograms per capita per day in Latin American countries.

Country	2012 MSW per capita (kg/capita/day)	MSW per capita (2025 projections)
Argentina	1.22	1.85
Bolivia	0.33	0.70
Brazil	1.03	1.60
Chile	1.08	1.50
Colombia	0.95	1.50
Ecuador	1.13	1.50
Guatemala	2.00	2.00
Mexico	1.24	1.75
Peru	1.00	1.40
Venezuela	1.14	1.15
LAC	1.09	1.56
World	1.19	1.42
Lower-income	0.6	0.86
Lower middle income	0.78	1.26
Upper middle income	1.16	1.59
High income	2.13	2.06

The authors created this table with the database from Hoornweg, D., & Bhada-Tata, P. (2012). What a waste: A global Review of solid waste management. Knowledge papers no. 15. World Bank. https://openknowledge.worldbank.org/handle/10986/17388.

[5] Data were extracted from the report 'What a Waste: A Global Review of Solid Waste Management' (Hoornweg & Bhada-Tata, 2012). Often, the MSW generation reported by the countries corresponds to an earlier year.

Table 6.6 Municipal solid waste collection rates in Latin American countries.

Country	Collection (%)
Brazil	83
Colombia	98
Ecuador	81
Guatemala	72
Mexico	91
Peru	74
LAC	10.62–100
Lower-income	10.62–55
Lower-middle-income	50.20–95
Upper-middle-income	50–100
High-income	76–100

The authors created this table with the Hoornweg, D., & Bhada-Tata, P. (2012). What a waste: A global Review of solid waste management. Knowledge papers no. 15. World Bank. https://openknowledge.worldbank.org/handle/10986/17388.

The municipal solid waste collection rates (Table 6.6) are highly heterogenous in the LAC region, ranging from 10.62% to 100%. All the largest countries in the region[6] have collection rates above 70%, which puts them within their income group range but far from universal coverage. In addition, solid waste management institutions in the region are plagued with bureaucracy, poor governance and the inability to collect revenue from services rendered (Hettiarachchi et al., 2018), which must be overcome to improve their performance.

Recycling and material reuse is fundamental in controlling the need to dispose of solid waste. Unfortunately, recycling is gaining traction in developed Europe, but it is still incipient in most LAC countries. To the best of our knowledge, there is no reliable database that tracks recycling rates in this region. UNEP (2018) provides some estimates of recycling rates for selected LAC countries that range from less than 2% in Brazil to approximately 17% in Colombia.

6.6 Biological resources and biodiversity

Biological resources are a crucial indicator of ecosystems' health and sustainability. Preserving biodiversity should be amongst the top priorities of humankind, as it provides raw materials essential to many sectors of the economy, prevents desertification and soil erosion and stabilises Earth's climate. Ecosystems are deeply interconnected communities of living organisms, and disruptive human interventions may generate a disequilibrium that ultimately leads to its collapse. Human activities cause pressure on biological resources both directly and indirectly. The overexploitation of resources, the

[6] No data were available for Argentina, Bolivia, Chile and Venezuela.

splitting of habitats by human constructions and the chemical poisoning of land and water fit the former category. In contrast, the anthropogenic climate change generated by man fits the latter.

The LAC region harbours a rich biological diversity, with around 60% of its global terrestrial life (UNEP-WCMC, 2016). It encompasses several biomes, extending from wetlands and coastal ecosystems to deserts, tropical forests, extensive savannah grasslands and high-altitude Andean habitats (UNEP-WCMC, 2016). However, LAC flourishing ecosystems face various threats from human actions. With the rise in urbanisation and infrastructure development, population and economic growth lead to habitat loss that may disrupt ecosystems and cause species extinction. The overexploitation of natural resources, on which the region is highly dependent, resulted, in some cases, in vegetation removal and soil and water contamination (UNEP-WCMC, 2016). Furthermore, the expansion of agriculture is the primary driver of deforestation in the region (Carter et al., 2018).

In the last decade, the region implemented several measures to protect ecosystems and avoid their decay. For example, it expanded protected areas, devised control mechanisms to prevent illegal wildlife trade, implemented recovery programmes for targeted species and fostered a range of low-carbon, sustainable development approaches (UNEP-WCMC, 2016). However, the region still needs to enhance environmental regulation and enforcement to convey adequate protection for wildlife and the environment (UNEP-WCMC, 2016).

The LA region has one of the world's largest and most diverse forest areas, representing nearly 24% of the world total (World Bank Data, 2021). However, it also experienced a higher deforestation rate than any other region (Armenteras et al., 2017). In addition, economic development and the increasing demand for raw materials led to a sizable reduction in tree coverage in LAC (Cuaresma & Heger, 2019).

Fig. 6.10 shows that, in 1990, the LAC region had a higher fraction of forest area relative to the total land area (53.31%) than either the world (31.62%) or OECD countries (32.31%). However, it was also the region where deforestation was more severe until 2017: it saw a decrease in forest area higher than 6% throughout the period, while the world[7] experienced a 1% decrease and OECD a slight increase. Amongst the LAC countries depicted, Brazil (70.46%) and Colombia (58.55%) had the highest percentage of forest area relative to their territories, but they sustained considerable losses (10.75% in Brazil and 4.88% in Colombia). Argentina had the lowest tree cover in the region (12.86% in 1990), which decreased to 10.52% in 2018 (Table 6.7).

Protected areas signal the willingness of governments to safeguard the environment. However, they are not perfect indicators because sites may be protected for other reasons beyond biodiversity preservation (e.g., cultural and historical sites), and law enforcement is crucial to guaranteeing adequate protection.

The LAC region has a protected area, a percentage of total terrestrial area, which exceeds the world and OECD averages. Among the countries in the region, Venezuela shows the highest engagement in protecting the environment: more than half its

[7] The latest year available for the world is 2016.

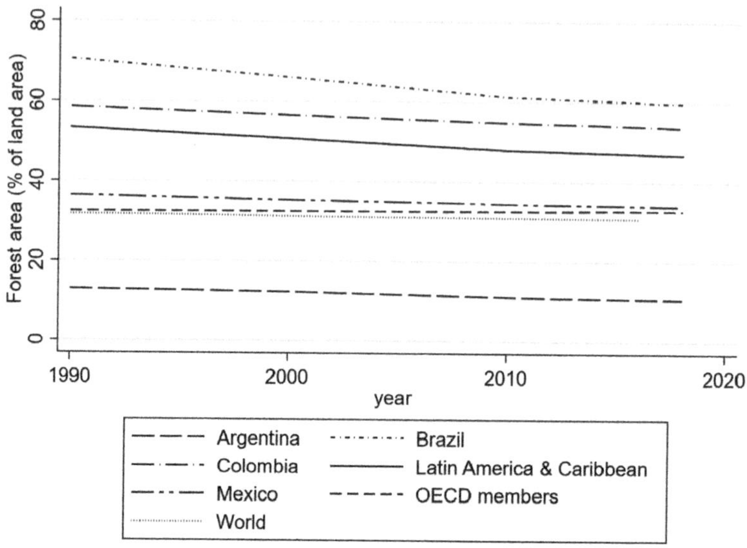

Figure 6.10 Forest area as a percentage of the total land area between 1990 and 2018 in the Latin America and Caribbean region.
The authors created this figure with the World Bank Open Data database (2021). http://www.worldbank.org/.

Table 6.7 Terrestrial protected areas as a percentage of total land area in 2018 in LA countries.

Country	Protected area (%)
Argentina	8.81
Bolivia	30.87
Brazil	29.42
Chile	18.49
Colombia	14.81
Ecuador	21.69
Guatemala	20.05
Mexico	14.50
Peru	21.31
Venezuela	54.14
LAC	23.46
OECD	15.09
World	14.73

The authors created this table with the World Bank Open Data database (2021). http://www.worldbank.org/.

territory is protected. Most countries have a fraction of protected areas at least as high as the world average, except for Mexico (14.53%) and Argentina (8.81%).

Protected areas are fundamental in preserving habitats and avoiding species extinction. The Red List Index (IUCN, 2021) is an indicator of ecosystems' health that focuses on the conservation status of significant species groups. The index comprises 0 and 1: the lowest value characterises the situation where all the species have gone extinct, and the highest represents the least concerning status (no species is expected to face extinction in the near future).

Fig. 6.11 shows that the performance of all countries and regions in protecting species has deteriorated over the years. In 1990, the LAC region had a higher fraction of endangered species than the world average (the Red List Index was 0.78 and 0.82 in the LAC region and the world, respectively). However, the world experienced a more pronounced decrease in the RLI than the LAC region, and both regions ended in 2018 with similar RLI values. Brazil (0.91) and Argentina (0.87) showed a good performance in protecting species in 1990 and managed to sustain their RLI values throughout the period. On the contrary, Colombia (0.78) and Mexico (0.73) started in a worse position than the world average and saw a reduction in the RLI scores to 0.73 and 0.67, respectively.

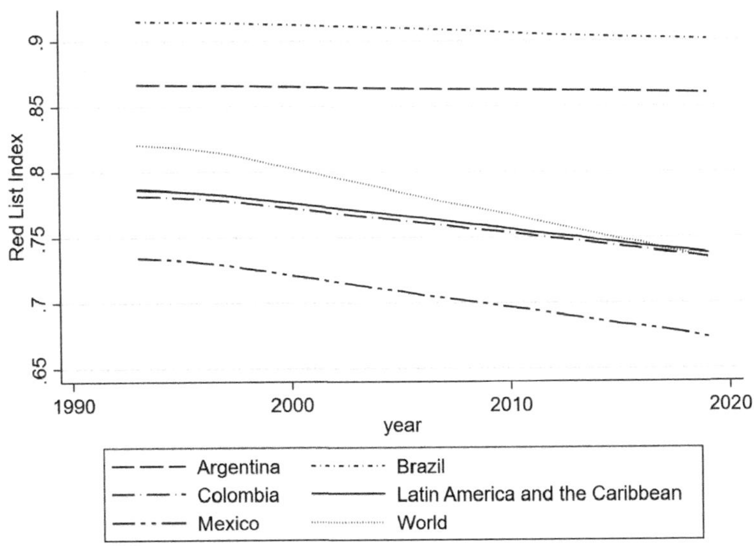

Figure 6.11 Red List Index between 1990 and 2018 in the Latin America and Caribbean region. (For interpretation of the references to colour in this figure legend, the reader is referred to the web version of this article.)
The authors created this figure with the Our World in Data database (2021). https://ourworldindata.org/.

6.7 Conclusion

In this chapter, we showed that the overall environmental performance of LA countries is slightly below the world average but has improved modestly over the past decade.

Regarding greenhouse gases, the LAC region compares favourably to the world and OECD averages, especially in CO_2 emissions, but it continues to exhibit a worrying increasing emissions trend. There is a clear need for more stringent regulation and enforcement if the region is to achieve the UN sustainable development goals. Renewable energy sources could play a leading role in the reduction of Kyoto gases. In the nineties, LA's share of renewable energy in total final energy consumption was above the world average, and it possesses exceptional conditions to produce energy from renewable sources. However, this share has been falling for several years. Renewable energy requires a considerable upfront investment compared to traditional fossil fuel sources, and the frail financial system of most LAC countries may compromise its development. Furthermore, these countries should gradually phase out the explicit and implicit subsidies to fossil fuels and divert them to the support of renewable energies.

The extent of the prevalence of toxic pollutants and their impacts in LA is broadly unknown. The region lacks widespread measurement instruments, and the data generated are often unreliable and non-comparable across countries (Furley et al., 2018). Thus, as a first step, countries in LAC should improve the data gathered about hazardous pollutants to diagnose the problem properly. Then, they can pass adequate regulations to control the spread of these contaminants into the environment and minimise their impacts.

LA has plentiful water resources, and the water stress level in the region is low. However, basic sanitation is far from universal. The institutions that manage these facilities are often poorly managed and bureaucratic and so are the solid waste management ones. The region should seek international support to solve these problems and adopt the best management practices from other countries. It should also revise the regulations to facilitate collecting revenues from services rendered, expanding the coverage of basic sanitation and waste management services.

LA is a haven of biodiversity, but deforestation may cause the loss of this status. It is faced with the conundrum of harbouring the largest rainforest on the planet and having scarce economic resources to protect it. Thus, it should seek international financial support to protect this global public good and technical assistance to design adequate regulations to ensure its preservation.

References

Acheampong, A. O. (2018). Economic growth, CO_2 emissions and energy consumption: What causes what and where? *Energy Economics, 74*, 677–692. https://doi.org/10.1016/j.eneco.2018.07.022

Aghahosseini, A., Bogdanov, D., Barbosa, L. S. N. S., & Breyer, C. (2019). Analysing the feasibility of powering the Americas with renewable energy and inter-regional grid interconnections by 2030. *Renewable and Sustainable Energy Reviews, 105*(January), 187–205. https://doi.org/10.1016/j.rser.2019.01.046

Arce, V. J. C., Raudales, R., Trubey, R., King, D. I., Chandler, R. B., & Chandler, C. C. (2009). Measuring and managing the environmental cost of coffee production in Latin America. *Conservation & Society, 7*(2), 141–144. https://www.jstor.org/stable/26392971.

Armenteras, D., Espelta, J. M., Rodríguez, N., & Retana, J. (2017). Deforestation dynamics and drivers in different forest types in Latin America: Three decades of studies (1980–2010). *Global Environmental Change, 46*(June), 139–147. https://doi.org/10.1016/j.gloenvcha.2017.09.002

Bezerra, M. O., Vollmer, D., Acero, N., Marques, M. C., Restrepo, D., Mendoza, E., Coutinho, B., Encomenderos, I., Zuluaga, L., Rodríguez, O., Shaad, K., Hauck, S., González, R., Hernandéz, F., Montelongo, R., Torres, E., & Serrano, L. (2021). Operationalizing integrated water resource management in Latin America: Insights from application of the freshwater health index. *Environmental Management, 69*. https://doi.org/10.1007/s00267-021-01446-1

Bravo, M. A., Son, J., De Freitas, C. U., Gouveia, N., & Bell, M. L. (2016). Air pollution and mortality in São Paulo, Brazil: Effects of multiple pollutants and analysis of susceptible populations. *Journal of Exposure Science and Environmental Epidemiology, 26*(2), 150–161. https://doi.org/10.1038/jes.2014.90

Carter, S., Herold, M., Avitabile, V., De Bruin, S., De Sy, V., Kooistra, L., & Rufino, M. C. (2018). Agriculture-driven deforestation in the tropics from 1990–2015: Emissions, trends and uncertainties. *Environmental Research Letters, 13*(1). https://doi.org/10.1088/1748-9326/aa9ea4

Costa, A. F., Hoek, G., Brunekreef, B., & Ponce de Leon, A. C. M. (2017). Effects of NO_2 exposure on daily mortality in São Paulo, Brazil. *Environmental Research, 159*(2), 539–544. https://doi.org/10.1016/j.envres.2017.08.041

Cuaresma, J. C., & Heger, M. (2019). Deforestation and economic development: Evidence from national borders. *Land Use Policy, 84*(January), e347–e353. https://doi.org/10.1016/j.landusepol.2018.12.039

De Barbosa, L. S. N. S., Bogdanov, D., Vainikka, P., & Breyer, C. (2017). Hydro, wind and solar power as a base for a 100% renewable energy supply for South and Central America. *PLoS ONE, 12*(3), 1–28. https://doi.org/10.1371/journal.pone.0173820

Deng, Q., Alvarado, R., Toledo, E., & Caraguay, L. (2020). Greenhouse gas emissions, non-renewable energy consumption, and output in South America: The role of the productive structure. *Environmental Science and Pollution Research, 27*(13), 14477–14491. https://doi.org/10.1007/s11356-020-07693-9

EPA. (2021). *Environmental protection agency*. https://www.epa.gov/.

Furley, T. H., Brodeur, J., Silva de Assis, H. C., Carriquiriborde, P., Chagas, K. R., Corrales, J., Denadai, M., Fuchs, J., Mascarenhas, R., Miglioranza, K. S. B., Miguez Caramés, D. M., Navas, J. M., Nugegoda, D., Planes, E., Rodriguez-Jorquera, I. A., Orozco-Medina, M., Boxall, A. B. A., Rudd, M. A., & Brooks, B. W. (2018). Toward sustainable environmental quality: Identifying priority research questions for Latin America. *Integrated Environmental Assessment and Management, 14*(3), 344–357. https://doi.org/10.1002/ieam.2023

Griffith-Jones, S., Spratt, S., Andrade, R., & Griffth-Jones, E. (2017). *Investment in renewable energy, fossil fuel prices and policy implications for Latin America and the Caribbean* (pp. 1–53). Financing for Development, CEPAL. https://doi.org/10.1017/CBO9781107415324.004

Gütschow, J., Jeffery, L., & Gieseke, R. (2019). *The PRIMAP-hist national historical emissions time series (1850–2016). V. 2.0.* GFZ Data Services. https://doi.org/10.5880/PIK.2019.001

Hanif, I. (2017). Economics-energy-environment nexus in Latin America and the caribbean. *Energy, 141*, 170–178. https://doi.org/10.1016/j.energy.2017.09.054

Hettiarachchi, H., Ryu, S., Caucci, S., & Silva, R. (2018). Municipal solid waste management in Latin America and the Caribbean: Issues and potential solutions from the governance perspective. *Recycling, 3*(2), 19. https://doi.org/10.3390/recycling3020019

Hoornweg, D., & Bhada-Tata, P. (2012). *What a waste: A global Review of solid waste management. Knowledge papers no. 15*. World Bank. https://openknowledge.worldbank.org/handle/10986/17388.

Hoornweg, D., & Bhada-Tata, P. (2012). *What a Waste : A Global Review of Solid Waste Management. Urban development series;knowledge papers no. 15*. Washington, DC: World Bank. https://openknowledge.worldbank.org/handle/10986/17388.

Intergovernmental Panel on Climate Change. (2014). Climate change 2014 mitigation of climate change. In *Climate change 2014 mitigation of climate change*. https://doi.org/10.1017/cbo9781107415416

IUCN. (2021). *The IUCN red list of threatened species*. Version 2021-1 https://www.iucnredlist.org.

Jardón, A., Kuik, O., & Tol, R. S. J. (2017). Economic growth and carbon dioxide emissions: An analysis of Latin America and the Caribbean. *Atmosfera, 30*(2), 87–100. https://doi.org/10.20937/ATM.2017.30.02.02

Johnson, D. L., Ambrose, S. H., Bassett, T. J., Bowen, M. L., Crummey, D. E., Isaacson, J. S., Johnson, D. N., Lamb, P., Saul, M., & Winter-Nelson, A. E. (1997). Meanings of environmental terms. *Journal of Environmental Quality, 26*(3), 581–589. https://doi.org/10.2134/jeq1997.00472425002600030002x

Khan, A. M., Zahir Khan, M., Zaman, K., & Naz, L. (2014). Global estimates of energy consumption and greenhouse gas emissions. *Renewable and Sustainable Energy Reviews, 29*, 336–344. https://doi.org/10.1016/j.rser.2013.08.091

Koengkan, M., Fuinhas, J. A., & Silva, N. (2021). Exploring the capacity of renewable energy consumption to reduce outdoor air pollution death rate in Latin America and the Caribbean region. *Environmental Science and Pollution Research, 28*(2), 1656–1674. https://doi.org/10.1007/s11356-020-10503-x

Liu, C., Chen, R., Sera, F., Vicedo-Cabrera, A. M., Guo, Y., Tong, S., Coelho, M. S. Z. S., Saldiva, P. H. N., Lavigne, E., Matus, P., Valdes Ortega, N., Osorio Garcia, S., Pascal, M., Stafoggia, M., Scortichini, M., Hashizume, M., Honda, Y., Hurtado-Díaz, M., Cruz, J., & Kan, H. (2019). Ambient particulate air pollution and daily mortality in 652 cities. *New England Journal of Medicine, 381*(8), 705–715. https://doi.org/10.1056/nejmoa1817364

NACAA. (2021). *National association of clean air agencies*. http://www.4cleanair.org/.

OECD. (2020). *Environment at a glance 2020*. OECD Publishing. https://doi.org/10.1787/4ea7d35f-en

OECD. (2021a). *OECD data*. https://data.oecd.org/.

OECD. (2021b). *Material consumption (indicator)*. https://doi.org/10.1787/84971620-en

O'Neill, M. S., Bell, M. L., Ranjit, N., Cifuentes, L. A., Loomis, D., Gouveia, N., & Borja-Aburto, V. H. (2008). Air pollution and mortality in Latin America: The role of education. *Epidemiology, 19*(6), 810–819. https://doi.org/10.1097/EDE.0b013e3181816528

Our World in Data. (2021). https://ourworldindata.org/.

Pablo-Romero, M. D. P., & De Jesús, J. (2016). Economic growth and energy consumption: The energy-environmental kuznets curve for Latin America and the caribbean. *Renewable and Sustainable Energy Reviews, 60*, 1343–1350. https://doi.org/10.1016/j.rser.2016.03.029

Pearson, T. R. H., Brown, S., Murray, L., & Sidman, G. (2017). Greenhouse gas emissions from tropical forest degradation: An underestimated source. *Carbon Balance and Management, 12*(1). https://doi.org/10.1186/s13021-017-0072-2

Peña-Guzmán, C., Ulloa-Sánchez, S., Mora, K., Helena-Bustos, R., Lopez-Barrera, E., Alvarez, J., & Rodriguez-Pinzón, M. (2019). Emerging pollutants in the urban water cycle in Latin America: A review of the current literature. *Journal of Environmental Management, 237*(December 2018), 408–423. https://doi.org/10.1016/j.jenvman.2019.02.100

Pope, C. A., Coleman, N., Pond, Z. A., & Burnett, R. T. (2020). Fine particulate air pollution and human mortality: 25+ years of cohort studies. *Environmental Research, 183*(August 2019), 108924. https://doi.org/10.1016/j.envres.2019.108924

Pope, C. A., & Dockery, D. W. (2013). Air pollution and life expectancy in China and beyond. *Proceedings of the National Academy of Sciences of the United States of America, 110*(32), 12861–12862. https://doi.org/10.1073/pnas.1310925110

Rees, W. E. (1992). Ecological footprints and appropriated carrying capacity: What urban economics leaves out. *Environment and Urbanisation, 4*(2), 121–130. https://doi.org/10.1177/095624789200400212

Reichert, G., Hilgert, S., Fuchs, S., & Azevedo, J. C. R. (2019). Emerging contaminants and antibiotic resistance in the different environmental matrices of Latin America. *Environmental Pollution, 255*(Part 1), 113140. https://doi.org/10.1016/j.envpol.2019.113140

Román-Collado, R., & Morales-Carrión, A. V. (2018). Towards a sustainable growth in Latin America: A multiregional spatial decomposition analysis of the driving forces behind CO_2 emissions changes. *Energy Policy, 115*(January), 273–280. https://doi.org/10.1016/j.enpol.2018.01.019

San Miguel, J. A. S. R. (2018). Water management in Europe and Latin America. *Management of Environmental Quality, 29*(2), 348–367. https://doi.org/10.1108/MEQ-05-2017-0044

Shahsavari, A., & Akbari, M. (2018). Potential of solar energy in developing countries for reducing energy-related emissions. *Renewable and Sustainable Energy Reviews, 90*(June 2017), 275–291. https://doi.org/10.1016/j.rser.2018.03.065

UN Environment (Ed.). (2019). *Global environment Outlook – GEO-6: Healthy planet, healthy people*. Cambridge University Press. https://doi.org/10.1017/9781108627146

UN General Assembly. (2015). *Transforming our world: The 2030 agenda for sustainable development, 21 October 2015, A/RES/70/1*. https://www.refworld.org/docid/57b6e3e44.html.

UNEP. (2018). *A third of urban waste ends up in open dumpsites or environment in Latin America and the Caribbean*. https://www.unep.org/ru/node/23696.

UNEP-WCMC. (2016). *The state of biodiversity in Latin America and the Caribbean*. In Unep-Wcmc. http://www.unep-wcmc.org/system/comfy/cms/files/files/000/000/734/original/Biodiversity_Review_LAC.pdf%0Apapers2://publication/uuid/AD0C28F3-9F24-4E1E-B3AE-91DC1B6D408-.

United Nations Environment Programme. (2019). *Emissions gap report 2019*. UNEP.

Van Ruijven, B. J., Daenzer, K., Fisher-Vanden, K., Kober, T., Paltsev, S., Beach, R. H., Calderon, S. L., Calvin, K., Labriet, M., Kitous, A., Lucena, A. F. P., & Van Vuuren, D. P. (2016). Baseline projections for Latin America: Base-year assumptions, key drivers and greenhouse emissions. *Energy Economics, 56*, 499–512. https://doi.org/10.1016/j.eneco.2015.02.003

Vodonos, A., Awad, Y. A., & Schwartz, J. (2018). The concentration-response between long-term PM 2.5 exposure and mortality; A meta-regression approach. *Environmental Research, 166*(June), 677–689. https://doi.org/10.1016/j.envres.2018.06.021

Wendling, Z. A., Emerson, J. W., de Sherbinin, A., & Esty, D. C. (2020). *2020 Environmental performance index*. Yale Center for Environmental Law & Policy.

World Bank Open Data. (2021). http://www.worldbank.org/.

World Wide Fund for Nature. (2020). *Climate change impacts in Latin America*. https://www.wwfca.org/en/our_work/climate_change_and_energy/climate_change_impacts_la/.

Causes of environmental degradation in the Latin American and Caribbean region

7

7.1 Introduction

What is environmental degradation? It is a multidimensional concept that includes all undesired changes in the environment. These undesired changes include the effect on the ecological footprint, deforestation, exposure to ambient $PM_{2.5}$ pollution, the damage to ecosystems, loss or extinction of wild species, depletion of natural resources or even agriculture and methane emissions. In addition, several behavioral patterns influence the environment—the level of urban solid residues, recycling, or sectoral composition of economies—for example, the percentage of people working in the tertiary sector. Extensive damage in soil, water and air is among the most common example of environmental degradation. In this chapter, we will concentrate on the loss of air quality.

An in-depth analysis of environmental degradation in the abstract is a task for tens of thousands of pages. Here we will focus on the environmental degradation closely related to several variables that the literature has identified as causing environmental degradation through harmful gas emissions.

Gas emissions have become central in the worries about our planet's health. Among them, the most concerning ones are global warming and the climate change associated with it. Some international initiatives have taken form, and the goals were and are ambitious. The best known is the United Nations Conference on Environment and Development (the Earth Summit) in 1992, the Kyoto Protocol—signed in 1997 and the Paris Agreement—signed in 2016. At the heart of these initiatives is an attempt to stabilise the greenhouse gas concentrations in the atmosphere to prevent anthropogenically induced alterations to Earth's climate. Along with international initiatives, multiple initiatives are performed at national, regional and local levels to curb environmental damage that is manifested in climate change.

Sometimes, the literature can seem to be confusing for people having their first contact with environmental concerns. There is an apparent contradiction in the fight for stopping environmental damage that the reader ought to be presented with. On the one hand, we have only one planet, and the Earth is a constant (we cannot alter its size). On the other hand, we have an economic and social dynamic that evolves, grows and demands more and more limited resources. Economists and other social scientists are used to transforming variables to identify the relationships better. Among the most common transformations to work with are per capita values or measures of intensity. These variables are sensitive to the relative growth between the components used in their construction. For example, we can have a per capita variable decreasing, but its absolute value increasing because the population grows faster than the variable

Obesity Epidemic and the Environment. https://doi.org/10.1016/B978-0-323-99339-5.00002-9

in per capita terms. A similar situation can occur, for example, with a variable, say, energy intensity (the quantity of energy required per unit of output). The total energy increases, but the energy intensity decreases because the output grows faster than the energy. The readers must be aware that we can only mitigate the environmental damage and not reverting the situation more often than not. For example, when the goal is to achieve carbon neutrality by 2050, we aggravate carbon dioxide (CO_2) in the atmosphere.

The analysis that we will develop seeks to determine whether economic and social variables cause environmentally harmful emissions. Gas emissions associated with human activities are, in general, well-identified. However, the same is not the case for whether one variable causes another or only has a statistical association. This distinction is crucial from the point of view of public policy decision-making. If one variable causes another, then it is on this that the authorities must act to achieve the desired results. In other words, the action must be on the disease and not on the symptoms. We apply the Granger causality notion, i.e., causality implies precedence over time.

We must be cautious in our conclusions. The empirical analysis that we will carry out will be done in a bivariate context. We will only focus on the causality of emissions' economic and social variables (a unidirectional causality). Often causality is a joint causality, i.e., to detect it, we have to consider the set of variables that interact with each other (endogeneity) to identify causality. These causalities, if any, are conditional on the model of description of reality that we have adopted. Finally, we recall the remark from Altman and Bland (1995): absence of evidence is not evidence of absence.

This first section assesses the Latin America and Caribbean (LAC) environmental degradation in the context of other world regions. We will compare the situation in seven regions of the world (WLD): LAC, Europe and Central Asia (ECS), East Asia and the Pacific (EAS), Central Europe and the Baltics (CEB), Middle East and North Africa (MEA), North America (NAC) and South Asia (SAS). The analysis will be focused on carbon dioxide, nitrous oxide, methane and greenhouse gas emissions. However, due to the scarcity of data on small particulates, only a brief explanation is made.

7.1.1 Total CO_2 emissions

The total annual CO_2 emissions varies greatly among the regions from 1960 to 2016 (see Fig. 7.1). Europe and Central Asia are the greatest polluters, followed by East Asia and Pacific and North America. In contrast, the LAC region was a minor contributor to the world total CO_2 emissions.

The annual CO_2 emissions of the LAC region grew consistently from 1960 to 2016 (see Fig. 7.2). Nevertheless, at least, two phases are discernible in this evolution. From 1960 to 1980, CO_2 emissions grew steeply, with a peak in 1970, reaching 9.87%. After 1980, CO_2 emissions evolved in a less pronounced way.

Mexico and Brazil were the highest CO_2 emitters among LAC countries, followed by Argentina and Venezuela (see Fig. 7.3). In contrast, Dominica and St. Vincent and Grenadines were the lowest CO_2 emitters.

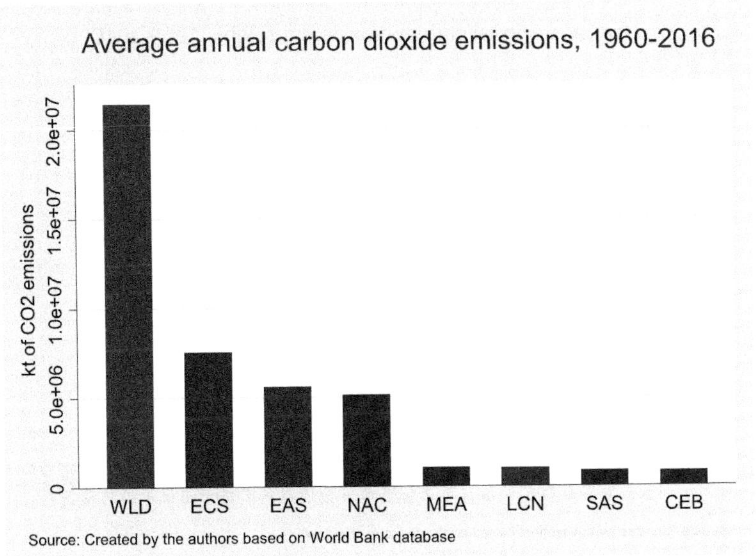

Figure 7.1 CO_2 emissions in the LAC region and other regions of the world. *CEB*, Central Europe and the Baltics; *EAS*, East Asia and Pacific; *ECS*, Europe and Central Asia; *LAC*, Latin America and the Caribbean; *MEA*, Middle East and North Africa; *NAC*, North America; *SAS*, South Asia; *WLD*, world.

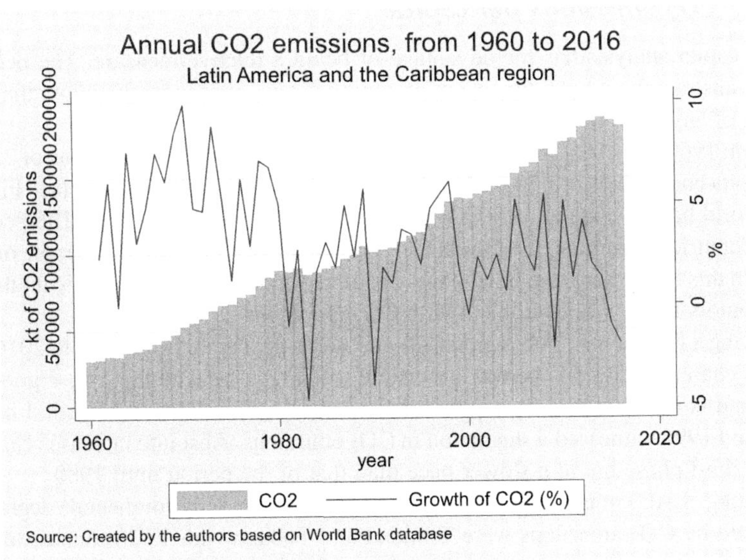

Figure 7.2 Total CO_2 emissions in the LAC region by year.

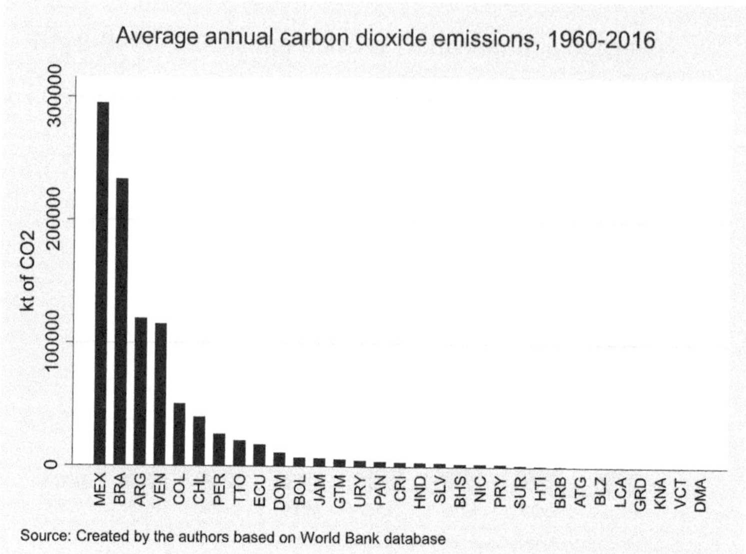

Figure 7.3 Average annual CO_2 emissions per capita in LAC countries. *ARG*, Argentina; *ATG*, Antigua and Barbuda; *BHS*, the Bahamas; *BLZ*, Belize; *BOL*, Bolivia; *BRA*, Brazil; *BRB*, Barbados; *CHL*, Chile; *COL*, Colombia; *CRI*, Costa Rica; *DMA*, Dominica; *DOM*, Dominican Republic; *ECU*, Ecuador; *GRD*, Grenada; *GTM*, Guatemala; *HND*, Honduras; *HTI*, Haiti; *JAM*, Jamaica; *KNA*, St. Kitts and Nevis; *LCA*, St. Lucia; *MEX*, Mexico; *NIC*, Nicaragua; *PAN*, Panama; *PER*, Peru; *PRY*, Paraguay; *SLV*, El Salvador; *SUR*, Suriname; *TTO*, Trinidad and Tobago; *URY*, Uruguay; *VCT*, St. Vincent and Grenadines; *VEN*, RB Venezuela.

7.1.2 CO_2 emissions per capita

A per capita analysis has the advantage of being a relative measure. The per capita values abstract the whole dimension of countries and, consequently, allow the contribution of any inhabitant to the environmental damage to be identified.

From 1960 to 2016, the LAC region had average annual CO_2 emissions of 2.31 (kt) per capita compared to 18.70 for North America and 0.67 for South Asia (see Fig. 7.4). The world had average annual CO_2 emissions of 4.10 (kt) per capita. The people of LAC countries are among the lowest CO_2 polluters, only surpassed by people of South Asia. In this big picture, the inhabitants of North America were by significant the most contributors to the world problem with CO_2 emissions.

Taking a look at the LAC region CO_2 emissions per capita from 1960 to 2016, three phases characterising this period are visible (see Fig. 7.5). The first phase goes up to 1980 and is characterised by rapid increases in CO_2 emissions. The period between 1981 and 1996 witnessed a stagnation in CO_2 emissions. Absolute increases characterise the third phase but at a slower pace than that of the period until 1980.

Among LAC inhabitants that most contribute to environmental degradation provoked by CO_2 emissions were Trinidad and Tobago with 17.18 tonnes, followed

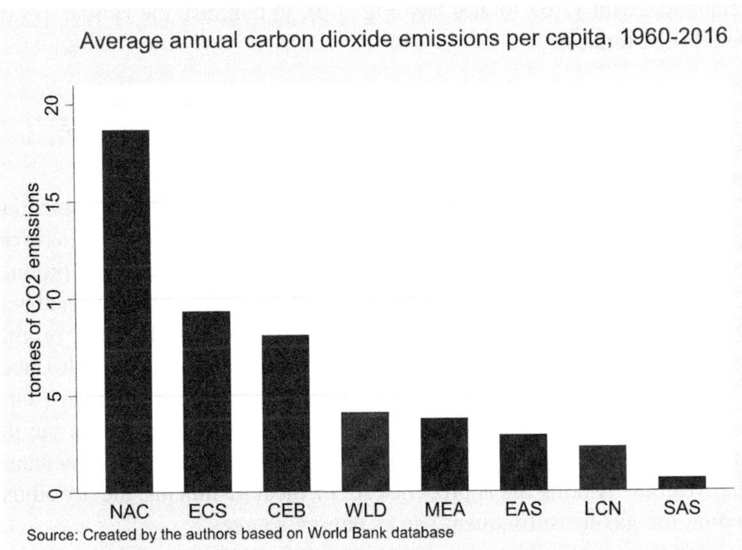

Figure 7.4 CO_2 emissions per capita in the LAC region and other regions of the World. *CEB*, Central Europe and the Baltics; *EAS*, East Asia and Pacific; *ECS*, Europe and Central Asia; *LCN*, Latin America and the Caribbean; *MEA*, Middle East andNorth Africa; *NAC*, North America; *SAS*, South Asia; *WLD*, world.

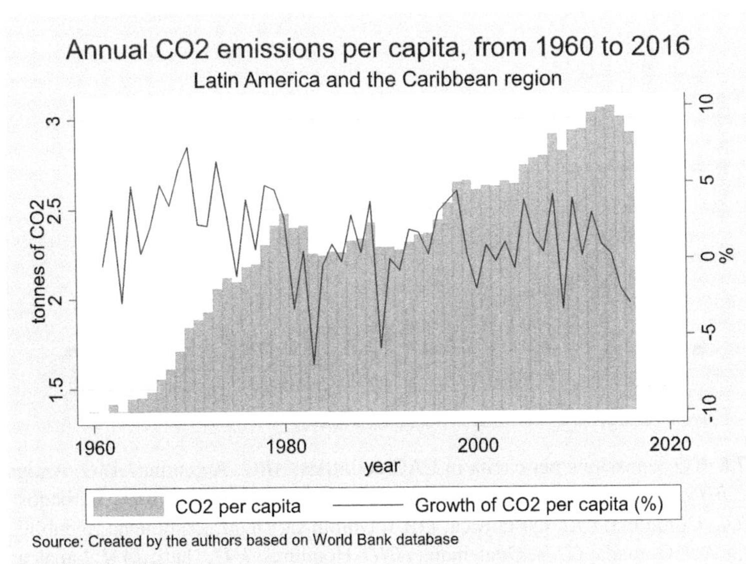

Figure 7.5 CO_2 emissions per capita in the LAC region by year.

by the Bahamas with 12.62 tonnes (see Fig. 7.6). In contrast, the Haitian people only contributed 0.14 tonnes.

7.1.3 Total methane, nitrous oxide and greenhouse gas emissions

Methane is a greenhouse gas that has substantial global warming potential. This potential is 34 times the potential of CO_2 over long time spans. Atmospheric methane can increase induced by climate change, mainly by increasing methane production in natural ecosystems. This phenomenon can produce climate change feedback (e.g., Dean et al., 2018). The bulk of Earth's methane emissions has a biogenic origin, being produced by anaerobic respiration. Ruminants (such as cattle, buffalo, goats and sheep), chickens and pigs burp methane. Anthropogenic annual methane emissions to the atmosphere account for about 37% of methane emissions. Given the threat of methane emissions, substantial efforts are on track to mitigate/reduce methane originating in livestock. Among the approaches are medical treatments, dietary adjustments and trapping the gas for subsequent use as energy sources.

An essential part of nitrous oxide emissions is of human origin (about 40%). Agriculture and industrial activity are the major contributors to the increase in

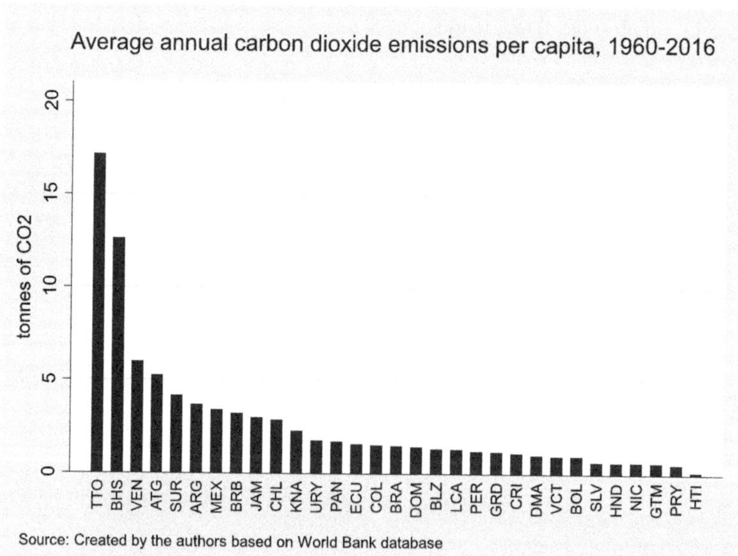

Figure 7.6 CO_2 emissions per capita in LAC countries. *ARG*, Argentina; *ATG*, Antigua and Barbuda; *BHS*, the Bahamas; *BLZ*, Belize; *BOL*, Bolivia; *BRA*, Brazil; *BRB*, Barbados; *CHL*, Chile; *COL*, Colombia; *CRI*, Costa Rica; *DMA*, Dominica; *DOM*, Dominican Republic; *ECU*, Ecuador; *GRD*, Grenada; *GTM*, Guatemala; *HND*, Honduras; *HTI*, Haiti; *JAM*, Jamaica; *KNA*, St. Kitts and Nevis; *LCA*, St. Lucia; *MEX*, Mexico; *NIC*, Nicaragua; *PAN*, Panama; *PER*, Peru; *PRY*, Paraguay; *SLV*, El Salvador; *SUR*, Suriname; *TTO*, Trinidad and Tobago; *URY*, Uruguay; *VCT*, St. Vincent and Grenadines; *VEN*, RB Venezuela.

emissions, mainly from developing economies. Nitrous oxide ranks third in the long-lived greenhouse gas emissions being a critical contributor to global warming. Nitrous oxide emissions are the most critical source of ozone layer depletion (e.g., Ravishankara et al., 2009).

The greenhouse gases present in Earth's atmosphere are composed (by order of abundance) of water vapor, carbon dioxide, methane, nitrous oxide, ozone, chlorofluorocarbons and hydrofluorocarbons. The anthropogenic sources of carbon dioxide emissions come mainly from the combustion of fossil fuels (mostly from coal, petroleum and natural gas). Deforestation and land-use changes also contribute to carbon dioxide emissions.

The data available to analyse methane, nitrous oxide and greenhouse gas emissions cover 1970–2012. These gases are measured in CO_2 equivalent to allow comparability among them and with CO_2 emissions. More precisely, the raw data of methane emissions are measured in kt of CO_2 equivalent, nitrous oxide emissions in 1000 metric tons of CO_2 equivalent and total greenhouse gas emissions in kt of CO_2 equivalent.

Some caution is required to analyse greenhouse gas emissions (Fig. 7.7), given that greenhouse gases include CO_2, methane and nitrous oxide emission.

In the LAC region, methane emissions increased steeply until 1980, but their growth was downward from 1970 to 2012 (see Fig. 7.8).

The pattern of nitrous oxide emissions was very similar to that of methane (see Figs 7.8 and 7.9) in the LAC region.

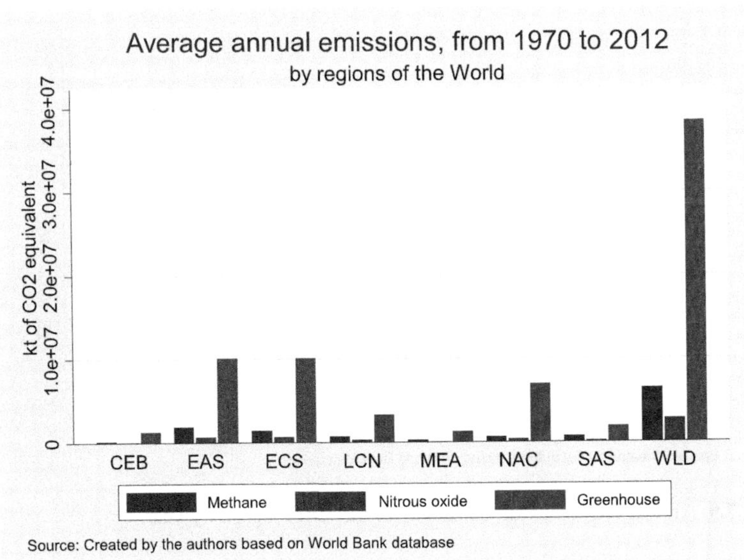

Figure 7.7 Average methane, nitrous oxide and greenhouse gas emissions in the LAC region and other regions of the world. *CEB*, Central Europe and the Baltics; *EAS*, East Asia and Pacific; *ECS*, Europe and Central Asia; *LCN*, Latin America and Caribbean; *MEA*, Middle East and North Africa; *NAC*, North America; *SAS*, South Asia; *WLD*, world.

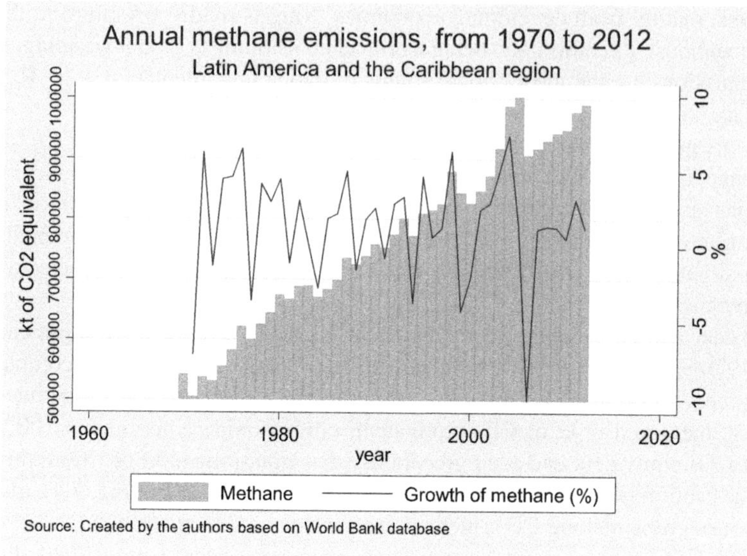

Figure 7.8 Annual methane emissions in the LAC region by year.

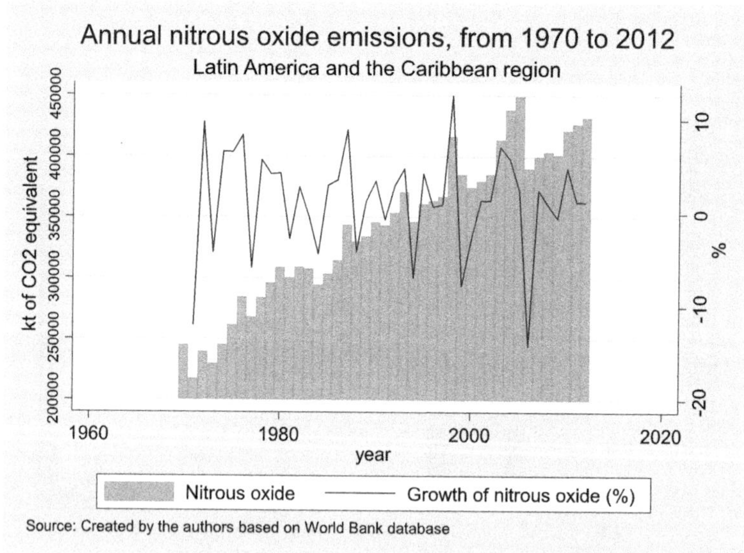

Figure 7.9 Annual nitrous oxide emissions in the LAC region by year.

The annual greenhouse gas emissions in the LAC region increased from 1970 to 2012 (see Fig. 7.10). It is discernible that after the last years of the 20th century, the volatility of emissions increased remarkably.

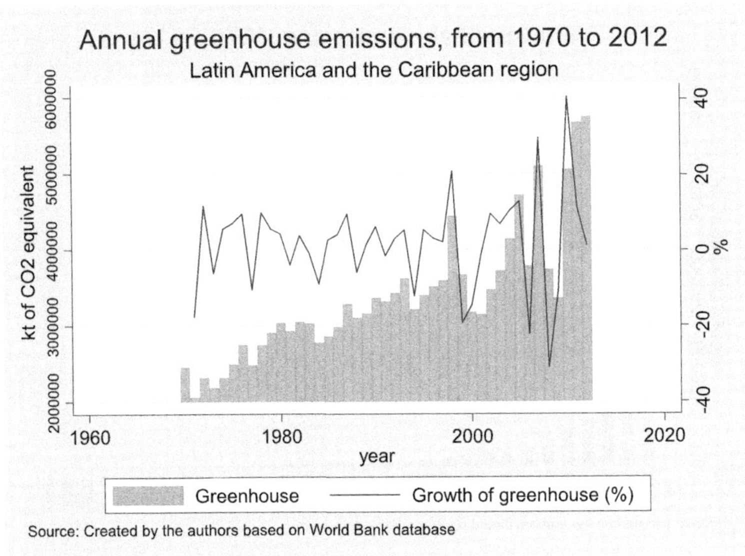

Figure 7.10 Annual greenhouse emissions in the LAC region by year.

Brazil was the major emitter of methane, a great distance ahead of Argentina and Mexico. Grenada was the lowest emitter of methane (see Fig. 7.11).

The behavior of nitrous oxide emissions was very similar to methane emissions (see Figs 7.11 and 7.12). Again, Brazil was the major emitter of nitrous oxide, a great distance ahead of Argentina and Mexico. St. Kitts and Nevis was the lowest emitter of nitrous oxide.

Brazil leads greenhouse emissions by far. Mexico, followed by Argentina, were also important greenhouse emitters (see Fig. 7.13).

7.1.4 Methane, nitrous oxide and greenhouse gas emissions per capita

North America was a colossal polluter in terms of per capita emissions of methane, nitrous oxide and greenhouse gases (see Fig. 7.14). Central Europe and the Baltics and Europe and Central Asia had more or less half the value of North America's position. LAC was slightly above the world per capita emissions. The Middle East and North Africa rank below everyone.

Methane emissions per capita in the LAC region revealed a tendency to decrease from 1970 to 2012 (see Fig. 7.15). Nevertheless, the years 2004 and 2005 registered a significant increase in emissions.

Nitrous oxide emissions behave in line with the pattern of methane emissions (see Figs 7.15 and 7.16).

Greenhouse gas emissions per capita have a behavior that does not allow any clear trend to be perceived (see Fig. 7.17), except an increase in volatility after the mid-2000s.

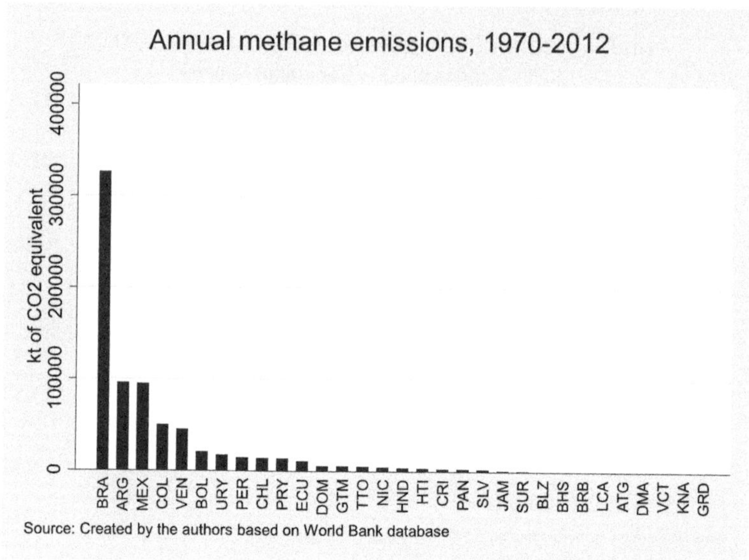

Figure 7.11 Annual methane emissions in LAC countries. *ARG*, Argentina; *ATG*, Antigua and Barbuda; *BHS*, the Bahamas; *BLZ*, Belize; *BOL*, Bolivia; *BRA*, Brazil; *BRB*, Barbados; *CHL*, Chile; *COL*, Colombia; *CRI*, Costa Rica; *DMA*, Dominica; *DOM*, Dominican Republic; *ECU*, Ecuador; *GRD*, Grenada; *GTM*, Guatemala; *HND*, Honduras; *HTI*, Haiti; *JAM*, Jamaica; *KNA*, St. Kitts and Nevis; *LCA*, St. Lucia; *MEX*, Mexico; *NIC*, Nicaragua; *PAN*, Panama; *PER*, Peru; *PRY*, Paraguay; *SLV*, El Salvador; *SUR*, Suriname; *TTO*, Trinidad and Tobago; *URY*, Uruguay; *VCT*, St. Vincent and Grenadines; *VEN*, RB Venezuela.

Uruguay was the biggest methane emitter per capita among LAC countries. Trinidad and Tobago was the next largest polluter. St. Lucia ranks last among methane polluters (Fig. 7.18).

Uruguay was the largest nitrous oxide emitter among LAC countries (see Fig. 7.19). Paraguay and Bolivia were the next largest polluters. Grenada is the best, with a very low level of nitrous oxide emissions per capita.

Trinidad and Tobago perform poorly in greenhouse emissions per capita (see Fig. 7.20). Bolivia was the next largest polluter, and Haiti was the country which was responsible for the least greenhouse emissions.

7.1.5 $PM_{2.5}$ air pollution by region

Fine particles with a diameter of 2.5 μm or less are designated $PM_{2.5}$. Exposure to $PM_{2.5}$ has caused millions of deaths worldwide, mainly from heart disease and stroke, lung cancer and chronic lung disease and respiratory infections (e.g., Health Effects Institute, 2018), and is a leading risk factor for premature death (e.g., Undark, 2018).

Data on $PM_{2.5}$ started in 1990 with a periodicity of 5 years until 2010, when it began annually. South Asia was the world region where human beings were most exposed to the dangers of small particulates (see Fig. 7.21). The Middle East and North

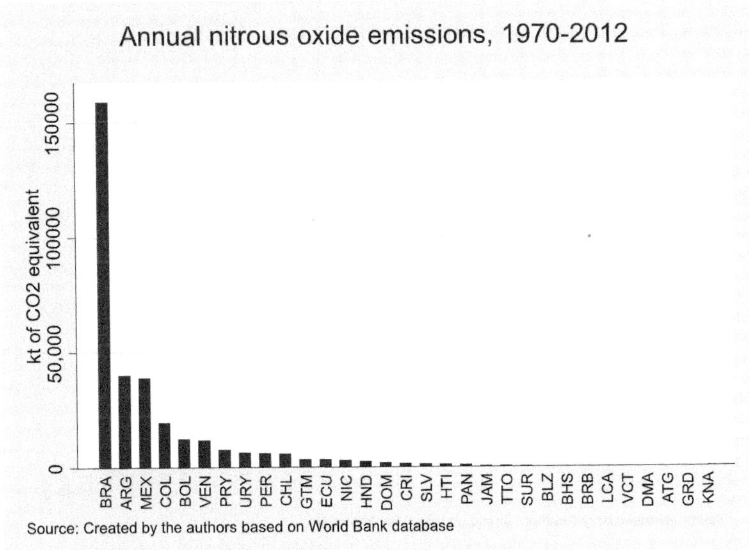

Figure 7.12 Annual nitrous oxide emissions in LAC countries. *ARG*, Argentina; *ATG*, Antigua and Barbuda; *BHS*, the Bahamas; *BLZ*, Belize; *BOL*, Bolivia; *BRA*, Brazil; *BRB*, Barbados; *CHL*, Chile; *COL*, Colombia; *CRI*, Costa Rica; *DMA*, Dominica; *DOM*, Dominican Republic; *ECU*, Ecuador; *GRD*, Grenada; *GTM*, Guatemala; *HND*, Honduras; *HTI*, Haiti; *JAM*, Jamaica; *KNA*, St. Kitts and Nevis; *LCA*, St. Lucia; *MEX*, Mexico; *NIC*, Nicaragua; *PAN*, Panama; *PER*, Peru; *PRY*, Paraguay; *SLV*, El Salvador; *SUR*, Suriname; *TTO*, Trinidad and Tobago; *URY*, Uruguay; *VCT*, St. Vincent and Grenadines; *VEN*, RB Venezuela.

Africa and East Asia and the Pacific are the next in exposure close to the world levels. LAC ranks as the place where people are exposed to the lowest levels of $PM_{2.5}$.

Fig. 7.22 complements the picture revealed by Fig. 7.21. Thus, during the period 1990–2017, the exposure to $PM_{2.5}$ remained barely stable inside each region.

7.2 Gas emissions and their determinants

Carbon dioxide, methane, nitrous oxide and greenhouse gas emissions are caused by nature and human activities. Factors essential for our analysis are the human determinants of these emissions and how they can degrade the environment.

7.2.1 Carbon dioxide emissions

Carbon dioxide emissions are the most important anthropogenic greenhouse gas (e.g., Intergovernmental Panel on Climate Change). Carbon dioxide emissions come mainly from the burning of fossil fuels and cement manufacture. Land use, land-use change and forestry, along with the emissions from international shipping or bunker fuels, are the other essential sources of CO_2 emissions.

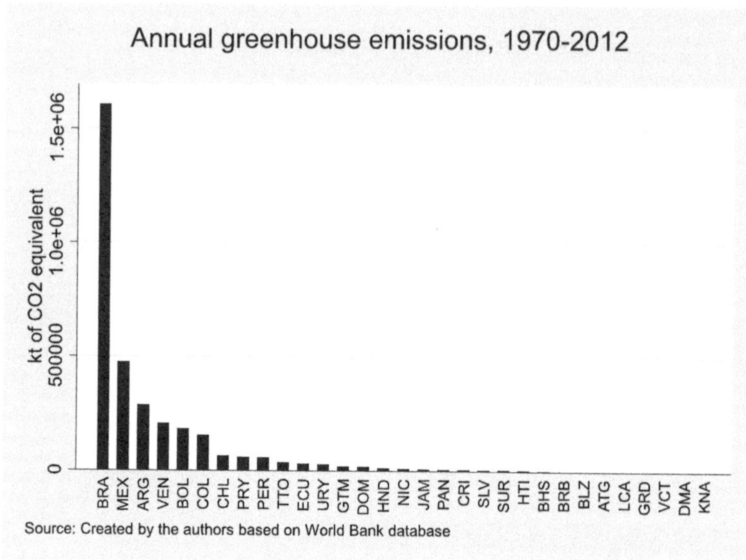

Figure 7.13 Annual greenhouse emissions in LAC countries. *ARG*, Argentina; *ATG*, Antigua and Barbuda; *BHS*, the Bahamas; *BLZ*, Belize; *BOL*, Bolivia; *BRA*, Brazil; *BRB*, Barbados; *CHL*, Chile; *COL*, Colombia; *CRI*, Costa Rica; *DMA*, Dominica; *DOM*, Dominican Republic; *ECU*, Ecuador; *GRD*, Grenada; *GTM*, Guatemala; *HND*, Honduras; *HTI*, Haiti; *JAM*, Jamaica; *KNA*, St. Kitts and Nevis; *LCA*, St. Lucia; *MEX*, Mexico; *NIC*, Nicaragua; *PAN*, Panama; *PER*, Peru; *PRY*, Paraguay; *SLV*, El Salvador; *SUR*, Suriname; *TTO*, Trinidad and Tobago; *URY*, Uruguay; *VCT*, St. Vincent and Grenadines; *VEN*, RB Venezuela.

7.2.2 Nitrous oxide emissions

Nitrous oxide (N_2O) is a significant depleter of stratospheric ozone and consequently deserves policymakers' attention. About 40% of N_2O emissions came from human activity, and agriculture and industry are among their primary sources. Most N_2O emissions come from microorganisms (bacteria and fungi) in lands and seas. Soils supporting vegetation are an essential source of N_2O, representing about 60% of natural emissions. The remaining natural sources of N_2O are oceans (about 35%) and the chemical reactions in the atmosphere (about 5%).

The most important sources of anthropogenic emissions of N_2O are: (1) agricultural fertilisers and livestock dung (about 42%); (2) excess and leakage of fertilisers (about 25%); (3) biomass burning (about 10%); (4) burning of fossil fuel and industrial activity (about 10%); (5) biological degradation of other nitrogen-containing atmospheric emissions (about 9%); and (6) human sewage (about 5%). The use of nitrogen fertilisers and manure in agriculture boosts N_2O emissions via naturally occurring bacteria.

7.2.3 Methane emissions

Methane (CH_4) emissions have been among the significant contributors to the rise in greenhouse gas concentrations in the atmosphere. Methane emissions have a natural

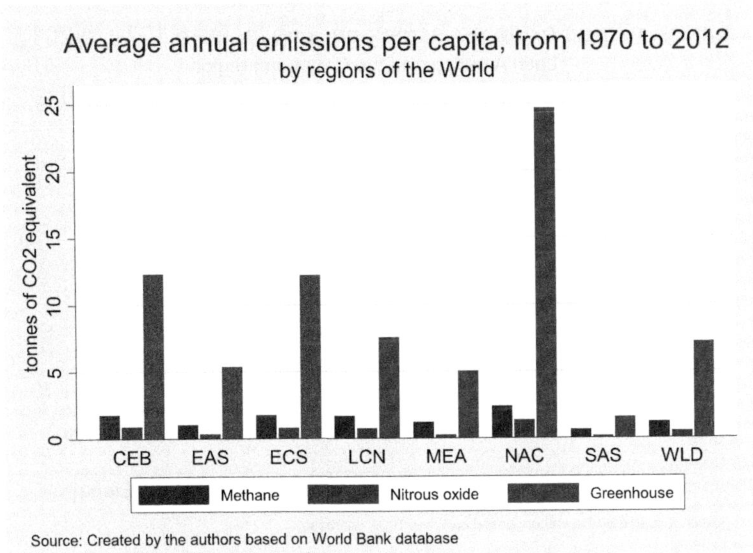

Figure 7.14 Average methane, nitrous oxide and greenhouse gas emissions per capita in the LAC region and other world regions. *CEB*, Central Europe and the Baltics; *EAS*, East Asia and Pacific; *ECS*, Europe and Central Asia; *LCN*, Latin America and Caribbean; *MEA*, Middle East and North Africa; *NAC*, North America; *SAS*, South Asia; *WLD*, world.

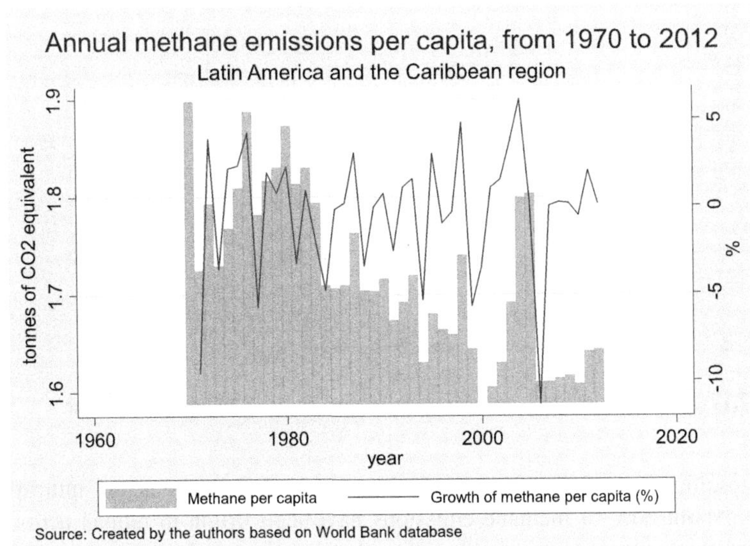

Figure 7.15 Total methane emissions in the LAC region by year.

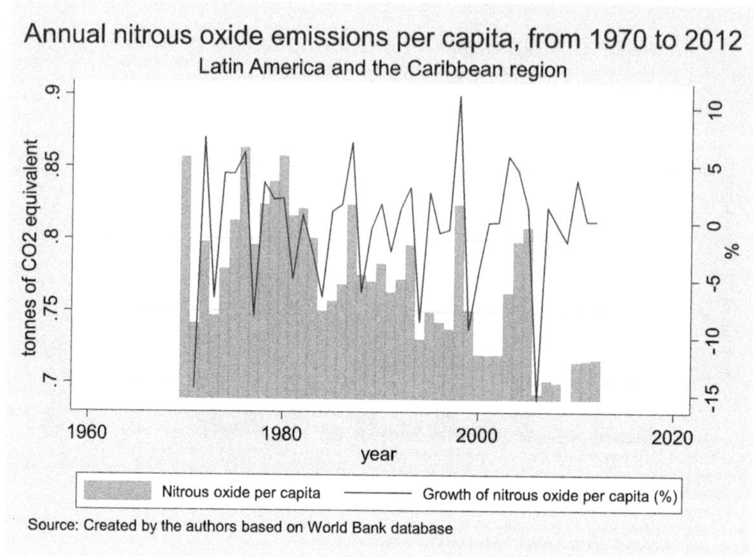

Figure 7.16 Nitrous oxide emissions per capita in the LAC region by year.

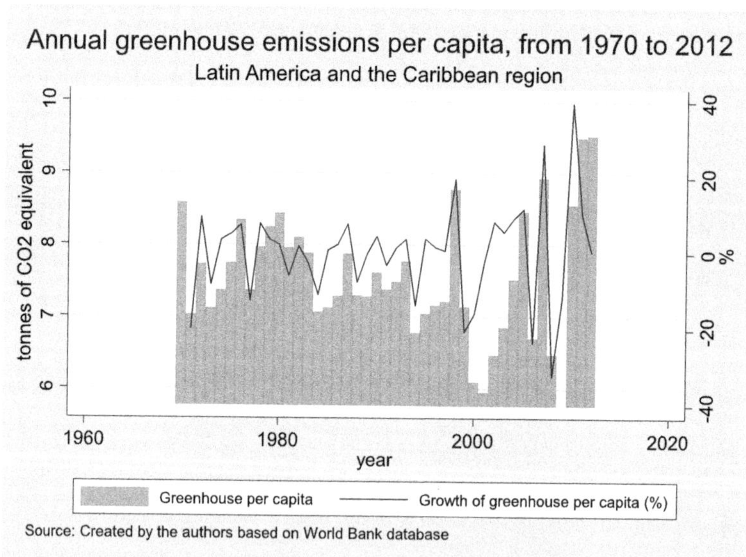

Figure 7.17 Annual greenhouse emissions per capita in the LAC region by year.

origin, resulting from the biological activity of natural ecosystems and anthropogenic activity. About 60% of methane emissions have their origin in human actions. The most important anthropogenic sources of methane are livestock (ruminants), representing about 30% of total emissions (mostly related to enteric fermentation). Waste

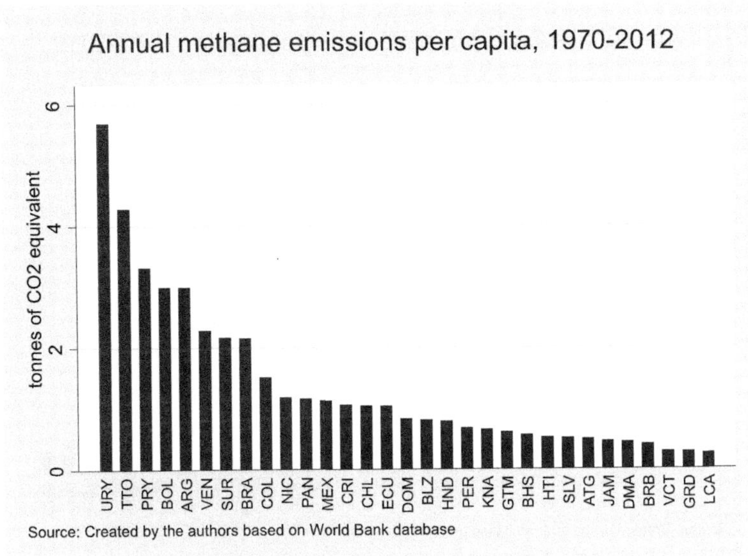

Figure 7.18 Annual methane emissions per capita in LAC countries. *ARG*, Argentina; *ATG*, Antigua and Barbuda; *BHS*, the Bahamas; *BLZ*, Belize; *BOL*, Bolivia; *BRA*, Brazil; *BRB*, Barbados; *CHL*, Chile; *COL*, Colombia; *CRI*, Costa Rica; *DMA*, Dominica; *DOM*, Dominican Republic; *ECU*, Ecuador; *GRD*, Grenada; *GTM*, Guatemala; *HND*, Honduras; *HTI*, Haiti; *JAM*, Jamaica; *KNA*, St. Kitts and Nevis; *LCA*, St. Lucia; *MEX*, Mexico; *NIC*, Nicaragua; *PAN*, Panama; *PER*, Peru; *PRY*, Paraguay; *SLV*, El Salvador; *SUR*, Suriname; *TTO*, Trinidad and Tobago; *URY*, Uruguay; *VCT*, St. Vincent and Grenadines; *VEN*, RB Venezuela.

(landfills and wastewater treatment) follows in the rank of the most significant emitters (about 18% of total emissions). Agriculture for food (with emphasis on traditional rice production) and biomass are also significant sources of methane (about 15% of total emissions).

Since methane emissions are also of natural origin, it is crucial to consider the human influence on the evolution of these sources of emissions. The most important natural source of methane is the wetlands (about 75% of natural emissions). A group of other sources follows this primary source, namely: (1) leakages from near-surface hydrocarbon and clathrate hydrate deposits; (2) volcanic discharges; (3) wildfires; (4) termite discharges; and (5) wild ruminant mammals. The necessity of mitigation/reduction of methane emissions is a source of concern. To achieve this goal, the capture and use of gas as a source of renewable energy are a benefit.

7.2.4 Greenhouse gas emissions

Greenhouse gas emissions result from various human activities that contribute to climate change, mainly by their influence on the Earth's atmosphere. Greenhouse gas is composed mostly of CO_2, nitrous oxide and methane emissions. Carbon dioxide emissions from burning fossil fuels (primarily coal and oil) and natural gas are the

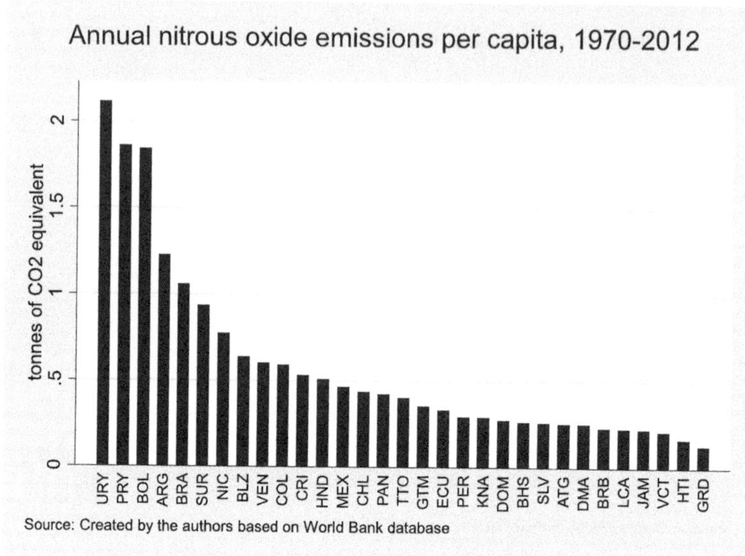

Figure 7.19 Annual nitrous oxide emissions per capita in LAC countries. *ARG*, Argentina; *ATG*, Antigua and Barbuda; *BHS*, the Bahamas; *BLZ*, Belize; *BOL*, Bolivia; *BRA*, Brazil; *BRB*, Barbados; *CHL*, Chile; *COL*, Colombia; *CRI*, Costa Rica; *DMA*, Dominica; *DOM*, Dominican Republic; *ECU*, Ecuador; *GRD*, Grenada; *GTM*, Guatemala; *HND*, Honduras; *HTI*, Haiti; *JAM*, Jamaica; *KNA*, St. Kitts and Nevis; *LCA*, St. Lucia; *MEX*, Mexico; *NIC*, Nicaragua; *PAN*, Panama; *PER*, Peru; *PRY*, Paraguay; *SLV*, El Salvador; *SUR*, Suriname; *TTO*, Trinidad and Tobago; *URY*, Uruguay; *VCT*, St. Vincent and Grenadines; *VEN*, RB Venezuela.

main components of greenhouse gas emissions. Deforestation and changes in land use also play an essential role in CO_2 emissions. Agriculture, through the use of fertilisers and manure, is responsible for higher N_2O emissions. Anthropogenic methane emissions come primarily from agriculture, gas venting and leakage from the fossil fuel industry. Another source of greenhouse gas is chlorofluorocarbons (CFCs) in refrigeration systems and CFCs and halogens in fire suppression systems and manufacturing processes.

7.2.5 Variables with explanatory power for gas emissions

Given that we are studying the influence of human behavior on the emissions of dangerous gases, we will focus our attention on economic and social variables that can cause these emissions. Among the most critical variables with potential explanatory power over gas emissions, we can identify GDP, energy consumption, urbanisation, globalisation, economic complexity and population. Many other variables can exert an influence on gas emissions. However, studying these influences requires data and formulations beyond the scope which we established for this chapter.

The former subsections identified the sources of human and natural emissions. Most environmental damage has its origin in economic activity and social behavior. The dimension of the intervention required to handle the challenges put to humankind requires correct identification of whether these variables cause gas emissions.

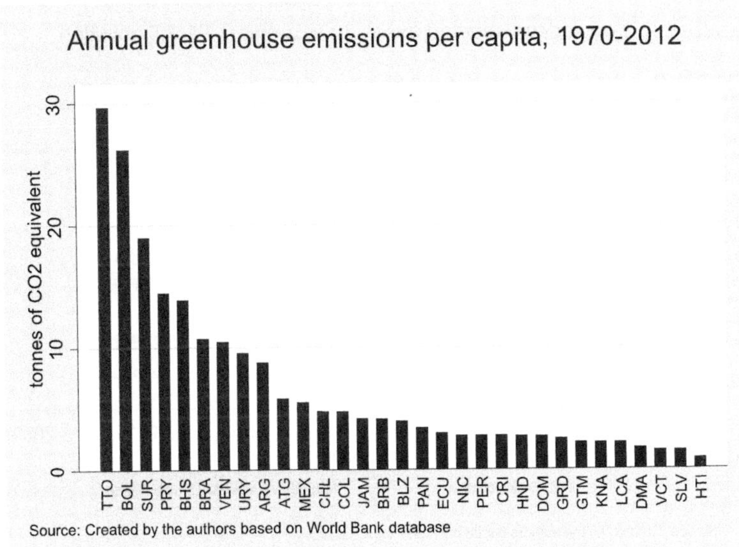

Figure 7.20 Annual greenhouse emissions per capita in LAC countries. *ARG*, Argentina; *ATG*, Antigua and Barbuda; *BHS*, the Bahamas; *BLZ*, Belize; *BOL*, Bolivia; *BRA*, Brazil; *BRB*, Barbados; *CHL*, Chile; *COL*, Colombia; *CRI*, Costa Rica; *DMA*, Dominica; *DOM*, Dominican Republic; *ECU*, Ecuador; *GRD*, Grenada; *GTM*, Guatemala; *HND*, Honduras; *HTI*, Haiti; *JAM*, Jamaica; *KNA*, St. Kitts and Nevis; *LCA*, St. Lucia; *MEX*, Mexico; *NIC*, Nicaragua; *PAN*, Panama; *PER*, Peru; *PRY*, Paraguay; *SLV*, El Salvador; *SUR*, Suriname; *TTO*, Trinidad and Tobago; *URY*, Uruguay; *VCT*, St. Vincent and Grenadines; *VEN*, RB Venezuela.

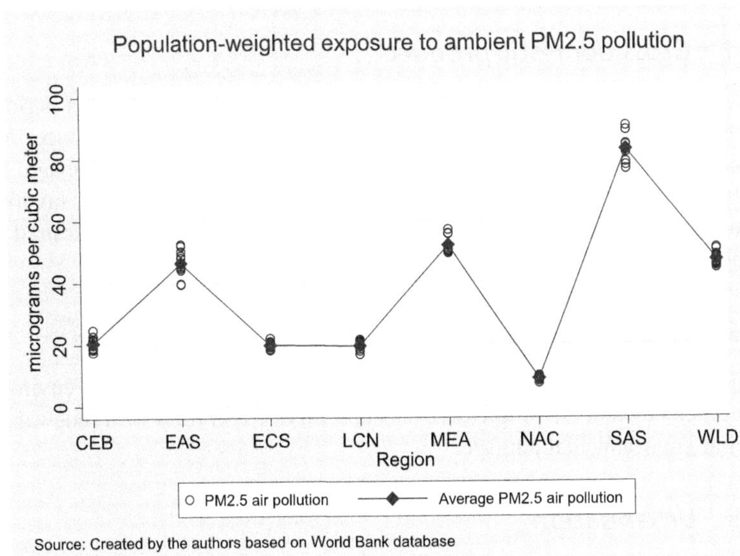

Figure 7.21 $PM_{2.5}$ air pollution by region. *CEB*, Central Europe and the Baltics; *EAS*, East Asia and Pacific; *ECS*, Europe and Central Asia; *LCN*, Latin America and Caribbean; *MEA*, Middle East and North Africa; *NAC*, North America; *SAS*, South Asia; *WLD*, world.

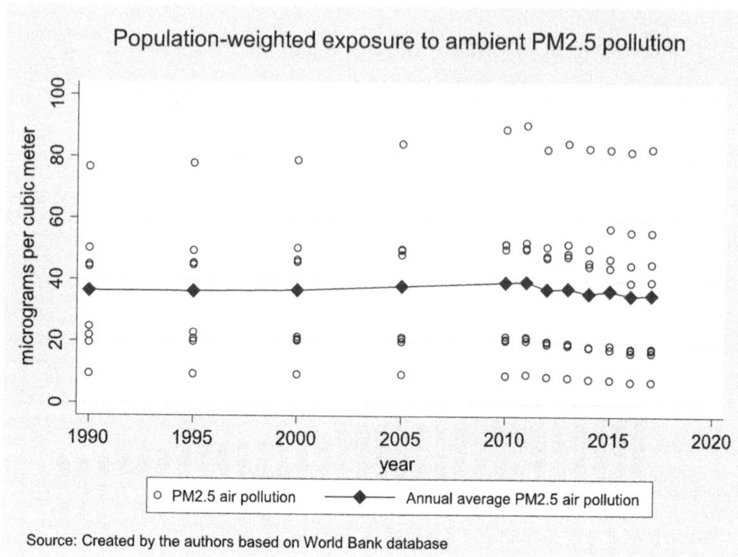

Figure 7.22 $PM_{2.5}$ air pollution by year (regions).

Economics has its own rules, and it is not prudent to trust that the behavior of free markets will move humankind to timely solutions. Nevertheless, markets can be a precious help when correctly framed. The great challenge remains and will continue to reconcile the actions necessary to reverse environmental damage without ruining economic growth.

7.2.5.1 Gross domestic product

GDP is an aggregate measure of the goods and services produced in a period. It is necessary to consume factors of production (among them land) to produce. Accordingly, economic growth is expected to contribute to environmental damage. Policymakers should intervene in the compositions of the parts of GDP, favoring the substitution of high-damage for low-damage goods to mitigate the undesired effects of economic growth on the environment.

7.2.5.2 Consumption of energy

Consumption of energy mainly from fossil sources is pernicious for the environment. Energy efficiency and the progressive substitution of fossil fuels with renewable ones helps limit environmental damage.

7.2.5.3 Urbanisation

Urbanisation is the phenomenon of concentration of people in cities. By its nature, this is a relative variable and is compatible with very different situations. Most of the time, the variable is the percentage of the total population that live in cities. Consequently, urbanisation can be growing because numbers of people living in cities are growing faster than the total population (which is compatible with the growing nonurban population). The urban population can also be increasing (either in absolute and relative terms) because people migrate from rural areas to urban ones.

7.2.5.4 Globalisation

Globalisation is the worldwide process of increasing interaction (integration) among persons, businesses and governments. Globalisation manifests itself in several ways, with the most visible being economic globalisation. Social and political globalisation also plays a significant role in the behavior of humankind. Globalisation has been a strong force behind the international specialisation that has influenced transportation and increased gas emissions. Globalisation also affects people's social behavior, leading to consumer standards that can influence gas emissions.

7.2.5.5 Economic complexity

As economies evolve to become more sophisticated, their production systems change in ways that can mitigate or otherwise aggravate the environment. More complex economies are also expected to be more efficient. These efficiency gains are a good outcome. The other side of the coin is that complexity is related to economies of scale. It leads to increased production levels, and more production exerts pressure on the environment.

7.2.5.6 Population growth

More people increase pressure on our planet. Indeed, contrary to several other dimensions, the Earth does not change. The capacity of ecosystems to regenerate is limited and below the actual burden exerted on them. Population growth is expected to contribute to evidence of environmental damage.

7.3 Data and method

7.3.1 Data

We use statistical information for the period from 1960 to 2016 for 31 LAC countries. Antigua and Barbuda (ATG), Argentina (ARG), the Bahamas (BHS), Barbados (BRB), Belize (BLZ), Bolivia (BOL), Brazil (BRA), Chile (CHL), Colombia (COL), Costa Rica (CRI), Dominica (DMA), Dominican Republic (DOM), Ecuador (ECU), El Salvador (SLV), Grenada (GRD), Guatemala (GTM), Haiti (HTI), Honduras (HND), Jamaica (JAM), Mexico (MEX), Nicaragua (NIC), Panama (PAN), Paraguay (PRY), Peru (PER), St. Kitts and Nevis (KNA), St. Lucia (LCA), St. Vincent and Grenadines (VCT), Suriname (SUR), Trinidad and Tobago (TTO),

Uruguay (URY) and Venezuela (VEN). The raw data variables used to perform this investigation are:

- Carbon dioxide emissions (**CO_2**) from the burning of fossil fuels and cement manufacture, in kilotons (kt) per capita. These include carbon dioxide produced during the consumption of solid, liquid and gas fuels and gas flaring retrieved from World Bank Open Data (2021).
- Nitrous oxide (**N_2O**) emissions (1000 metric tons of CO_2 equivalent), retrieved from World Bank Open Data (2021). Nitrous oxide emissions are emissions from agricultural biomass burning, industrial activities and livestock management.
- Methane (**CH_4**) emissions (kt of CO_2 equivalent), retrieved from World Bank Open Data (2021). Methane emissions are those stemming from human activities such as agriculture and industrial methane production.
- Total greenhouse gas (**GHG**) emissions (kt of CO_2 equivalent), retrieved from World Bank Open Data (2021). Total greenhouse gas emissions are composed of CO_2 totals excluding short-cycle biomass burning (such as agricultural waste burning and savanna burning) but including other biomass burning (such as forest fires, postburn decay, peat fires and decay of drained peatlands), all anthropogenic CH_4 sources, N_2O sources and F-gases (HFCs, PFCs and SF6).
- Gross domestic production (**Y**) in constant 2010 US$, retrieved from World Bank Open Data (2021). GDP at purchaser's prices is the sum of gross value added by all resident producers in the economy plus any product taxes and minus any subsidies not included in the value of the products. It is calculated without making deductions for depreciation of fabricated assets or depletion and degradation of natural resources. Data are in constant 2010 USD. Dollar figures for GDP are converted from domestic currencies using 2010 official exchange rates. An alternative conversion factor is used for a few countries where the official exchange rate does not reflect the rate effectively applied to actual foreign exchange transactions.
- Globalisation index (**G**), de facto, retrieved from the KOF Globalisation Index (KOF Globalization Index, 2021). Globalisation index (de facto) is the mean of economic globalisation (de facto), social globalisation (de facto) and political globalisation (de facto). Economic globalisation (de facto) is composed of trade globalisation (de facto) (trade in goods, trade in services and trade partner diversity) and financial globalisation (de facto) (foreign direct investment, portfolio investment, international debt, international reserves and international income payments). Social globalisation (de facto) is composed of interpersonal globalisation (de facto) (international voice traffic, transfers, international tourism, international students and migration), informational globalisation (de facto) (used internet bandwidth, international patents and high technology exports) and cultural globalisation (de facto) (trade in cultural goods, trade-in personal services, international trademark, McDonald's restaurants and IKEA stores). Political globalisation (de facto) is composed of embassies, UN peacekeeping missions and international NGOs.
- Economic globalisation index (**EG**), de facto, retrieved from the KOF Globalisation Index (KOF Globalization Index, 2021). This variable measures trade and financial globalisation. Trade globalisation is determined based on trade in goods and services, and financial globalisation includes foreign investment in various categories.
- Social globalisation index (**SG**), de facto, retrieved from the KOF Globalization Index (2021). This variable measures interpersonal contact flows of information and cultural proximity. Interpersonal contact is measured within the de facto segment concerning international telephone connections, tourist numbers and migration. Flows of information are determined within the de facto segment concerning international patent applications, international students and trade in high-technology goods. Cultural proximity is measured in the de facto

segment via trade in cultural goods, international trademark registrations and the number of McDonald's restaurants and IKEA stores.

- Urban population (% of the total population) (**U**), retrieved from World Bank Open Data (2021). Urban population refers to people living in urban areas as defined by national statistical offices. The data are collected and smoothed by the United Nations Population Division.
- Energy use (kg of oil equivalent per capita) (**Epc**), retrieved from World Bank Open Data (2021). Energy use refers to the use of primary energy before transformation to other end-use fuels, equal to indigenous production plus imports and stock changes, minus exports and fuels supplied to ships and aircraft engaged in international transport.
- Population-weighted exposure to ambient $\mathbf{PM_{2.5}}$ pollution is defined as the average level of exposure of a nation's population to concentrations of suspended particles measuring less than 2.5 microns in aerodynamic diameter, capable of penetrating deep into the respiratory tract and causing severe damage to health. Exposure is calculated by weighting to mean annual concentrations of $PM_{2.5}$ by population in both urban and rural areas. This variable was retrieved from World Bank Open Data (2021).
- Economic complexity index (**ECI**), retrieved from Observatory of Economic Complexity (OEC) (2021). The economic complexity index (ECI) measures the relative knowledge intensity of an economy. ECI measures the knowledge intensity of an economy by considering the knowledge intensity of the products it exports.
- Total population, retrieved from World Bank Open Data (2021). Total population is based on the de facto definition of population, which counts all residents regardless of legal status or citizenship. The values shown are midyear estimates.

The variables carbon dioxide (CO_2) emissions, nitrous oxide (N_2O) emissions, methane (CH_4) emissions and total greenhouse gas (GHG) emissions were also converted to per capita (by dividing them by the total population). These new variables use the suffix "pc" to design per capita ones (CO_2pc, N_2Opc, CH_4pc and GHGpc). The per capita value allows disparities to be controlled for population growth over time and within countries. This analysis is proper, especially when the countries have very different population growth rates. The variable energy use was retrieved per capita. To allow the relationship of energy use (E) with the other variables, the variable Epc was multiplied by the population.

The variables were converted to natural logarithms and the first differences between these variables in natural logs were taken to facilitate the analysis. The natural logs smooth the variables that have been subject to growth over time. It also has the advantage of its linear relationship becoming elasticities. The first differences of variables in natural logs give the compound growth rate between the two differentiated moments.

7.3.2 Method

To analyse the environmental degradations of LAC countries, we use pairwise Granger causality, which was computed, using Dumitrescu & Hurlin' (2012) approach (see Box 7.1). This econometric methodology allows it to be identified if the past values of one variable have explanatory power on contemporaneous values of another variable (i.e., Granger causality). The Granger causality will be compared with the correlations between variables.

Box 7.1 Pairwise Granger causality

The Granger (1969) causality tests were first performed in the context of time series. The Granger causality test is computed running a bivariate regression.

In the presence of longitudinal data (i.e., panel data), some adaptations of the original approach have to be done.

In the presence of panel data, the bivariate estimations are as follows – Eq. (7.1).

$$y_{i,t} = \alpha_{0,i} + \alpha_{1,i}y_{i,t-1} + \cdots + \alpha_{k,i}y_{i,t-k} + \beta_{1,i}x_{i,t-1} + \cdots + \beta_{k,i}x_{i,t-k} + e_{i,t}$$
$$x_{i,t} = \gamma_{0,i} + \gamma_{1,i}x_{i,t-1} + \cdots + \gamma_{k,i}x_{i,t-k} + \delta_{1,i}y_{i,t-1} + \cdots + \delta_{k,i}y_{i,t-k} + \mu_{i,t} \quad (7.1)$$

There are two main approaches to estimate Eq. (7.1). The first is to estimate an OLS pooled but does not allow lagged values of one individual to join the lagged values of the next individual, i.e., an OLS preserving some panel data features. In this approach, the estimated coefficients are common to all individuals (see Eq. 7.2).

$$\alpha_{0,i} = \alpha_{0,j},\ \alpha_{1,i} = \alpha_{1,j}, \cdots, \alpha_{k,i} = \alpha_{k,j}, \beta_{1,i} = \beta_{1,j}, \cdots, \beta_{1,i} = \beta_{1,j},\ \forall i,j$$
$$\gamma_{0,i} = \gamma_{0,j},\ \gamma_{1,i} = \gamma_{1,j}, \cdots, \gamma_{k,i} = \gamma_{k,j}, \delta_{1,i} = \delta_{1,j}, \cdots, \delta_{1,i} = \delta_{1,j},\ \forall i,j \quad (7.2)$$

The second is the causality approach of Dumitrescu and Hurlin (2012), who assume that all coefficients across the cross-sections are different (see Eq. 7.3).

$$\alpha_{0,i} \neq \alpha_{0,j},\ \alpha_{1,i} \neq \alpha_{1,j}, \cdots, \alpha_{k,i} \neq \alpha_{k,j}, \beta_{1,i} \neq \beta_{1,j}, \cdots, \beta_{1,i} \neq \beta_{1,j},\ \forall i,j$$
$$\gamma_{0,i} \neq \gamma_{0,j},\ \gamma_{1,i} \neq \gamma_{1,j}, \cdots, \gamma_{k,i} \neq \gamma_{k,j}, \delta_{1,i} \neq \delta_{1,j}, \cdots, \delta_{1,i} \neq \delta_{1,j},\ \forall i,j \quad (7.3)$$

Dumitrescu and Hurlin's (2012) approach is well suited to work with the heterogeneous panel. The Dumitrescu–Hurlin test is computed in two steps. First, the Granger causality test statistics were estimated for each individual. The second is to compute the average of the test statistics calculated in the first step. This statistic is designed as the $W_{\widetilde{bar}}$ statistic. Unbalanced panels distort the accuracy of the $W_{\widetilde{bar}}$ statistic. To surpass this limitation, a standardised version of the $W_{\widetilde{bar}}$ statistic was created, which is weighted for the unbalanced panels, called the $Z_{\widetilde{bar}}$ statistic, and follows a standard normal distribution.

LAC countries share some characteristics that simultaneously manifest themselves, giving rise to the phenomena of cross-sectional dependence (or contemporaneous correlation among crosses). One way to assess the presence of cross-sectional dependence is to use Pesaran's CD test (2004).

7.4 Empirical results

We will divide the analysis into the relationships between the growth of variables (first differences of natural logarithms) and the relationships between variables in natural logarithms. These analyses assess different behaviors between explanatory variables and the gas emissions in LAC countries.

7.4.1 Short-run analysis

Table 7.1 shows the summary statistics of the first differences of variables and the cross-sectional dependence CD test of Pesaran (2004). Some conclusions arise from the analysis of summary statistics. First, the volatility of gas emissions (measured by the standard deviation) is very high, with enormous values of minimum and maximum variables. Second, economic, social and demographic variables reveal both high volatility and range of values. Third, all variables reveal statistically significant cross-sectional dependence, supporting the fact that LAC countries share common features. The very high absolute value can see the intensity of the shared common features of the correlations for the changes in population growth (0.782) and urbanisation (0.717). Economic growth and economic growth per capita also reveal synchronisation. Among the gas emissions, greenhouse gas is the most integrated.

Table 7.2 reveals the correlations between gas emissions. The correlations between variables are high except with carbon dioxide emissions. The correlations between variables and variables per capita are very similar, indicating that population growth does not distort the relationships between the variables. The strongest association is between methane and nitrous oxide emissions.

Table 7.3 shows the correlation between gas emissions and the economic, social and demographic variables. In general, the correlations are low except between energy consumption and carbon dioxide emissions and to a lesser extent between economic growth and carbon dioxide emissions.

Table 7.4 reveals Granger's causalities between variables and CO_2 emissions. Economic growth, energy consumption, urbanisation and population growth Granger-cause CO_2 emissions. However, for CO_2 per capita, only economic growth per capita and energy consumption per capita were revealed to be statistically significant.

Table 7.5 presents pairwise Granger causalities for nitrous oxide emissions. Only economic growth per capita seems to cause nitrous oxide emissions per capita and is statistically significant at 7%.

Table 7.6 reveals that the growth in social globalisation Granger-causes growth in methane and methane percapita. Growth of urbanisation and population Granger-causes methane emissions per capita.

Table 7.7 shows variables Granger causality on greenhouse emissions. Economic growth and economic growth per capita Granger-cause greenhouse emissions and

Table 7.1 Descriptive statistics and cross-sectional dependence.

Variables	Obs	Mean	Std. dev	Min	Max	CD-test	*P*-value	Corr	Abs(corr)
dlCO_2	1736	0.0417277	0.1584846	−1.710081	1.564441	4.51	0.000	0.032	0.136
dlCO_2pc	1736	0.0258881	0.1580848	−1.724198	1.545911	4.48	0.000	0.032	0.136
dlN2O	1302	0.0119263	0.1412826	−1.543172	1.711207	7.08	0.000	0.051	0.167
dlN2Opc	1302	−0.0030571	0.1414084	−1.549496	1.687589	6.99	0.000	0.050	0.167
dlCH4	1302	0.0125921	0.1004949	−1.234615	1.343765	3.03	0.002	0.022	0.146
dlCH4pc	1302	−0.0023914	0.100877	−1.25831	1.320146	2.43	0.015	0.017	0.143
dlGHG	1203	0.021331	0.1742306	−1.331709	1.797539	5.64	0.000	0.044	0.188
dlGHGpc	1203	0.0058783	0.1745246	−1.355404	1.781315	5.76	0.000	0.045	0.188
dlY	1714	0.0329143	0.0432664	−0.3075962	0.2322159	16.14	0.000	0.150	0.439
dlYpc	1714	0.0167496	0.0428449	−0.3375354	0.2186651	17.12	0.000	0.155	0.445
dlE	945	0.0290334	0.0613135	−0.3583775	0.3980103	4.28	0.000	0.049	0.376
dlEpc	945	0.0117516	0.0612193	−0.3701725	0.3817601	4.57	0.000	0.051	0.377
dlU	1829	0.0228345	0.0153539	−0.0360765	0.0662975	67.18	0.000	0.716	0.717
dlG	1423	0.0079966	0.0441265	−0.2000711	0.2588634	5.29	0.000	0.056	0.146
dlEG	1423	0.0118811	0.0864497	−0.5154891	1.024626	3.61	0.000	0.038	0.130
dlSG	1423	0.0088202	0.0344791	−0.2185407	0.1977379	27.60	0.000	0.294	0.298
dlECI	1054	−0.0003247	0.0560387	−0.4187156	0.3264169	10.99	0.000	0.116	0.241
dlPOP	1829	0.0155089	0.0098812	−0.0210915	0.050559	73.52	0.000	0.782	0.782

Obs denotes the number of observations in the model; Std. dev denotes the standard deviation; Min and Max denote minimum and maximum, respectively; CD test denotes Pesaran's CD test (2004); Corr denotes the average value of the off-diagonal elements correlation matrix between the cross-sectional units; Abs(corr) denotes the average absolute value of the off-diagonal elements correlation matrix between the cross-sectional units; "l" denotes natural logarithm; "d" denotes first differences; to avoid losing observations when we apply the logarithmic transformation, we add three to all observations of the economic complexity variable (raw data range from −2.5 to 2.5); for the variables dlGHG and dlGHGpc, cross-sectional dependence was computed excluding Antigua and Barbuda, the Bahamas, Barbados and St.Kitts and Nevis, as they only share short time spans with other countries.

Table 7.2 Pairwise correlations between gas emissions.

	dlCO_2	**dlN2O**	**dlCH4**	**dlGHG**		**dlCO_2pc**	**dlN2Opc**	**dlCH4pc**	**dlGHGpc**
dlCO_2	1				**dlCO2pc**	1			
dlN2O	**0.0492**	1			**dlN2Opc**	**0.0506**	1		
dlCH4	0.0450	**0.7900**	1		**dlCH4pc**	**0.0483**	**0.7902**	1	
dlGHG	**0.1304**	**0.6153**	**0.6291**	1	**dlGHGpc**	**0.1334**	**0.6164**	**0.6308**	1

Statistical significance levels below 10% are marked in bold.

Table 7.3 Pairwise correlations between gas emissions and economic and social variables.

	$dlCO_2$	dlN2O	dlCH4	dlGHG		$dlCO_2pc$	dlN2Opc	dlCH4pc	dlGHGpc
dlY	**0.2270**	**0.0465**	**0.0522**	**0.0721**	**dlYpc**	**0.2162**	**0.0512**	**0.0642**	**0.0799**
dlE	**0.4652**	0.0162	0.0259	**0.1183**	**dlEpc**	**0.4611**	0.0135	0.0282	**0.1188**
dlU	**0.0700**	0.0240	0.0160	−0.0015	**dlU**	0.0207	−0.0256	**−0.0536**	−0.0411
dlG	**0.0456**	−0.0331	−0.0426	0.0021	**dlG**	**0.0447**	−0.0335	−0.0430	0.0020
dlEG	**0.0683**	−0.0100	0.0016	−0.0043	**dlEG**	**0.0664**	−0.0115	−0.0004	−0.0055
dlSG	0.0366	0.0060	−0.0198	0.0268	**dlSG**	0.0365	0.0071	−0.0182	0.0280
dlECI	−0.0358	−0.0214	−0.0164	−0.0037	**dlECI**	−0.0379	−0.0222	−0.0174	−0.0045
dlPOP	**0.0716**	0.0190	0.0043	−0.0055	**dlPOP**	0.0090	**−0.0462**	**−0.0871**	**−0.0583**

Statistical significance levels below 10% are marked in bold.

Table 7.4 Pairwise Granger causality—carbon dioxide emissions.

Excluded	W-bar	Z-bar	*P*-value	Z-bar tilde	*P*-value	Lags (BIC)
Dependent variable: **dlCO_2**						
dlY	3.6636	10.4867	**0.0000**	9.6354	**0.0000**	1
dlE	2.3780	4.4654	**0.0000**	3.9200	**0.0000**	1
dlU	1.7465	2.9391	**0.0500**	2.5948	**0.0500**	1
dlG	0.7088	−1.1465	0.2500	−1.2317	0.1600	1
dlEG	1.0693	0.2728	0.8500	0.0696	0.9500	1
dlSG	1.4199	1.6530	0.1200	1.3351	0.2200	1
dlECI	0.6415	−1.1337	0.2400	−1.1787	0.1800	1
dlPOP	1.8387	3.3020	**0.0400**	2.9333	**0.0400**	1
Dependent variable: **dlCO_2pc**						
dlYpc	3.4161	9.5122	**0.0000**	8.7264	**0.0000**	1
dlEpc	2.2491	4.0475	**0.0100**	3.5386	**0.0100**	1
dlU	1.1437	0.5656	0.6600	0.3807	0.7300	1
dlG	0.7177	−1.1113	0.2800	−1.1995	0.1700	1
dlEG	1.0590	0.2323	0.8500	0.0325	0.9700	1
dlSG	1.3962	1.5598	0.1700	1.2496	0.2300	1
dlECI	0.6536	−1.0953	0.2400	−1.1431	0.1700	1
dlPOP	1.1747	0.6879	0.5600	0.4948	0.6500	1

H_0: independent variable does not Granger-cause-dependent variable; H_1: independent variable Granger-cause-dependent variable for at least one panelvar (country); *P*-values computed using 100 bootstrap replications; seed(20210331); the BIC criterion was used identify the optimal lag order; due to the lack of observations for the variable dlE and dlEpc, the countries Antigua and Barbuda, the Bahamas, Barbados, Belize, Dominica, Grenada, St. Kitts and Nevis, St. Lucia, St. Vincent and Grenadines and Suriname were excluded from the panel.

Table 7.5 Pairwise Granger causality—nitrous oxide emissions.

Excluded	W-bar	Z-bar	*P*-value	Z-bar tilde	*P*-value	Lags (BIC)
Dependent variable: **dlN2O**						
dlY	1.3752	1.4770	0.1100	1.1427	0.2800	1
dlE	0.7899	−0.6807	0.5800	−0.7842	0.4900	1
dlU	0.6934	−1.2071	0.3200	−1.2949	0.2800	1
dlG	1.3225	1.2696	0.3000	0.9544	0.3900	1
dlEG	1.1129	0.4447	0.6900	0.2052	0.8100	1
dlSG	0.7557	−0.9619	0.3800	−1.0722	0.3200	1
dlECI	1.1793	0.5669	0.6200	0.3553	0.7100	1
dlPOP	0.9095	−0.3563	0.7100	−0.5222	0.6100	1
Dependent variable: **dlN2Opc**						
dlYpc	1.4193	1.6508	**0.0700**	1.3006	0.1500	1
dlEpc	0.7465	−0.8215	0.4800	−0.9117	0.3600	1
dlU	0.7410	−1.0198	0.4000	−1.1247	0.3200	1
dlG	1.2790	1.0983	0.4000	0.7988	0.4800	1
dlEG	1.1010	0.3975	0.7500	0.1624	0.8900	1
dlSG	0.7939	−0.8114	0.5000	−0.9355	0.3900	1

Continued

Table 7.5 Pairwise Granger causality—nitrous oxide emissions.—cont'd

Excluded	W-bar	Z-bar	*P*-value	Z-bar tilde	*P*-value	Lags (BIC)
dlECI	1.1798	0.5684	0.6400	0.3567	0.7800	1
dlPOP	1.1142	0.4496	0.6700	0.2097	0.8000	1

H_0: independent variable does not Granger-cause-dependent variable; H_1: independent variable Granger-causes-dependent variable for at least one panelvar (country); *P*-values computed using 100 bootstrap replications; seed(20210331); the BIC criterion was used identify the optimal lag order; due to the lack of observations for the variable dlE and dlEpc, the countries Antigua and Barbuda, the Bahamas, Barbados, Belize, Dominica, Grenada, St. Kitts and Nevis, St. Lucia, St. Vincent and Grenadines and Suriname were excluded from the panel.

Table 7.6 Pairwise Granger causality – methane emissions.

Excluded	W-bar	Z-bar	*P*-value	Z-bar tilde	*P*-value	Lags (BIC)
Dependent variable: **dlCH4**						
dlY	1.2083	0.8200	0.4400	0.5460	0.6200	1
dlE	1.0650	0.2106	0.8100	0.0230	0.9900	1
dlU	1.1062	0.4180	0.7400	0.1810	0.9200	1
dlG	1.1733	0.6824	0.5900	0.4211	0.7800	1
dlEG	1.1254	0.4936	0.7200	0.2496	0.8100	1
dlSG	1.7048	2.7746	**0.0500**	2.3211	**0.0500**	1
dlECI	1.1940	0.6135	0.5900	0.3976	0.7500	1
dlPOP	0.9924	−0.0300	0.9900	−0.2259	0.9000	1
Dependent variable: **dlCH4pc**						
dlYpc	1.3299	1.2988	0.2500	0.9808	0.3200	1
dlEpc	1.1088	0.3525	0.6400	0.1516	0.7900	1
dlU	1.6110	2.4053	**0.1000**	1.9858	0.1100	1
dlG	1.1518	0.5977	0.6500	0.3441	0.8300	1
dlEG	1.1348	0.5309	0.6900	0.2835	0.7900	1
dlSG	1.7781	3.0634	**0.0300**	2.5834	**0.0300**	1
dlECI	1.2324	0.7351	0.5500	0.5080	0.7000	1
dlPOP	1.7537	2.9673	**0.0500**	2.4961	**0.0500**	1

H_0: independent variable does not Granger-cause-dependent variable; H_1: independent variable Granger-cause-dependent variable for at least one panelvar (country); *P*-values computed using 100 bootstrap replications; seed(20210331); the BIC criterion was used identify the optimal lag order; due to the lack of observations for the variable dlE and dlEpc, the countries Antigua and Barbuda, the Bahamas, Barbados, Belize, Dominica, Grenada, St. Kitts and Nevis, St. Lucia, St. Vincent and Grenadines and Suriname were excluded from the panel.

Table 7.7 Pairwise Granger causality—total greenhouse gas emissions.

Excluded	W-bar	Z-bar	*P*-value	Z-bar tilde	*P*-value	Lags
Dependent variable: **dlGHG**						
dlY	2.1083	4.3634	**0.0000**	3.7640	**0.0002**	1
dlE	1.1887	0.6113	0.5410	0.3860	0.6995	1
dlU	0.9201	−0.3147	0.7530	−0.4844	0.6281	1
dlG	1.4413	1.7374	**0.0823**	1.3792	0.1678	1

Table 7.7 Pairwise Granger causality—total greenhouse gas emissions.—cont'd

Excluded	W-bar	Z-bar	*P*-value	Z-bar tilde	*P*-value	Lags
dlEG	1.2210	0.8700	0.3843	0.5914	0.5542	1
dlSG	1.1190	0.4686	0.6394	0.2269	0.8205	1
dlECI	0.6227	−1.1931	0.2328	−1.2431	0.2138	1
dlPOP	0.9694	−0.1206	0.9040	−0.3082	0.7580	1
Dependent variable: **dlGHGpc**						
dlYpc	2.1186	4.4039	**0.0000**	3.8008	**0.0001**	1
dlEpc	1.0737	0.2387	0.8113	0.0485	0.9613	1
dlU	0.7237	−1.0877	0.2767	−1.1864	0.2355	1
dlG	1.4143	1.6310	0.1029	1.2826	0.1996	1
dlEG	1.1759	0.6924	0.4887	0.4302	0.6671	1
dlSG	1.0635	0.2500	0.8026	0.0284	0.9773	1
dlECI	0.6372	−1.1472	0.2513	−1.2014	0.2296	1
dlPOP	0.7855	−0.8444	0.3985	−0.9654	0.3343	1

H_0: independent variable does not Granger-cause-dependent variable; H_1: independent variable Granger-cause-dependent variable for at least one panelvar (country); the lag order was fixed at one lag due to short time span available to do the analysis; due to the lack of observations for the variable dlE and dlEpc, the countries Antigua and Barbuda, the Bahamas, Barbados, Belize, Dominica, Grenada, St. Kitts and Nevis, St. Lucia, St. Vincent and Grenadines and Suriname were excluded from the panel; the short time span does not allow the use of bootstrap to compute more accurate statistical levels of significance.

greenhouse emissions per capita. Growth in globalisations only causes the growth of greenhouse emissions.

7.4.2 Long-run analysis

The long-run analysis reveals a denser relationship/causality than that revealed by variables in first-differences, as shown in the following tables. Table 7.8 shows Pesaran's (2007) CIPS panel unit root testing of the second generation. The test indicates that social globalisation is stationary. The remaining variables seem to have a unit root or at least to be borderline I(0)/I(1).

Only the correlations between nitrous emissions per capita and carbon dioxide emissions per capita have a low (statistically not significant) correlation in natural logarithms. As a rule, correlations between variables per capita reveal much lower correlations.

Table 7.9 presents the pairwise correlations between gas emissions and variables. With few exceptions, economic, social and demographic variables are statistically associated with gas emissions, both in logs or logs per capita. Without exception, all variables have statistically significant correlations with greenhouse emissions. It should be noted that we have negative and positive associations between the variables and the gas emissions. There is a visible disparity between the relationships between variables and variables per capita. This supports that population growth interferes with relationships (Table 7.10).

Table 7.8 Panel unit root test (CIPS).

Variables	CIPS (Zt-bar)			Variables	CIPS (Zt-bar)		
	Lags	Without trend	With trend		Lags	Without trend	With trend
lCO_2	1	−2.146**	−1.095	lYpc	1	−0.695	0.808
	2	−0.213	0.746		2	−1.438*	0.170
lN2O	1	−0.716	1.865	lE	1	n.a.	n.a.
	2	1.258	3.654		2	n.a.	n.a.
lCH4	1	−1.669**	0.166	lEpc	1	n.a.	n.a.
	2	−0.040	0.979		2	n.a.	n.a.
lGHG	1	−3.761***	−1.330*	lU	1	−3.835***	0.336
	2	−1.644**	0.648		2	−3.363***	0.918
lCO2pc	1	−1.726**	1.517	lG	1	−2.129**	−0.900
	2	−0.103	3.238		2	−1.275	−0.119
lN2Opc	1	1.743	2.179	lEG	1	−2.442***	−1.249
	2	4.736	6.175		2	−1.519*	−0.217
lCH4pc	1	1.082	0.055	lSG	1	−4.053***	−3.644***
	2	2.495	1.007		2	−3.364***	−2.919***
lGHGpc	1	−2.926***	0.285	lECI	1	−2.833***	−1.183
	2	−0.267	3.090		2	−2.169**	−0.984
lY	1	−2.512***	−0.328	lPOP	1	−5.961***	−6.930***
	2	−2.527***	−0.645		2	−1.787**	0.226

***, ** and * denote statistical significance at 1%, 5% and 10% level, respectively; panel unit root test (CIPS) assumes that cross-sectional dependence is in the form of a single unobserved common factor and H_0: series is I(1); n.a. denotes “not available.”

Table 7.9 Pairwise correlations between gas emissions.

	ICO_2	IN2O	ICH4	IGHG		ICO_2pc	IN2Opc	ICH4pc	IGHGpc
ICO_2	1				ICO2pc	1			
IN2O	**0.8831**	1			IN2Opc	0.0395	1		
ICH4	**0.9101**	**0.9874**	1		ICH4pc	**0.2433**	**0.8477**	1	
IGHG	**0.9489**	**0.9688**	**0.9757**	1	IGHGpc	**0.6096**	**0.7030**	**0.7556**	1

Statistical significance levels below 5% are marked in bold.

Table 7.10 Pairwise correlations between gas emissions and variables.

	lCO_2	lN2O	lCH4	lGHG		lCO_2pc	lN2Opc	lCH4pc	lGHGpc
lY	**0.9640**	**0.9099**	**0.9252**	**0.9410**	lYpc	**0.8212**	−0.0497	**0.0871**	**0.4580**
lE	**0.9627**	**0.8512**	**0.9051**	**0.9148**	lEpc	**0.9069**	0.0492	**0.4087**	**0.5475**
lU	**0.9232**	**0.9694**	**0.9715**	**0.9531**	lU	**0.0548**	**0.4485**	**0.5389**	**0.2864**
lG	**0.2875**	0.0192	0.0316	**0.1406**	lG	**0.6289**	**0.0856**	**0.1206**	**0.3405**
lEG	**−0.3457**	**−0.5382**	**−0.5393**	**−0.4399**	lEG	**0.4315**	**−0.2051**	**−0.2582**	**0.0857**
lSG	**−0.2971**	**−0.6129**	**−0.5875**	**−0.4842**	lSG	**0.5685**	**−0.3160**	**−0.2821**	**0.0590**
lECI	**0.2028**	**0.1263**	**0.1493**	**0.0919**	lECI	**0.1852**	−0.0523	−0.0130	**−0.1461**
lPOP	**0.8978**	**0.9662**	**0.9658**	**0.9416**	lPOP	−0.0193	**0.4165**	**0.5010**	**0.2346**

Statistical significance levels below 5% are marked in bold.

Table 7.11 reveals the Granger causalities tests between variables and CO_2 emissions. Although the causality test statistics are not straightforward reading, we can get an idea of the type of causality they refer to. Given that all variables are in natural logs, the acceptance of Granger's causalities refers to the coefficients between the independent and the dependent variable that are one atypical kind of elasticity. Therefore, it can be concluded that a percentage change in GDP and energy consumption, both lagged once, causes a percentage change in CO_2 emissions, and this conclusion is equally valid for per capita values. The population also cause CO_2 emissions.

Nitrous oxide emissions are caused by all variables, except economic globalisation on nitrous oxide emissions per capita (see Table 7.12), contrasting with the CO_2 emissions analysed above.

Table 7.13 reveals the pairwise Granger causality of variables in methane emissions. More often than not, variables Granger-cause methane emissions and methane emissions per capita. The exceptions are globalisation on methane emissions, economic globalisation and economic complexity on methane emissions per capita.

Table 7.14 shows the pairwise Granger causality of variables on greenhouse gas emissions. All variables Granger-cause greenhouse gas emissions.

Temporal cross-correlations between gas emissions (t) and economic and social variables lagged once (t-1) suggest that past values of GDP, energy consumption,

Table 7.11 Pairwise Granger causality—carbon dioxide emissions.

Excluded	**W-bar**	**Z-bar**	***P*-value**	**Z-bar tilde**	***P*-value**	**Lags (BIC)**
Dependent variable: **lCO_2**						
lY	5.5535	17.9272	**0.0000**	16.6015	**0.0000**	1
lE	4.5439	11.4836	**0.0100**	10.3538	**0.0100**	1
lU	3.6706	10.5143	0.0600	9.6771	0.0600	1
lG	1.8025	3.1594	0.6800	2.7264	0.6800	1
lEG	1.9127	3.5932	0.4000	3.1250	0.4000	1
lSG	1.7636	3.0062	0.6800	2.5856	0.6800	1
lECI	1.2226	0.7039	0.5900	0.5288	0.5288	1
lPOP	3.7315	10.7539	**0.0500**	9.9010	**0.0500**	1
Dependent variable: **lCO2pc**						
lYpc	4.2837	12.9278	**0.0000**	11.9315	**0.0000**	1
lEpc	3.1804	7.0653	**0.0000**	6.3117	**0.0000**	1
lU	2.7840	7.0235	0.1800	6.4164	0.1800	1
lG	2.0983	4.3241	0.4300	3.7965	0.4300	1
lEG	2.3060	5.1419	0.2100	4.5479	0.2100	1
lSG	2.4566	5.7347	0.2300	5.0926	0.2300	1
lECI	1.3391	1.0722	0.5100	0.8708	0.5300	1
lPOP	2.6228	6.3890	0.2100	5.8237	0.2100	1

H_0: independent variable does not Granger-cause dependent variable; H_1: independent variable Granger-causes-dependent variable for at least one panelvar (country); *P*-values computed using 100 bootstrap replications; seed (20210331); the BIC criterion was used identify the optimal lag order; due to the lack of observations for the variable lE and lEpc, the countries Antigua and Barbuda, the Bahamas, Barbados, Belize, Dominica, Grenada, St. Kitts and Nevis, St. Lucia, St. Vincent and Grenadines and Suriname were excluded from the panel.

Table 7.12 Pairwise Granger causality—nitrous oxide emissions.

Excluded	W-bar	Z-bar	*P*-value	Z-bar tilde	*P*-value	Lags (BIC)
Dependent variable: **lN2O**						
lY	6.7069	22.4681	**0.0000**	20.2634	**0.0000**	1
lE	6.2165	16.9033	**0.0000**	15.1872	**0.0000**	1
lU	6.9419	23.3932	**0.0000**	21.1057	**0.0000**	1
lG	2.7109	6.7356	**0.0400**	5.9390	**0.0400**	1
lEG	2.8998	7.4794	**0.0100**	6.6162	**0.0100**	1
lSG	2.6008	6.3025	**0.0800**	5.5447	**0.0800**	1
lECI	2.4087	4.4547	**0.0700**	3.9003	**0.0700**	1
lPOP	6.7201	22.5200	**0.0000**	20.3107	**0.0000**	1
Dependent variable: **lN2Opc**						
lYpc	3.0064	7.8990	**0.0200**	6.9983	**0.0200**	1
lEpc	2.5034	4.8717	**0.0700**	4.2607	**0.0700**	1
lU	5.6616	18.3527	**0.0000**	16.5163	**0.0000**	1
lG	2.7722	6.9773	**0.0600**	6.1591	**0.0600**	1
lEG	2.1213	4.4146	0.1500	3.8257	0.1500	1
lSG	4.1671	12.4689	**0.0000**	11.1592	**0.0000**	1
lECI	2.1973	3.7863	**0.0600**	3.2918	**0.0600**	1
lPOP	5.6675	18.3758	**0.0000**	16.5374	**0.0000**	1

H_0: independent variable does not Granger-cause-dependent variable; H_1: independent variable Granger-cause-dependent variable for at least one panelvar (country); *P*-values computed using 100 bootstrap replications; seed(20210331); the BIC criterion was used identify the optimal lag order; due to the lack of observations for the variable lE and lEpc, the countries Antigua and Barbuda, the Bahamas, Barbados, Belize, Dominica, Grenada, St. Kitts and Nevis, St. Lucia, St. Vincent and Grenadines and Suriname were excluded from the panel.

Table 7.13 Pairwise Granger causality—methane emissions.

Excluded	W-bar	Z-bar	*P*-value	Z-bar tilde	*P*-value	Lags (BIC)
Dependent variable: **lCH4**						
lY	5.9979	19.6769	**0.0000**	17.7220	**0.0000**	1
lE	5.8647	15.7635	**0.0000**	14.1522	**0.0000**	1
lU	5.5094	17.7536	**0.0000**	15.9709	**0.0000**	1
lG	2.3374	5.2655	0.1100	4.6005	0.1100	1
lEG	2.5753	6.2021	**0.0300**	5.4532	**0.0300**	1
lSG	4.4493	13.5800	**0.0000**	12.1708	**0.0000**	1
lECI	2.8596	5.8805	**0.0200**	5.1985	**0.0200**	1
lPOP	62.0502	56.8828	**0.0100**	11.8001	**0.0100**	12
Dependent variable: **lCH4pc**						
lYpc	5.3755	17.2262	**0.0000**	15.4907	**0.0000**	1
lEpc	2.3777	4.4642	**0.0500**	3.8907	**0.0500**	1
lU	3.5316	9.9668	**0.0200**	8.8810	**0.0200**	1
lG	2.5837	6.2350	**0.0500**	5.4832	**0.0500**	1
lEG	2.2603	4.9619	0.1300	4.3240	0.1300	1

Table 7.13 Pairwise Granger causality—methane emissions.—cont'd

Excluded	W-bar	Z-bar	*P*-value	Z-bar tilde	*P*-value	Lags (BIC)
lSG	4.2899	12.9524	**0.0000**	11.5994	**0.0000**	1
lECI	2.1097	3.5093	0.1300	3.0396	0.1300	1
lPOP	63.9538	59.0463	**0.0000**	12.3101	**0.0000**	12

H_0: independent variable does not Granger-cause-dependent variable; H_1: independent variable Granger-cause-dependent variable for at least one panelvar (country); *P*-values computed using 100 bootstrap replications; seed (20210331); the BIC criterion was used identify the optimal lag order; due to the lack of observations for the variable lE and lEpc, the countries Antigua and Barbuda, the Bahamas, Barbados, Belize, Dominica, Grenada, St. Kitts and Nevis, St. Lucia, St. Vincent and Grenadines and Suriname were excluded from the panel.

Table 7.14 Pairwise Granger causality—total greenhouse gas emissions.

Excluded	W-bar	Z-bar	*P*-value	Z-bar tilde	*P*-value	Lags
Dependent variable: **lGHG**						
lY	7.7618	26.6212	**0.0000**	24.0448	**0.0000**	1
lE	9.1872	26.5295	**0.0000**	23.9293	**0.0000**	1
lU	8.1133	28.0050	**0.0000**	25.3048	**0.0000**	1
lG	2.5699	6.1806	**0.0000**	5.4336	**0.0000**	1
lEG	3.2399	8.8186	**0.0000**	7.8356	**0.0000**	1
lSG	2.9003	7.4814	**0.0000**	6.6180	**0.0000**	1
lECI	2.4703	4.6494	**0.0000**	4.0777	**0.0000**	1
lPOP	8.0480	27.7479	**0.0000**	25.0707	**0.0000**	1
Dependent variable: **lGHGpc**						
lYpc	4.3881	13.3391	**0.0000**	11.9514	**0.0000**	1
lEpc	3.3109	7.4881	**0.0000**	6.6368	**0.0000**	1
lU	3.5867	10.1839	**0.0000**	9.0787	**0.0000**	1
lG	2.4954	5.8872	**0.0000**	5.1665	**0.0000**	1
lEG	2.6348	6.4362	**0.0000**	5.6664	**0.0000**	1
lSG	3.2079	8.6925	**0.0000**	7.7207	**0.0000**	1
lECI	2.1124	3.5176	**0.0004**	3.0471	**0.0023**	1
lPOP	3.6171	10.3035	**0.0000**	9.1876	**0.0000**	1

H_0: independent variable does not Granger-cause-dependent variable; H_1: independent variable Granger-cause-dependent variable for at least one panelvar (country); the lag order was fixed at one lag due to short-time span available to do the analysis; due to the lack of observations for the variable lE and lEpc, the countries Antigua and Barbuda, the Bahamas, Barbados, Belize, Dominica, Grenada, St. Kitts and Nevis, St. Lucia, St. Vincent and Grenadines and Suriname were excluded from the panel; the short time span does not allow the use of bootstrap to compute more accurate statistical levels of significance.

urbanisation, and population are related (at least statistically associated) to gas emissions (see Table A.1 in the appendix). If we consider per capita values, only GDP and energy consumption have a very high temporal cross-correlation with carbon dioxide emissions. There is also evidence of negative cross-correlations of economic and social globalisation on gas emissions. Nevertheless, this former evidence almost disappears for greenhouse emissions when assessing per capita values.

The Granger causality between nonstationary variables should be cautiously analysed if they are not cointegrated. Table 7.15 shows Westerlund's (2007) cointegration test. The Westerlund (2007) cointegration approach is based on four statistics. The statistics Gt and Ga are groups mean tests, and the statistics Pt and Pa are panel mean tests. The statistics Ga and Pa are normalised by T (number of observations). This normalisation may cause the test to reject the null too frequently, especially when the number of lags is comparably large (Westerland, 2007). The tests have the null hypothesis of no cointegration. The Ga and Gt test statistics test H_0 of no cointegration versus H_1 of at least one cross being cointegrated. The rejection of H_0 reveals evidence of cointegration of at least one of the cross-sectional units. The statistics Pa and Pt test pool information over all the cross-sectional units to test H_0 of panel no cointegration versus H_1 of panel cointegration. The rejection of H_0 reveals evidence of panel cointegration (i.e., the panel is cointegrated as a whole). Westerlund (2007) suggested that cross-sectional dependence Pt and Gt tests have the most negligible size distortions. Bootstrapping the four tests allow *P*-values to be obtained that are robust to cross-sectional dependence.

We can conclude that there is more evidence of panel cointegration than of group mean regarding carbon dioxide emissions. CO_2 is cointegrated with GDP, energy consumption, urbanisation and population. The results are also similar in per capita.

We can conclude that there is more evidence of panel cointegration than the group means regarding nitrous dioxide emissions (see Table 7.16). N_2O is cointegrated with GDP, energy consumption, urbanisation, globalisation (global, economic and social), economic complexity and population. The results are also similar for values per capita.

We can conclude that there is more evidence of panel cointegration than the group means regarding methane emissions (see Table 7.17). CH_4 is cointegrated with GDP, energy consumption, urbanisation and population. The analysis with per capita values also reveals evidence of cointegration with social globalisation.

We can conclude that there is more evidence of panel cointegration than the group means regarding greenhouse gas emissions. Greenhouse gas emissions are cointegrated with all variables. The results are also similar in per capita values (Table 7.18).

7.5 Discussion

The relationship between economic and social variables and gas emissions does not behave in the same way for all. Since the origin of the gases was associated with different contexts, it was expected that their relationship with the variables would also be different. However, the results of variables and the variables expressed per capita reveal different results demonstrating that the increasing population changes the relationships between variables.

Table 7.15 Westerlund cointegration test—carbon dioxide emissions.

	Value	Z-value	*P*-value	Value	Z-value	*P*-value	Value	Z-value	*P*-value	Value	Z-value	*P*-value
	lCO_2 and lY			**lCO_2 and lEG**			**lCO_2pc and lYpc**			**lCO_2pc and lEG**		
Gt	−2.321	−3.369	**0.000**	−1.386	2.428	0.992	−2.269	−3.046	**0.001**	−1.726	0.317	0.624
Ga	−10.507	−3.441	**0.000**	−4.041	3.172	0.999	−9.715	−2.632	**0.004**	−6.316	0.845	0.801
Pt	−13.392	−5.364	**0.000**	−7.666	0.397	0.654	−11.768	−3.729	**0.000**	−9.570	−1.519	**0.064**
Pa	−9.631	−6.773	**0.000**	−3.856	0.469	0.681	−7.823	−4.506	**0.000**	−5.830	−2.006	**0.022**
	lCO_2 and lE[a]			**lCO_2 and lSG**			**lCO_2pc and lEpc[a]**			**lCO_2pc and lSG**		
Gt	−2.203	−2.219	**0.013**	−1.527	1.555	0.940	−1.906	−0.671	0.251	−1.960	−1.129	0.129
Ga	−10.570	−2.953	**0.002**	−5.368	1.815	0.965	−8.057	−0.788	0.215	−8.286	−1.170	0.121
Pt	−10.513	−3.745	**0.000**	−8.344	−0.286	0.388	−8.617	−1.837	**0.033**	−9.771	−1.721	**0.043**
Pa	−10.093	−6.194	**0.000**	−4.266	−0.045	0.482	−6.940	−2.863	**0.002**	−5.268	−1.302	**0.097**
	lCO_2 and lU			**lCO_2 and lECI[b]**			**lCO_2pc and lU**			**lCO_2pc and lECI[b]**		
Gt	−2.338	−3.472	**0.000**	−1.707	0.331	0.630	−2.389	−3.791	**0.000**	−1.790	−0.060	0.476
Ga	−6.159	1.005	0.843	−3.781	2.620	0.996	−5.660	1.516	0.935	−5.093	1.597	0.945
Pt	−12.501	−4.467	**0.000**	−6.626	−0.488	0.313	−12.901	−4.869	**0.000**	−7.260	−1.125	0.130
Pa	−6.435	−2.765	**0.003**	−2.634	1.525	0.936	−6.436	−2.766	**0.003**	−4.041	0.181	0.572
	lCO_2 and lG			**lCO_2 and lPOP**			**lCO_2pc and lG**			**lCO_2pc and lPOP**		
Gt	−1.510	1.659	0.951	−2.271	−3.058	**0.001**	−1.822	−0.274	0.392	−2.271	−3.058	**0.001**
Ga	−5.661	1.515	0.935	−6.256	0.907	0.818	−7.405	−0.269	0.394	−5.599	1.578	0.943
Pt	−7.764	0.298	0.617	−10.618	−2.573	**0.005**	−9.284	−1.231	0.109	−10.481	−2.435	**0.007**
Pa	−4.574	−0.431	0.333	−5.908	−2.104	**0.018**	−6.011	−2.233	**0.013**	−5.542	−1.645	**0.050**

Model with constant; to compute Westerland cointegration test continuous time-series are required. The Stata command xtwest developed by Persyn and Westerlund (2008) was used. [a] ATG, BHS, BLZ, BRB, DMA, GRD, KNA, LCA, and VCT were excluded. [b] TTO and VEN were excluded.

Table 7.16 Westerlund cointegration test—nitrous dioxide emissions.

	Value	Z-value	*P*-value	Value	Z-value	*P*-value	Z-value	*P*-value	Value	Z-value	*P*-value	Value
		N2O and lY			**lN2O and lEG**			**lN2Opc and lYpc**			**lN2Opc and lEG**	
Gt	−2.303	−3.255	**0.001**	−1.647	0.807	0.790	−2.041	−1.635	**0.051**	−1.834	−0.350	0.363
Ga	−11.880	−4.846	**0.000**	−6.526	0.630	0.736	−9.072	−1.974	**0.024**	−7.657	−0.527	0.299
Pt	−11.751	−3.712	**0.000**	−8.964	−0.909	0.182	−10.895	−2.852	**0.002**	−9.794	−1.744	**0.041**
Pa	−8.843	−5.785	**0.000**	−6.032	−2.259	**0.012**	−8.178	−4.950	**0.000**	−6.670	−3.060	**0.001**
		lN2O and lE[a]			**lN2O and lSG**			**lN2Opc and lEpc[a]**			**lN2Opc and lSG**	
Gt	−2.169	−2.041	**0.021**	−2.066	−1.790	**0.037**	−1.949	−0.897	0.185	−2.237	−2.844	**0.002**
Ga	−11.774	−3.991	**0.000**	−8.069	−0.947	0.172	−8.893	−1.508	**0.066**	−10.167	−3.093	**0.001**
Pt	−9.216	−2.440	**0.007**	−12.084	−4.048	**0.000**	−8.733	−1.954	**0.025**	−13.334	−5.305	**0.000**
Pa	−8.283	−4.282	**0.000**	−8.114	−4.871	**0.000**	−7.463	−3.416	**0.000**	−9.852	−7.051	**0.000**
		lN2O and lU			**lN2O and lECI**			**lN2Opc and lU**			**lN2Opc and lECI**	
Gt	−2.467	−4.274	**0.000**	−1.681	0.481	0.685	−2.493	−4.436	**0.000**	−1.888	−0.547	0.292
Ga	−6.715	0.437	0.669	−6.960	0.150	0.559	−6.328	0.833	0.798	−8.387	−1.022	0.153
Pt	−11.525	−3.485	**0.000**	−7.782	−1.316	**0.094**	−11.572	−3.533	**0.000**	−8.072	−1.608	**0.054**
Pa	−5.572	−1.682	**0.046**	−6.358	−2.144	**0.016**	−5.466	−1.549	**0.061**	−7.540	−3.334	**0.000**
		lN2O and lG			**lN2O and lPOP**			**lN2Opc and lG**			**lN2Opc and lPOP**	
Gt	−1.743	0.212	0.584	−2.739	−5.956	**0.000**	−2.001	−1.384	**0.083**	−2.739	−5.956	**0.000**
Ga	−7.441	−0.305	0.380	−6.167	0.997	0.841	−8.651	−1.544	**0.061**	−5.700	1.475	0.930
Pt	−8.878	−0.823	0.205	−12.675	−4.642	**0.000**	−10.109	−2.061	**0.020**	−12.686	−4.653	**0.000**
Pa	−5.885	−2.075	**0.019**	−7.042	−3.526	**0.000**	−7.006	−3.481	**0.000**	−7.088	−3.583	**0.000**

Model with constant; to compute Westerland cointegration test, continuous time-series are required. [a] ATG, BHS, BLZ, BRB, DMA, GRD, KNA, LCA, and VCT were excluded.

Table 7.17 Westerlund cointegration test—methane emissions.

	Value	Z-value	*P*-value	Value	Z-value	*P*-value	Z-value	*P*-value	Value	Value	Z-value	*P*-value
	lCH4 and lY			**lCH4 and lEG**			**lCH4pc and lYpc**			**lCH4pc and lEG**		
Gt	−2.166	−2.406	**0.008**	−1.230	3.392	1.000	−2.037	−1.610	**0.054**	−1.612	1.023	0.847
Ga	−9.901	−2.822	**0.002**	−4.390	2.815	0.998	−7.684	−0.554	0.290	−6.444	0.714	0.762
Pt	−12.248	−4.213	**0.000**	−5.299	2.777	0.997	−9.767	−1.717	**0.043**	−6.200	1.870	0.969
Pa	−9.821	−7.011	**0.000**	−2.527	2.136	0.984	−6.438	−2.769	**0.003**	−2.945	1.612	0.947
	lCH4 and lE[a]			**lCH4 and lSG**			**lCH4pc and lEpc[a]**			**lCH4pc and lSG**		
Gt	−2.088	−1.623	**0.052**	−1.773	0.027	0.511	−1.984	−1.076	0.141	−1.799	−0.135	0.446
Ga	−9.375	−1.924	**0.027**	−5.987	1.181	0.881	−8.307	−1.004	0.158	−7.524	−0.391	0.348
Pt	−10.543	−3.775	**0.000**	−8.023	0.037	0.515	−9.076	−2.299	**0.011**	−9.396	−1.344	**0.090**
Pa	−10.783	−6.923	**0.000**	−4.255	−0.032	0.487	−7.953	−3.933	**0.000**	−5.618	−1.741	**0.041**
	lCH4 and lU			**lCH4 and lECI**			**lCH4pc and lU**			**lCH4pc and lECI**		
Gt	−2.186	−2.530	**0.006**	−1.347	2.143	0.984	−2.218	−2.729	**0.003**	−1.646	0.655	0.744
Ga	−5.378	1.805	0.965	−4.719	1.991	0.977	−5.214	1.972	0.976	−7.274	−0.108	0.457
Pt	−9.520	−1.469	**0.071**	−4.449	2.036	0.979	−9.754	−1.704	**0.044**	−4.074	2.414	0.992
Pa	−3.509	0.905	0.817	−2.555	1.688	0.954	−3.528	0.881	0.811	−2.581	1.662	0.952
	lCH4 and lG			**lCH4 and lPOP**			**lCH4pc and lG**			**lCH4pc and lPOP**		
Gt	−1.455	1.999	0.977	−2.350	−3.547	**0.000**	−1.811	−0.209	0.417	−2.350	−3.547	**0.000**
Ga	−5.237	1.949	0.974	−4.834	2.361	0.991	−7.544	−0.410	0.341	−4.336	2.871	0.998
Pt	−6.250	1.821	0.966	−10.091	−2.043	**0.021**	−7.262	0.802	0.789	−10.127	−2.079	**0.019**
Pa	−3.137	1.372	0.915	−4.280	−0.062	0.475	−3.795	0.546	0.708	−4.147	0.104	0.541

Model with constant; to compute Westerland cointegration test, continuous time-series are required. [a] ATG, BHS, BLZ, BRB, DMA, GRD, KNA, LCA, and VCT were excluded.

Table 7.18 Westerlund cointegration test—Greenhouse gas emissions.

	Value	Z-value	*P*-value	Value	Z-value	*P*-value	Z-value	*P*-value	Value	Value	Z-value	*P*-value
	lGHG and lY[a]			**lGHG and lEG[d]**			**lGHGpc and lYpc[a]**			**lGHGpc and lEG[d]**		
Gt	−2.152	−1.996	**0.023**	−1.734	0.231	0.591	−2.162	−2.053	**0.020**	−2.236	−2.444	**0.007**
Ga	−12.261	−4.510	**0.000**	−8.301	−1.021	0.154	−10.946	−3.351	**0.000**	−11.421	−3.770	**0.000**
Pt	−13.262	−6.356	**0.000**	−9.053	−2.123	**0.017**	−12.140	−5.228	**0.000**	−12.062	−5.149	**0.000**
Pa	−17.258	−14.074	**0.000**	−8.650	−4.774	**0.000**	−14.375	−10.959	**0.000**	−14.217	−10.788	**0.000**
	lGHG and lE[b]			**lGHG and lSG[d]**			**lGHGpc and lEpc[b]**			**lGHGpc and lSG[d]**		
Gt	−2.485	−3.433	**0.000**	−2.153	−2.005	**0.023**	−1.987	−1.018	0.154	−2.269	−2.625	**0.004**
Ga	−15.301	−6.533	**0.000**	−9.311	−1.911	**0.028**	−11.220	−3.265	**0.001**	−11.618	−3.943	**0.000**
Pt	−14.275	−8.011	**0.000**	−8.841	−1.910	**0.028**	−12.412	−6.137	**0.000**	−10.117	−3.193	**0.001**
Pa	−23.951	−19.362	**0.000**	−7.774	−3.828	**0.000**	−18.679	−14.186	**0.000**	−10.266	−6.520	**0.000**
	lGHG and lU[c]			**lGHG and lECI[e]**			**lGHGpc and lU[c]**			**lGHGpc and lECI[e]**		
Gt	−2.765	−5.269	**0.000**	−1.846	−0.324	0.373	−2.797	−5.441	**0.000**	−2.261	−2.284	**0.011**
Ga	−7.600	−0.403	0.343	−8.316	−0.915	0.180	−6.267	0.771	0.780	−11.009	−3.014	**0.001**
Pt	−13.618	−6.714	**0.000**	−7.609	−1.475	**0.070**	−13.991	−7.089	**0.000**	−10.965	−4.852	**0.000**
Pa	−11.248	−7.581	**0.000**	−7.330	−2.962	**0.002**	−10.395	−6.659	**0.000**	−14.007	−9.343	**0.000**
	lGHG and lG[d]			**lGHG and lPOP[a]**			**lGHGpc and lG[d]**			**lGHGpc and lPOP[a]**		
Gt	−1.739	0.207	0.582	−2.706	−4.955	**0.000**	−2.310	−2.840	**0.002**	−2.706	−4.955	**0.000**
Ga	−8.841	−1.496	**0.067**	−5.660	1.306	0.904	−12.178	−4.436	**0.000**	−5.543	1.409	0.921
Pt	−9.149	−2.219	**0.013**	−12.214	−5.302	**0.000**	−12.574	−5.664	**0.000**	−12.264	−5.353	**0.000**
Pa	−9.982	−6.213	**0.000**	−6.412	−2.357	**0.009**	−16.323	−13.064	**0.000**	−6.263	−2.196	**0.014**

Model with constant; to compute Westerland cointegration test continuous time-series are required. [a] ATG, BHS, BRB, JAM, LCA, SUR, TTO, and VCT were excluded. [b] ATG, BHS, BLZ, BRB, DMA, GRD, JAM, LCA, SUR, TTO, and VCT were excluded. [c] ATG, BHS, BRB, JAM, LCA, SUR, TTO, and VCT were excluded. [d] ATG, BRB, JAM, KNA, LCA, SUR, TTO, and VCT were excluded. [e] JAM and TTO were excluded.

The gas emissions analysis in LAC countries should include the potential influence of cross-sectional dependence and spurious regressions if the series are not cointegrated. In addition, possible bidirectional causality between emissions and the economic and social variables introducing feedback phenomena also should be considered to understand better the dynamics of interaction among gas emissions and economic and social variables.

The short-run analysis (first differences) imply the loss of one trend in nonstationary data. The dissimilarity of results between the analysis in first differences and logs (long run) support that the tendency in variables is essential in explaining gas emissions.

Changes in nitrous oxide and methane extensively share the same sources of emissions, so it is not strange that they are strongly correlated. However, not so expected is the strong correlation of methane with greenhouse emissions. Indeed, the most significant contributor to greenhouse emissions is carbon dioxide, not methane emissions.

When looking for the short-run dynamics of relationships, energy consumption is strongly correlated with CO_2 emissions, but CO_2 emissions are Granger-caused by economic growth and energy consumption. This discrepancy between correlation and causality is mainly due to economic growth requiring the use of energy. However, the efficiency gains in energy use have been consistent over time but not necessarily uniform. Two causalities deserve a short explanation: (1) methane emissions are caused by social globalisation and (2) greenhouse emissions are caused by economic growth. Social globalisation in LAC countries exerts influence over the diet of people generalising the consumption of meat. Production of meat changes from traditional to intensive farming that generates massive methane emissions as explained before. The causality of economic growth on greenhouse emissions results from increased income raising the demand for food from intensive farming, which increases methane and nitrous oxide emissions, and the increase in cement production and the rise in transportation, both huge emitters of CO_2.

The mitigation and control of carbon dioxide emissions require energy transition from fossil energy sources to renewables to be reinforced. Research and development policies for environmentally friendly technologies must be implemented to increase efficiency in energy use. People's behavior should also be oriented to the consumption of low energy-intensive goods and services. Policies can be used to provoke changes in the relative prices of substitute energy sources to encourage the consumption of less damaging ones. Policies of reforestation and preventing wildfires can also play an important role. Given that cement is a significant emitter, policies promoting the substitution of cement with less harmful materials should be considered one option.

The mitigation and control of nitrous dioxide emissions imply changing agricultural fertiliser to be less aggressive. The livestock dung should be treated to generate energy. The excess and leakage of fertilisers and biomass burning must be regulated to implement good practices. All economic and social variables cause nitrous oxide emissions except economic globalisation on per capita values. These results emphasise the importance of policymakers monitoring urbanisation, globalisation and economic complexity and mitigating their undesired environmental effects.

The significant anthropogenic source of methane emissions is ruminant livestock, so it is crucial to promote dietary changes to less damaging food. Policies to reduce waste generation (mainly solid residues and wastewater) and promote intensive packaging reuse (or suppression when possible) and recycling must be implemented. Policymakers also should promote active actions to change agricultural practices (especially in traditional rice production) to more sustainable ones and promote efficient use of biomass. The necessity of mitigation/reduction of methane emissions is a source of concern. To achieve this goal, capturing and utilising gas as a source of renewable energy are welcome. Policymakers should be aware that methane emissions are caused by all economic and social variables except globalisation and economic globalisation and economic complexity on per capita values.

Greenhouse emissions are mainly composed of carbon dioxide, nitrous oxide and methane emissions. Consequently, most of the policies used to mitigate/control carbon dioxide, nitrous oxide and methane emissions will be materialised in greenhouse emissions. Nevertheless, other sources of greenhouse gas are chlorofluorocarbons (CFCs) in refrigeration systems and CFCs and halogens in fire suppression systems and manufacturing processes should be under tight regulation. In LAC countries, policymakers should be aware of the complexity of fighting environmental damage as all economic and social variables cause greenhouse emissions.

The negative association between gas emissions and economic and social globalisation (except for per capita greenhouse emissions) should be better understood and exploited by policymakers of LAC countries.

7.6 Conclusion

Given that we have only one Earth, the absolute impacts are much more critical than per capita variables evolve. Nevertheless, per capita variables can more precisely signal situations where population growth blurs the picture. We should also bear in mind that variables retain much more information in their levels than their growth rates. Indeed, variables in differences lose any trend presiding the evolution of variables. We performed both analyses, and as expected, variables in levels preserve more visible relationships.

The causality between economic variables and environmental damage through gas emissions was analysed for the Latin American and Caribbean countries from 1960 to 2016, using the pairwise Granger causality tests. In addition, the causality from GDP, energy consumption, urbanisation, globalisation, economic complexity and population to carbon dioxide, nitrous oxide, methane and greenhouse gas emission was analysed.

From the short-run analysis (first differences), it can be concluded that nitrous emissions are strongly correlated with methane and greenhouse gases. Methane emissions are also strongly correlated with greenhouse gas emissions. Only energy consumption is strongly correlated with CO_2 emissions. CO_2 emissions are caused by economic growth and energy consumption. Methane emissions seem to be caused by social globalisation. Greenhouse emissions are caused by economic growth.

From the long-run analysis (natural logarithms), one can conclude that the pairwise correlations between emissions are robust, but slow down when measured in per capita values and show no statistical significance between carbon dioxide emissions and nitrous dioxide emissions. Contrary to the first differences, the correlations between emissions and the variables are mostly statistically significant. Economic and social globalisation were revealed to be predominantly negatively correlated with emissions, except per capita greenhouse emissions. CO_2 emissions are caused by GDP and energy consumption. All variables cause nitrous oxide emissions except economic globalisation on per capita values. Methane emissions are caused by all variables except globalisation, economic globalisation and economic complexity on per capita values. All variables cause greenhouse emissions. There is some evidence of variables being cointegrated. The cointegrations are notorious between economic and social variables and nitrous oxide and greenhouse emission, slightly less with carbon dioxide emissions and mostly absent with methane emissions.

Policymakers have to act on economic and social variables to mitigate/control/reduce their impact on gas emissions that damage the environment. Some variables contribute to circumventing gas emissions. Policy actions should be oriented to potentiate this effect. Most of the variables we analysed contribute to damaging the environment. In these variables, policymakers have to intervene in their composition to mitigate, control and reduce their impact on the environment. One of the first steps is to inventory the emissions of gases from goods and services and, in this way, allow the adoption of policies focused on environmental policy objectives. Indeed, the strategies to target an environmental sustainable anthropogenic behavior involve implementing measures combining promoting some behaviors or activities (e.g., granting subsidies and reducing taxation) and burdening the undesired ones. The disclosure of emissions, for example, on packaging labels, also contributes to intensifying environmental awareness and thus changing consumer behavior in a more environmentally friendly way.

The analysis is subject to some shortcomings that can influence the results and deserves further research. Among the most concerning are the possible influence of outliers, the panels being unbalanced, cross-sectional dependence and the limitations of working bivariate models. Further research should also consider the impacts of economic and social variables in the stocks of emissions.

Appendix

Table A.1 Temporal cross-correlations between gas emissions (t) and variables (t−1).

	lCO_{2t}	$lN2O_t$	$lCH4_t$	$lGHG_t$		lCO_2pc_t	$lN2Opc_t$	$lCH4pc_t$	$lGHGpc_t$
lY_{t-1}	**0.9640**	**0.9085**	**0.9240**	**0.9403**	$lYpc_{t-1}$	**0.8195**	−0.0499	**0.0873**	**0.4606**
lE_{t-1}	**0.9628**	**0.8512**	**0.9051**	**0.9147**	$lEpc_{t-1}$	**0.9048**	0.0471	**0.4081**	**0.5474**
lU_{t-1}	**0.9242**	**0.9692**	**0.9714**	**0.9533**	lU_{t-1}	**0.0471**	**0.4485**	**0.5392**	**0.2875**
lG_{t-1}	**0.2809**	0.0122	0.0251	0.1345	lG_{t-1}	**0.6306**	**0.0861**	**0.1217**	**0.3448**
lEG_{t-1}	**−0.3487**	**−0.5408**	**−0.5419**	**−0.4411**	lEG_{t-1}	**0.4322**	**−0.2053**	**−0.2580**	**0.0917**
lSG_{t-1}	**−0.3078**	**−0.6251**	**−0.5990**	**−0.4967**	lSG_{t-1}	**0.5721**	**−0.3180**	**−0.2824**	**0.0627**
$lECI_{t-1}$	**0.1961**	**0.1135**	**0.1413**	**0.0812**	$lECI_{t-1}$	**0.1798**	−0.0555	−0.0057	**−0.1478**
$lPOP_{t-1}$	**0.8989**	**0.9661**	**0.9658**	**0.9416**	$lPOP_{t-1}$	−0.0274	**0.4162**	**0.5011**	**0.2348**

Notes: Statistical significance levels below 5% are marked in bold.

References

Altman, D. G., & Bland, M. (1995). Absence of evidence is not evidence of absence. *BMJ, 311*(7003), 485. https://doi.org/10.1136/bmj.311.7003.485

Dean, J. F., Middelburg, J. J., Röckmann, T., Aerts, R., Blauw, L. G., Egger, M., Jetten, M. S. M., de Jong, A. E. E., & Meisel, O. H. (2018). Methane feedbacks to the global climate system in a warmer world. *Reviews of Geophysics, 56*(1), 207–250. https://doi.org/10.1002/2017RG000559

Dumitrescu, E.-I., & Hurlin, C. (2012). Testing for Granger non-causality in heterogeneous panels. *Economic Modelling, 29*, 1450–1460. https://doi.org/10.1016/j.econmod.2012.02.014

Granger, C. W. J. (1969). Investigating causal relations by econometric models and cross-spectral methods. *Econometrica, 37*(3), 424–438.

Health Effects Institute. (2018). *State of global air/2018 a special report on global exposure to air pollution and its disease burden.* https://www.stateofglobalair.org/sites/default/files/soga-2018-report.pdf.

Intergovernmental Panel on Climate Change. https://www.ipcc.ch/.

KOF Globalization Index. (2021). https://www.kof.ethz.ch/en/forecastsand-indicators/indicators/kof-globalisation-index.html.

Observatory of Economic Complexity (OEC). (2021). https://oec.world/en/resources/data/.

Persyn, D., & Westerlund, J. (2008). Error correction based cointegration tests for panel data. *Stata Journal, 8*(2), 232–241.

Pesaran, M. H. (2004). General diagnostic tests for cross-section dependence in panels. In *Cambridge working papers in economics No. 0435*. University of Cambridge. Faculty of Economics.

Pesaran, M. H. (2007). A simple panel unit root test in the presence of cross-section dependence. *Journal of Applied Econometrics, 22*(2), 265–312. https://doi.org/10.1002/jae.951

Ravishankara, A. R., Daniel, J. S., & Portmann, R. W. (2009). Nitrous oxide (N_2O): The dominant ozone-depleting substance emitted in the 21st century. *Science, 326*(5949), 123–125. https://doi.org/10.1126/science.1176985

Undark. (2018). *The weight of numbers: Air pollution and PM2.5.* https://undark.org/breathtaking/.

Westerlund, J. (2007). Testing for error correction in panel data. *Oxford Bulletin of Economics and Statistics, 69*(6), 709–748. https://doi.org/10.1111/j.1468-0084.2007.00477.x

World Bank Data (WBD). (2021). http://databank.worldbank.org/data/home.aspx.

The consequences of the obesity epidemic on food production: empirical evidence from Latin American and Caribbean countries

8

8.1 Introduction

Obesity in Latin American and Caribbean (LAC) countries has been a public health problem since the 1980s when countries went through macro-economic International Monetary Fund (IMF) programs. As a result, economies experienced economic growth, which drove up other phenomena like urbanisation and integration in the global economy. However, this economic growth has also generated an increasing trend of overweight and obesity, which ended up being considered an epidemic in LAC countries. Chapter 1 of this book describes the problem of obesity in these countries.

In the LAC region, 19% of the adult population in the LA region was obese in 2016, while this value was (9%) in 1990 (see Fig. 8.1 below).

Figure 8.1 Share of adults that are obese (%) in the LA region, between 1990 and 2016. Being overweight is defined as having a body mass index (BMI) greater than or equal to 25. BMI is a person's weight in kilograms divided by height in meters squared.
The authors created this figure with the Our World in Data (2021). Obesity. https://ourworldindata.org/obesity.

Obesity Epidemic and the Environment. https://doi.org/10.1016/B978-0-323-99339-5.00010-8

The increase of the share of adults that are obese is related to the rapid economic development, as mentioned before. When we approach major LAC economies (e.g., Argentina, Brazil, Chile, Colombia, Ecuador, Mexico, Peru and Venezuela), this trend of growth in the obesity problem can be identified (see Fig. 8.2 below).

The obesity problem reached 17% of the adult population in Argentina, while in Chile this was 17%, Mexico 16%, Venezuela 15%, Colombia 12%, Peru 10%, Brazil 10% and Ecuador 9% in 1990. However, in 2016, the obesity problem reached the following values: Mexico 30%, Chile 28%, Argentina 28%, Venezuela 26%, Colombia 22%, Brazil 22%, Peru 20% and Ecuador 20%. Exhibit 8.1 discusses the increase of overweight and obesity problems in the LAC region.

The myriad factors that explain obesity are presented in Chapter 2. Genetic factors and other individual characteristics may explain obesity. However, more importantly, obesity may be influenced by factors that are controllable and can be shaped by governments. Among all the determinants of obesity, we have highlighted economic growth, urbanisation, globalisation and poverty, as described in Chapter 3. It was shown that the obesity problem in LAC countries fits the phenomenon of nutritional transition well, and we also showed that obesity is interrelated with economic growth, urbanisation, globalisation and poverty. This means that these factors determine obesity, but obesity also contributes to economic growth, urbanisation, globalisation and poverty. Thus, there is a sustainable cycle that reinforces the obesity epidemic in this geographic region.

It was confirmed that obesity contributes to economic growth in Chapter 5. Not only is obesity positively associated with economic growth, but food production

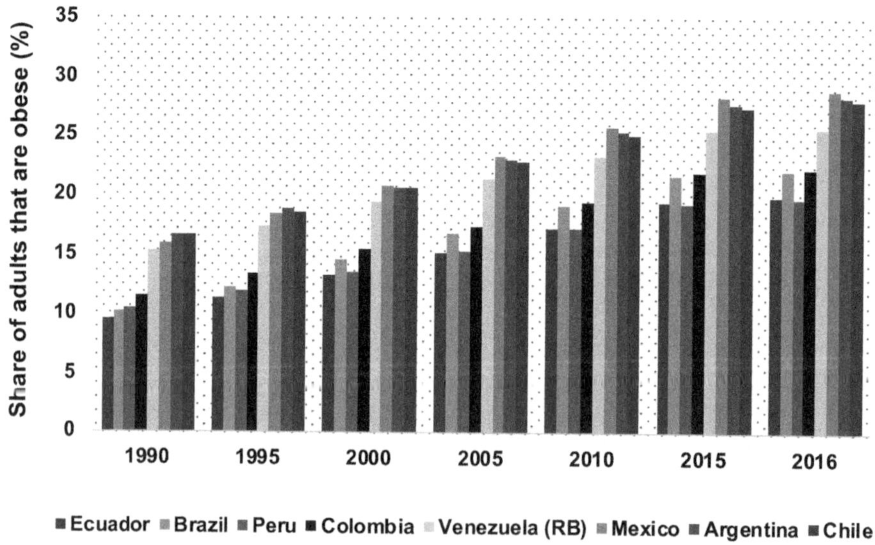

Figure 8.2 Share of adults that are obese (%) in leading LA economies, between 1990 and 2016. Being overweight is defined as having a body mass index (BMI) greater than or equal to 25. Obesity is characterised by a BMI greater than or equal to 30. BMI is a person's weight in kilograms divided by their height in meters squared.
The authors created this figure with the Our World in Data (2021). Obesity. https://ourworldindata.org/obesity.

Exhibit 8.1 Obesity and overweight problems in the LAC region

More than 300 million adults were overweight in the LAC region in 2014, and of these, more than 100 million were obese. Obesity and overweight are defined as abnormal or excessive fat accumulation that may impair health (Garcia-Garcia, 2021). In the world, 39% (2.0 billion) of the world's adult population (38% of men and 40% of women) were overweight and close to 13% (600 million, 11% of men and 15% of women) were obese in 2014. The global prevalence of obesity more than doubled between 1975 and 2014 (World Health Organization (WHO), 2020; International Food Policy Research Institute, 2016; and NCD-RisC, 2016).

Moreover, obesity has become a significant health challenge in the LAC region. Around 57% (302 million) of the adult population in the LAC region (54% men and 70% of women) are overweight, while 19% (100.8 million) are obese (15% in men and 24% in women) (Garcia-Garcia, 2021).

In other low- to middle-income countries, the overweight problem impacts 61% of women and 54% of men and obesity affects 24% of women and 15% of men. Indeed, the obesity problem is more prevalent in women than in men. In 14 LAC countries, the prevalence in females is greater than 20%. The highest prevalence of obesity problem in the adult population is found in El Salvador (33%) and Paraguay (30%) for women and in Uruguay (23%) and Chile (22%) for men (Ng et al., 2014).

Moreover, the prevalence of overweight and obesity in children in the LAC region is also high, with 16% of children impacted. It ranges from more than 12% for girls in Chile, Uruguay and Costa Rica, to less than 5% in Bolivia, Ecuador, Peru, Honduras and Guatemala. The highest prevalence of obesity in children is found in Chile (12%) and Mexico (11%) in boys and Uruguay (18%) and Costa Rica (12%) in girls (Ng et al., 2014).

Overweight and obesity are significant risks for non-communicable diseases such as cardiovascular disease (heart disease and stroke). Furthermore, the leading cause of death (30% of all causes) in the LAC region is diabetes, hypertension and chronic kidney disease (Garcia-Garcia, 2021).

also accompanies economic growth. One view of these findings is that food production feeds economic growth, but then economic growth feeds obesity.

In this chapter, we aim to close this triangle relationship and answer the research question: Does the epidemic of obesity determine the amount of food production in LAC countries? We use data from 16 LAC countries between 1990 and 2016 and estimate an ordinary least squares (OLS) model to answer this question. The main result of this estimation is the confirmation of the positive correlation between food production and obesity. In this way, the triangle relationship is empirically verifiable: food production increases with obesity, obesity increases with economic growth, economic growth is also favoured by the rise of obesity, and finally, economic growth contributes to the increase in food production. Exhibit 8.2 discusses food and nutrition security in the LAC region.

Exhibit 8.2 Food and nutrition security in the LAC region

In the world (2012–14), 805 million people are affected by hunger. It represents a fall of more than 100 million in the last 10 years and 209 million fewer than in 1990–92, according to the United Nations Food and Agriculture Organization (FAO) (2021). Indeed, in the LAC region, hunger affects more than 37 million people, which is 6% of the region's population, which is a significant advance from the 69 million (15.3%) who suffered from hunger in the 3 years between 1990 and 1992 (FAO, 2021).

Indeed, between 1990 and 2014, the LAC region, as a whole, reduced its proportion of the undernourished population by 60%, making it the only region in the world to achieve the goal of 'halving the percentage of people suffering from hunger' set for 2015 by the Millennium Development Goals (MDGs).

Therefore, the LAC region experience showed that to deal with significant challenges, in particular, extreme poverty and hunger, what is needed is a combination of economic growth, strong political commitment and decisive public action, in the form of public policies that have a high impact on the most vulnerable segments of the population (FAO, 2021).

Food production

The LAC region is one of the world's leading food-producing and exporting regions. It has enormous natural wealth, a flourishing agricultural industry and a family farming sector that is fundamental for the food security of its population. Moreover, the region produces sufficient food to meet the needs of all its inhabitants. However, the central problem concerning hunger in the region is not a lack of food but rather the problems that the poorest members of society face in gaining access to that food (FAO, 2021).

(a) **Family farming**: According to FAO (2021), farming accounts for a total share of food products consumed internally in the LAC region. On average, holdings run by small farmers represent over 80% of the total and provide between 30% and 40% of the region's agricultural GDP. Family farming also fosters employment in rural areas where pockets of poverty and food insecurity are at their worst;

(b) **Livestock**: According to FAO (2021), in the LAC region, the accelerated growth of livestock production has converted the region into the world's leading exporter of beef and poultry, accounting for about 45% of the region's agricultural GDP. However, this growth requires a sustainable approach if it is to avoid increasing pressure on the region's natural resources and environment;

(c) **Fisheries and aquaculture**: According to FAO (2021), in the LAC region, aquaculture's contribution to the regional economy has increased substantially in the last 10 years. It provides direct employment to more than 200,000 people and indirect employment to a further 500,000.

Family farming: A key sector for food security

According to FAO and Banco Interamericano de Desarrollo (BID, 2007), family farming in **Brazil** produced 67% of beans, 84% of cassava, 49% of maize and

Exhibit 8.2 Food and nutrition security in the LAC region—cont'd

52% of milk consumed in the country. In **Colombia**, family farming would appear to account for more than 30% of annual crop output, with a very high contribution of maize and beans. In **Ecuador**, this sector accounts for 64% of potato production, 85% of onion production and 70% of maize. Argentina produces 64% of pork and 33% of dairy cattle and represents 75% of the rural workforce while only occupying 17% of the sown area.

Vulnerable population

(a) **Indigenous peoples**: According to FAO (2021), in the LAC region, the indigenous peoples have contributed more than anyone to the domestication of agrobiodiversity that feeds humanity today, yet their levels of food insecurity are several times higher than those of the non-indigenous population.

(b) **Women and food security**: According to FAO (2021), in the LAC region, rural women account for more than half of food production, play an essential role in the conservation of biodiversity, and ensure food sovereignty and food security with the production of wholesome foods. However, they live under social, political and economic inequality conditions, receiving only 30% of land titles, 10% credit and 5% technical assistance. What is therefore needed is concerted public policy to promote gender parity in the region. Moreover, in the LAC region, 58 million women live in rural areas, 17 million are part of the economically active population, and 4.5 million are agricultural producers. The gender gap constitutes a real cost for society in terms of agricultural production, food security and economic growth. If women farmers enjoyed the same conditions as men, it would be possible to feed an additional 150 million people in the world;

(c) **Food prices**: According to FAO (2021), in the LAC region, higher food prices impact directly on family welfare, reducing purchasing power, and thus affecting both the quantity and the quality of food purchased by households, especially those that are poorest and most vulnerable, given that these spend between 60% and 70% of their income on food.

Rural employment

According to FAO (2021), the region's worst poverty and food insecurity pockets are found in rural areas. Regrettably, the rural labour market in the LAC region is predominantly informal and casual, without normal social security provisions. Reinforcing the institutional framework for employment is one of the keys to reducing poverty and improving income distribution in the rural areas of LAC countries. Therefore, absolute policy priority needs to generate an institutional framework for employers to protect workers' rights, encourage dignified, formal and social security and develop workers' skills through education and vocational training.

We also analyse the relationship between economic growth, food production, poverty and food production to complement our main aim. In the first case, we expect to find a positive relationship because economic growth is described as the increase in the value-added created in the country, reflecting an increase in food production.

In the second case, poverty reduction due to economic growth and development may increase food production, which would minimise food insecurity. In this way, we would find an inverse relationship between poverty and food production. However, there could be another way to hypothesise this relationship through obesity. Increasing levels of obesity are associated with increasing levels of poverty and increasing food production, so positive relationship would be expected between poverty and food production.

This chapter proceeds with the literature review; Section 8.3 describes data and the econometric method used; empirical findings are presented in Section 8.4. Finally, in Sections 8.5 and 8.6, results are discussed and the conclusion is presented.

8.2 Literature review

8.2.1 Obesity and food production

Obesity is a physical condition in the human body when the expenditure of calories is smaller than the intake, which happens by eating food. LAC countries are going through nutritional transition (Popkin, 2002), which may be described as a change in the diet patterns toward a 'Western' diet and toward an inactive lifestyle. The 'Western' diet includes high levels of saturated fats, sugar, salt, processed and refined foods, which strongly contribute to the accumulation of fat in the human body (Popkin & Gordon-Larsen, 2004).

The nutritional transition in LAC countries results from the economic growth these economies have experienced in the last 30 years. Economic growth in these countries began during the 1980s when they were subject to IMF stabilising programs. As a result, the economies soon were open to investment and international trade and became part of the global economy. At the same time, people were moving to urban centres looking for employment and better living conditions, even though many of them found 'urban' poverty.

Many people in LAC countries' 'Western' lifestyle meant the lack of physical exercise and easy access to and ingestion of unhealthy food. In large cities, women participated in the labour market and kept the family responsible for providing meals. The absence of a fresh food market nearby, together with the available short time to cook, made precooked meals and fast food easy options to feed the family.

Foreign investment has increased since the 1980s, and a large proportion of this investment was directed to agriculture and food transformation (Gereffi, 1990). Globalisation opened societies to fast-food chains and multinational supermarket chains that offer a wide range of processed foods (Fox et al., 2019). Processed, ultraprocessed and refined foods include ingredients such as flavourings and colourings to make them tasty. This food also accounts for free sugars, total fats, saturated fats, or an excess of sodium. Packing and marketing are also used to make these sorts of food appealing and make it a first-choice option when choosing food.

Indeed, between 2009 and 2014, the sales of ultraprocessed food and drink products grew by 8.3% in LAC countries, and the trend is expected to continue (PAHO, 2019, p. 2019). In 2014, it was estimated that more than 30% of total energy was obtained from eating ultraprocessed food, where more than 40% were free sugars; over 50% of total fat and saturated fat was ingested in the form of cookies, margarine and other spreadable oils and fats, sweet and savoury snacks and confectionery. Over 50% of sodium was obtained from ultraprocessed sauces and dressings. The diversity of sales in ultraprocessed food is large across LAC countries. For example, sales of these products are two or three times higher in countries like Chile and Mexico than in Colombia and Peru (PAHO, 2019, p. 2019).

Summing up, the nutritional transition in LAC countries has resulted in the generalisation of a diet based on processed and refined food, which is reflected in the increasing levels of obesity. On the other hand, overweight and obese people tend to choose these processed and refined foods produced and offered by international chains of restaurants and supermarkets. In this way, a positive relationship is expected between obesity and food production levels because of increased demand for industrialised food.

8.2.2 Economic growth and food production

Despite the rare empirical work on this topic, economic growth is intuitively associated with food production, as far as we are aware. On the one hand, the increasing population needs to be feed, which is the demand side of the market for food; on the other hand, economic growth accounts for the increase in production, which includes value created by (industrialised) agriculture and the food industry, the supply side of the market for food. Recently, Fukase and Martin (2020) found that increasing income per capita is a more critical driver of food demand than population growth. Thus, despite the different speeds of growth by demand and supply of food, what is expected to be observed is that economic growth increases income per capita and reduces poverty, which drives up the demand for food and thus food production.

From the nutritional transition perspective, economic growth comes together with urbanisation and globalisation, which forces a change in peoples' diet toward a Western type of diet. We have shown in Chapter 4 that the increase of body mass index is directly related to the urbanisation process in LAC countries, where 70% of the people were living in cities by 2000. Urbanisation recreated the food system, so food supply is carried out by supermarkets and fast-food restaurant chains, industrialised farming and food processors. Then, processed food is advertised, and demand is trapped into nice, good-looking food, with short cooking time, making easy and cheap meals. Different authors have noticed the increase of food consumption as economic and income growth, as people abandon food insecurity toward food abundance and excess caloric consumption (e.g., Gerbens-Leenes et al., 2010; Roskam et al., 2010; and Sullivan et al., 2008).

Globalisation is the other process taking place along with the economic growth of LAC countries. It is also associated with the nutrition transition (Hawkes, 2006) since international food companies affect food consumption patterns. Consequently, globalisation drives up the prevalence of obesity (Costa-Font & Mas, 2016), and it also plays a role in the nutritional transition observed in these countries.

Economic growth in LAC countries took off in the 1980s. It is associated with two relevant phenomena that explain the increased level of obesity in the region: urbanisation and globalisation. Chapter 5 showed these relations and showed that economic growth correlates with the increasing food production level.

8.2.3 Poverty and food production

Chapter 3 found that poverty is positively associated with obesity in LAC countries and negatively associated with economic growth. Poor people in massive urban centres continue to face food insecurity or access to unhealthy cheap food, which results in a large proportion of poor obese people. In Chapter 5, we found that economic growth is determined both by obesity and food production. As LAC countries experienced economic growth, they also experienced urbanisation and globalisation. These two processes motivated the change in diet, increasing the proportion of obese people and driving up food production.

The association between poverty and food production is generated by two mechanisms working together. On the one hand, the decrease in poverty could be determining food production because economic growth is helping to decrease poverty. On the other hand, the increase in the prevalence of obesity contributes to the increase of poverty and economic development, resulting in the joint increase of poverty and food production. It is possible these two mechanisms are taking place in LAC countries. As shown in Chapters 3 and 5, the speed of adjustment or the strength of the adjustment forces determines the relationship between poverty and food production.

Some empirical evidence tends to support that the increase of income in a country is associated with the rise in food consumption (e.g., Marques et al., 2018; Rask & Rask, 2011; and Skoufias et al., 2011), which implies the increase in food production, while the rise in income may be associated with a decrease in the proportion of poverty in the country. Bearing in mind the findings by Fukase and Martin (2020), it is to be expected that as poverty decreases, due to the increase in income per capita, the food production tends to increase, so food insecurity is overcome in low- to middle-income countries.

One possible nexus of poverty reduction and food production is the increase in food security: better food security contributes to economic growth and reduces poverty (Manap & Ismail, 2019). The reduction of poverty is associated with an increase in food production. The World Bank has already noticed this relation in rural areas. Stryker (1979) showed that increasing food production would reduce rural poverty. Updating these findings nowadays, one expects to find the same type of relation in broader settings, including at the country level. The following section will present the data and method this chapter will use to carry out this investigation.

8.3 Data and method

This section will be divided into two parts. The first will approach the group of countries and data/variables used in the chapter, while the second will show the method.

8.3.1 Data

This chapter will use annual data that were collected from 1991 to 2016 of 16 countries from the LAC region, i.e., **Argentina**, **Belize**, **Bolivia**, **Brazil**, **Chile**, **Colombia**, **Costa Rica**, **Dominican Republic**, **Ecuador**, **El Salvador**, **Honduras**, **Mexico, Panama**, **Paraguay**, **Peru** and **Uruguay**. The use of time-series between 1991 and 2016 is due to data availability until 2016 for the variable OBESITY for all countries selected. The variables that were chosen to perform this investigation will be shown in Table 8.1 below.

'Log' denotes variables in natural logarithms, 'Obs.' denotes the number of observations in the model, 'Std. Dev' denotes the Standard Deviation, and 'Min and Max' denote Minimum and Maximum. All variables are in natural logarithms to harmonise the interpretation of results and linearise the relationships between variables. These summary statistics were obtained from the command *sum* of **Stata 16.0.** The board below shows how to get the summary statistics of variables.

> ***How to do:***
>
> ****The summary statistics****
>
> *sum* l_logfood_prod l_logobesity l_loggdp_ppp l_logpoverty

This section presents countries from the LAC region that this chapter will focus on and the variables used. In the following subsection, we will present the method used to carry out the empirical investigation of this chapter.

8.3.2 Method

This section will show the method and strategy that this chapter will use. Ordinary least squares (OLS) will be computed and the OLS linear regression model can be represented by the following linear Eq. (8.1):

$$y = X\beta + \varepsilon, \tag{8.1}$$

where y is an $N \times 1$ vector of outcome observations, $\mathbf{X}$ is an $N \times p$ full rank fixed matrix of predictors, β is a $p \times 1$ vector of regression parameters, and ε is an $N \times 1$

Table 8.1 Description of variables and summary of statistics.

Description of variables	
Variables	**Source**
Cereal production (metric tons) per capita. This variable measures the production data on cereals related to crops harvested for dry grain only. Cereal crops harvested for hay or harvested green for food, feed, or silage and used for grazing are excluded. In this investigation, we called this variable **'FOOD_PROD.'**	World Bank Open Data (2021)
Share of adults that are obese (percent). Obesity is defined as having a body mass index (BMI) equal to or greater than 30. BMI is a person's weight in kilograms divided by their height in meters squared. In this investigation, we called this variable **'OBESITY.'**	Our World in Data (2021)
Gross domestic product (GDP) per capita based on purchasing power parity (PPP). This variable is converted to international dollars using purchasing power parity rates. An international dollar has the same purchasing power over GDP as the US dollar has in the United States. GDP at purchaser's prices is the sum of gross value added by all resident producers in the country plus any product taxes and minus any subsidies not included in the value of the products. It is calculated without deductions for the depreciation of fabricated assets or depletion and degradation of natural resources. Moreover, this variable is in the current 2017 international dollars. In this investigation, we called this variable **'GDP_PPP.'**	World Bank Open Data (2021)
The poverty gap at $5.50 a day (2011 PPP) is the mean shortfall in income or consumption from the poverty line of $5.50 a day (counting the non-poor as having zero shortfalls), expressed as a percentage of the poverty line. This measure reflects the depth of poverty as well as its incidence. In this investigation, we called this variable **'POVERTY.'**	World Bank Open Data (2021)

Summary statistics

Variables	**Obs.**	**Mean**	**Std. Dev**	**Min**	**Max**
LogFOOD_PROD	416	−1.6737	0.7839	−3.1690	0.2754
LogOBESITY	416	2.8113	0.2786	2.0918	3.3428
LogGDP_PPP	416	8.9337	0.5272	7.5976	10.1393
LogPOVERTY	316	2.5877	0.7352	−0.1053	3.9473

(Log) denotes variables in the natural logarithms.

Table 8.2 Preliminary tests.

Test	Finality
Variance inflation factor (VIF) test (Belsley et al., 1980)	This test verifies the presence of multicollinearity between the variables.
Cross-sectional dependence (CSD) test (Pesaran, 2004)	This test identifies the presence of cross-sectional dependence (CSD) in the panel data. The null hypothesis of this test is the presence of cross-section independence CD ~ N(0,1).
Levin-Lin-Chu unit-root test (LLC-test) (Levin et al., 2002)	This test verifies the presence of unit roots in the variables. The null hypothesis of the LLC-test is that all the panels contain a unit root.
Panel unit root test (CIPS) test (Pesaran, 2007)	This test verifies the presence of unit roots in the variables. The panel unit root test (CIPS) null hypothesis is that all series have a unit root.
Bias-corrected LM-based test (Born & Breitung, 2015)	This test verifies the presence of serial correlation in the fixed-effects panel model. The null hypothesis of this test is the non-presence of autocorrelation up to the second order.

vector of errors. Here, N is the total number of observations, and p is the number of predictors (including the intercept) in the regression equation. Indeed, we assume that the errors are normally distributed and $E(\varepsilon) = 0$, and $var(\varepsilon) = \Phi = dia\left[\sigma_i^2\right]$. Therefore, under the assumption of homoscedasticity, the elements in the error vector have constant variance σ^2, and then $\Phi = \sigma^2 I_N$, where I_N is an identity matrix of order N. The OLS estimation of regression coefficients is $\widehat{\beta} = (X'X)^{-1} X'y$, so

$$var\left(\widehat{\beta}\right) = \sum = (X'X)^{-1} X'\Phi X (X'X)^{-1} \tag{8.2}$$

However, before the realisation of the OLS model regression, it is necessary to verify the proprieties of the variables, which includes the presence of multicollinearity and cross-sectional dependence (CSD) in the panel data; order of integration of the variables; the presence of individual effects in the models; and the presence of serial correlation in the fixed-effects panel model. Table 8.2 below shows the tests that need to be applied before the OLS model regression.

After the OLS model regression, it is necessary to apply some postestimation tests (see Table 8.3 below).

These tests need to be applied to verify the presence of heteroscedasticity in the model. The estimation and testing procedures are carried out using **Stata 16.0.**

Table 8.3 Postestimation tests.

Test	Finality
The Wald test (Agresti, 1990)	This test verifies the global significance of the estimated models. The null hypothesis of the Wald test is that all the coefficients are equal to zero.
Breusch-Pagan/Cook-Weisberg test for heteroskedasticity (Breusch-Pagan, 1979; Cook-Weisberg, 1983)	This test identifies the presence of heteroscedasticity. Therefore, the null hypothesis of this test is the presence of homoscedasticity.

8.4 Empirical results

As mentioned before, this section will approach the empirical results of our investigation. Therefore, we will show the results from preliminary tests, OLS model regression and postestimation tests. To find the level of multicollinearity between the variables in our panel data, the VIF tests developed by Belsley et al. (1980) were calculated. Table 8.4 below shows the outcomes from the VIF test.

Table 8.4 VIF test.

Variables	VIF	1/VIF	Mean VIF
LogFOOD_PROD			
LogOBESITY	2.52	0.3974	2.67
LogGDP_PPP	3.25	0.3074	
LogPOVERTY	2.24	0.4466	

(Log) denotes variables in the natural logarithms.

The results from the VIF test show that the values are lower than the usually accepted benchmark of 10 in the case of the VIF values and six in the case of the mean VIF values. The results of the VIF test were obtained from the command *estat vif* in **Stata 16.0**. The board below shows how to carry out and obtain the results from the VIF test.

How to do:

**** The Variance Inflation Factor test****

reg l_logfood_prod l_logobesity l_loggdp_ppp l_logpoverty
estat vif

Table 8.5 Pesaran CD test.

Variables	CD-test	*P*-value	
LogFOOD_PROD	2.16	0.056	*
LogOBESITY	55.81	0.000	***
LogGDP_PPP	54.35	0.000	***
LogPOVERTY		N.A	

*** and * denote statistically significant at the 1% and 10% levels, respectively; (Log) denotes variables in the natural logarithms; NA denotes not available.

After realising the VIF test, it is necessary to find cross-sectional dependence (CSD) in the panel data, and the Pesaran CD-test developed by Pesaran (2004) was used in this investigation. The null hypothesis of this test is the non-presence of cross-sectional dependence CD $\sim$ N (0,1) for $N \rightarrow \infty$, and that T is sufficiently large. Table 8.5 below shows the results from the Pesaran CD-test.

The results of the CSD test show the presence of cross-section dependence for the variables LogFOOD_PROD, LogOBESITY and LogGDP_PPP. However, the variable LogPOVERTY could not be computed by the CSD test because the test requires strongly balanced data. Therefore, cross-sectional dependence can be a sign that the countries selected in our study share the same characteristics and shocks as mentioned in Fuinhas et al. (2017). The results of the Pesaran CD test were obtained from the command *xtcd* in **Stata 16.0**. The board below shows how to carry out and obtain the results from the Pesaran CD test.

How to do:

****The Pesaran CD-test****

xtcd l_logfood_prod l_logobesity l_loggdp_ppp l_logpoverty

In the presence of CSD, it is necessary to verify the order of integration of the variables. To this end, the panel unit root tests, such as the LLC test developed by Levin et al. (2002) and the CIPS test developed by Pesaran (2007), were calculated. Table 8.6 below shows the results from the unit root tests.

The results from the LLC-test obtained indicate that the variable LogGDP_PPP is on the borderline between I(0) and I(1) of the order of integration. In contrast, the variables LogFOOD_PROD and LogOBESITY are stationary. However, the variable LogPOVERTY could not be computed by the LLC test because the test requires strongly balanced data. Moreover, the results from the CIPS test obtained indicate that the variable LogFOOD_PROD is stationary, while the variables LogOBESITY and LogGDP_PPP are on the borderline between I(0) and I(1) of the order of integration. The CIPS test could not compute the variable LogPOVERTY because this test

Table 8.6 Unit root tests.

	Levin-Lin-Chu unit-root test (LLC-test)					Panel unit root test (CIPS) (Zt-bar)				
	Without trend			With trend		Without trend			With trend	
Variables	**Lags**	**Adjusted t**		**Adjusted t**		**Lags**	**Zt-bar**		**Zt-bar**	
LogFOOD_PROD	1	−2.1962	**	−2.1727	**	1	−3.262	***	−2.154	**
LogOBESITY	1	−18.6970	***	−3.6607	***	1	−0.445		−2.506	***
LogGDP_PPP	1	0.4364		−2.0103	**	1	−0.671		−1.401	*
LogPOVERTY	N.A					NA				

***, ** and * denote statistically significant at the 1%, 5% and 10% levels, respectively; (Log) denotes variables in the natural logarithms; N.A denotes not available.

requires strongly balanced data. The results of the LLC test and CIPS test were obtained from the commands *xtunitroot* and *multipurt* in **Stata 16.0**. The board below shows how to carry out and obtain the results from the LLC and CIPS tests.

How to do:

**** The Levin-Lin-Chu unit-root test****

```
xtunitroot llc l_logfood_prod
xtunitroot llc l_logfood_prod, trend

xtunitroot llc l_logobesity
xtunitroot llc l_logobesity, trend

xtunitroot llc l_loggdp_ppp
xtunitroot llc l_loggdp_ppp, trend

xtunitroot llc l_logpoverty
xtunitroot llc l_logpoverty, trend
```

****The CIPS-test****

```
multipurt l_logfood_prod l_logobesity l_loggdp_ppp l_logpoverty,lags(1)
```

After identifying the order of integration between the variables of our model, it is necessary to check serial correlation in the fixed-effects panel model. To this end, the bias-corrected LM-based test developed by Born and Breitung (2015) was computed. The null hypothesis of this test is the non-presence of autocorrelation up to the second order. Table 8.7 below shows the results from the bias-corrected LM-based test.

The results from the bias-correct LM-based test indicate the presence autocorrelation up to the second order in the fixed-effects panel model, where the null hypothesis can be rejected. However, the variable LogPOVERTY could not be computed by the bias-corrected LM-based test because this test requires strongly balanced data. The results of the bias-corrected LM-based test were obtained from the command *xtqptest* in

Table 8.7 Bias-corrected LM-based test.

Variables	LM (k)-stat	
LogFOOD_PROD	2.59	**
LogOBESITY	10.34	***
LogGDP_PPP	8.27	***
LogPOVERTY	N.A	

*** and ** denote statistically significant at the 1% and 5% levels, respectively; (Log) denotes variables in the natural logarithms; NA denotes not available.

Stata 16.0. The board below shows how to carry out and obtain the results from this test.

How to do:

**** The Bias-corrected LM-based test ****

```
xtqptest l_logfood_prod l_logobesity l_loggdp_ppp l_logpoverty, order(1)
```

After carrying out the preliminary tests, we can conduct the OLS model regression to assess the impact of obesity on food production in LAC and the postestimation tests (e.g., the Wald test and Breusch-Pagan/Cook-Weisberg test for heteroscedasticity). The null hypothesis of the Wald test is that all the coefficients are equal to zero. In contrast, the null hypothesis of the Breusch-Pagan/Cook-Weisberg test is the presence of homoscedasticity. Table 8.8 below shows the results of OLS model estimations and the postestimation tests to confirm the presence of heteroskedasticity in the models.

OLS and OLS robust estimations indicate that the obesity epidemic increases food production by 1.1501 and economic growth by 0.2867, while poverty reduces food production by −0.2556. The postestimation tests (e.g., the Wald test and Breusch-Pagan/Cook-Weisberg test) indicated that the time-fixed effects are needed, and heteroscedasticity is present in the models. The result from the postestimation tests is an indicator that the estimations that this investigation use are adequate.

Table 8.8 Pooled OLS model estimations and postestimation tests.

Independent variables	**Dependent variable (LogFOOD_PROD)**		
	OLS		**OLS robust**
Trend	−0.0483	***	***
LogOBESITY	1.1501	***	***
LogGDP_PPP	0.2867	**	***
LogPOVERTY	−0.2556	***	***
Constant	−6.0658	***	***
Obs	316		316
Postestimation tests for Pooled OLS and OLS robust			
Statistics	The Wald test	Breusch-Pagan/Cook-Weisberg test for heteroskedasticity	The Wald test
	F(3,311) = 36.37***	chi2(1) = 23.88***	F(3,311) = 52.89 ***

*** and ** denote statistically significant at the 1% and 5% levels, respectively; (Log) denotes variables in the natural logarithms.

The results of OLS model regression were obtained from the command *reg* and *reg* with option *robust* in **Stata 16.0**. Moreover, the results from the postestimation test, the Wald test, were obtained from the command *testparm*, while the Breusch-Pagan/Cook-Weisberg test was obtained from the command *hettest* in **Stata 16.0**. The board below shows how to carry out and obtain the results from the OLS model regressions and the results from the postestimation tests.

How to do:

****Pooled OLS****

reg l_logfood_prod trend l_logobesity l_loggdp_ppp l_logpoverty

****The Wald test****

testparm l_logfood_prod l_logobesity l_loggdp_ppp l_logpoverty

****Breusch-Pagan/Cook-Weisberg test****

reg l_logfood_prod trend l_logobesity l_loggdp_ppp l_logpoverty

hettest

****Pooled OLS Robust****

reg l_logfood_prod trend l_logobesity l_loggdp_ppp l_logpoverty, robust

****The Wald test****

testparm l_logfood_prod l_logobesity l_loggdp_ppp l_logpoverty

Fig. 8.3 below summarises the impact of independent variables on dependent ones. This figure was based on the results from the OLS model regressions.

In this section, we showed the empirical results of this investigation. The following section will show the discussion of the empirical results.

8.5 Discussion

The capacity of economic growth to increase food production could be related to the process of income transition from lower to high income caused by economic development. Therefore, this process encourages higher levels of food consumption, often unhealthy and with high energy-dense animal sources (e.g., Gerbens-Leenes et al., 2010;

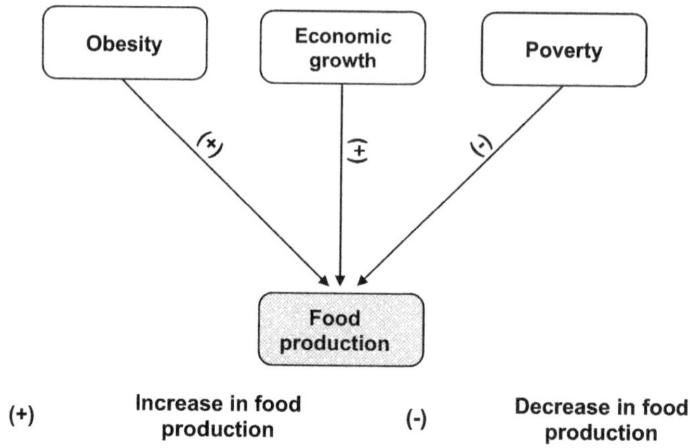

Figure 8.3 Summarises the impact of independent variables on dependent ones. Figure created by the authors.

and Roskam et al., 2010), as highlighted by the nutritional transition approach (e.g., Popkin, 2002; and Popkin & Gordon-Larsen, 2004).

Therefore, the increase in food consumption caused by income growth contributes to overweight and obesity levels increase. In LAC countries, the evidence that income growth increases the obesity problem was found in Chapter 3, where economic growth increases the obesity epidemic by 0.0041. This phenomenon occurred in LAC countries because the region registered high economic growth in the last 20 years caused by economic reforms and commodities boom, as mentioned earlier. Moreover, this rapid economic development reduced poverty in the region between 2000 and 2012 and influenced urbanisation and globalisation.

According to Costa-Font and Mas (2016), higher economic development reduces economic inequality and poverty. This reduction also influences the increase in overweight and obesity. Therefore, in countries with high economic growth but low income, food insecurity underscores the risk of overweight and obesity due to the high food consumption in the following stage of higher income. Sullivan et al. (2008) add that an excess of caloric consumption follows the nutritional deficits caused by food insecurity. Therefore, economic growth and poverty reduction could be associated with overweight and obesity increase, both with its contemporary and future effects on food consumption and production.

Indeed, the evidence that the reduction of poverty caused by economic development increases the obesity problem in LAC countries was found in Chapter 3, where the reduction of poverty increases the obesity epidemic by 0.0014. In this investigation, it was expected that reducing poverty caused by economic development would increase food production in LAC countries. However, we found that poverty decreases food production. This reduction could be related to the permanence of poverty in several countries of the LAC region. The process of economic development caused by rapid economic growth has not been enough to reduce extreme poverty in the lowest strata of society. That is, the distribution of income was not equitable in most countries from the region. The economies in the LAC region did not create effective policies for reducing poverty during the economic boom from 2002 to 2012. Another factor related to reducing food production by poverty is the end of the 'commodities prices' cycle from 2013. Many countries from the region saw a decrease in their commodity rents and a deterioration in their fiscal situation.

Consequently, many countries reduced the fund transfers for income transfer programs and cut the number of people served by social programs to ensure fiscal equilibrium. Another consequence of the end in the cycle of 'commodities prices' was a reduction in public investment and consequently reduced jobs creation. Therefore, all these factors reversed the downward trend in poverty and consequently reflected food consumption and production.

Moreover, other factors have also contributed to obesity in LAC countries and food production, such as urbanisation and economic and social globalisation. In Chapter 4, it was found that the process of urbanisation increases obesity by 0.0739, while the economic and social globalisation increases this problem by 0.0534 and 0.1112 in LAC countries, respectively.

Therefore, the increase in the obesity epidemic caused by economic development, poverty reduction, urbanisation and globalisation will positively affect food consumption and production (Koengkan et al., 2021, pp. 1−7; and Koengkan & Fuinhas, 2021). The same authors also add that the overweight or obesity epidemic increases the consumption of processed foods from multinational food corporations, fast-food chains and multinational supermarket chains and food production on farms. This will impact the multinational food corporations and farm production positively to attend to the processed foods demand. Furthermore, this increase will impact economic activity. The evidence that food production increases economic growth in LAC countries was found in Chapter 5. This section showed the possible explanations for the impact of the obesity epidemic, economic development and poverty on food production. The following section will give the main conclusions of this chapter.

8.6 Conclusion

This chapter approached the impact of the prevalence of obesity on food production. This chapter used data for 16 countries from the LAC region, i.e., Argentina, Belize, Bolivia, Brazil, Chile, Colombia, Costa Rica, Dominican Republic, Ecuador, El Salvador, Honduras, Mexico, Panama, Paraguay, Peru and Uruguay for the period between 1991 and 2016. This study used the method of the ordinary least-squares (OLS) model.

The results from the preliminary tests, such as the VIF-test, indicated low multicollinearity between the variables. The values are lower than the usually accepted benchmark of 10 in the VIF values and six in the mean VIF values. The results of the CSD-test showed the presence of cross-section dependence for the variables LogFOOD_PROD, LogOBESITY and LogGDP_PPP. However, the variable

LogPOVERTY could not be computed by the CSD-test because the test requires strongly balanced data. Therefore, cross-sectional dependence can signify that the countries selected in our study share the same characteristics and shocks.

The results from the LLC-test obtained indicated that the variable LogGDP_PPP is on the borderline between I(0) and I(1) of the order of integration. In contrast, the variables LogFOOD_PROD and LogOBESITY are stationary. However, the variable LogPOVERTY could not be computed by the LLC-test because the test requires strongly balanced data. Moreover, the results from the CIPS-test obtained indicated that the variables LogFOOD_PROD is stationary, while the variables LogOBESITY and LogGDP_PPP are on the borderline between I(0) and I(1) of the order of integration. The CIPS-test could not compute the variable LogPOVERTY because this test requires strongly balanced data. The results from the bias-correct LM-based test indicate the presence of autocorrelation up to the second-order in the fixed-effects panel model, where the null hypothesis can be rejected. However, the variable LogPOVERTY could not be computed by the bias-corrected LM-based test because this test requires strongly balanced data.

OLS and OLS robust estimations indicated that the obesity epidemic increases food production by 1.1501 and economic growth by 0.2867, while poverty reduces food production by −0.2556. The post-estimation tests (e.g., the Wald test and Breusch-Pagan/Cook-Weisberg test) indicated that the time fixed-effects are needed in the model and heteroscedasticity is present. The result from the post-estimation tests is an indicator that the estimations used in this investigation are adequate.

Therefore, the main finding was the role of obesity as a determinant factor of food production, reflecting the nutritional transition and economic growth happening in these countries. As a result, people tend to choose processed and refined food, feeding the obesity pandemic produced by the food industry and processors. This increase in food production contributes to economic growth and the rise of income per capita.

The complementary finding presented in this chapter was the negative relationship between poverty and food production, i.e., as poverty reduces, food production increases. This finding supports the idea that food production ensures food security and reduces the prevalence of poverty. As the share of people facing poverty reduces, more people have enough income to buy food consistently. This increased demand for food corresponds to an increase in food production. This chapter complements previous chapters and helps in understanding the effects of pandemic obesity on food production in LAC countries.

References

Agresti, A. (1990). *Categorical data analysis*. John Wiley and Sons. ISBN 0-471-36093-7.

Belsley, D. A., Kuh, E., & Welsch, R. E. (1980). *Regression diagnostics: Identifying influential data and sources of collinearity*. Wiley. https://doi.org/10.1002/0471725153

Born, B., & Breitung, J. (2015). Testing for serial correlation in fixed-effects panel data models. *Econometric Reviews, 35*(7), 1290–1316. https://doi.org/10.1080/07474938.2014.976524

Breusch, T. S., & Pagan, A. R. (1979). A simple test for heteroscedasticity and random coefficient variation. *Econometrica, 47*, 1287–1294. https://doi.org/10.2307/1911963

Cook, R. D., & Weisberg, S. (1983). Diagnostics for heteroscedasticity in regression. *Biometrika, 70*, 1–10. https://doi.org/10.2307/2335938

Costa-Font, J., & Mas, N. (2016). 'Globesity'? The effects of globalisation on obesity and caloric intake. *Food Policy, 64*, 121–132. https://doi.org/10.1016/j.foodpol.2016.10.001

FAO and BID. (2007). *Políticas para la agricultura familiar en américa latina y el caribe.* https://idbdocs.iadb.org/wsdocs/getdocument.aspx?docnum=35555762.

Food and Agriculture Organization of the United Nations (FAO). (2021). *FAO regional office for Latin America and the Caribbean: Food and nutrition security in Latin America and the Caribbean.* http://www.fao.org/americas/priorities/seguridad-alimentaria/en/.

Fox, A., Feng, W., & Asal, V. (2019). What is driving global obesity trends? Globalisation or 'modernisation'? *Globalization and Health, 15*(32), 1–16. https://doi.org/10.1186/s12992-019-0457-y

Fuinhas, J. A., Marques, A. C., & Koengkan, M. (2017). Are renewable energy policies upsetting carbon dioxide emissions? The case of Latin America countries. *Environmental Science and Pollution Research, 24*(17), 15044–15054. https://doi.org/10.1007/s11356-017-9109-z

Fukase, E., & Martin, W. (2020). Economic growth, convergence, and world food demand and supply. *World Development, 132*, 104954. https://doi.org/10.1016/j.worlddev.2020.104954

Garcia-Garcia, G. (2021). *Obesity and overweight populations in Latin America.* The Lancet Kidney Campaign. https://www.thelancet.com/campaigns/kidney/updates/obesity-and-overweight-populations-in-latin-america.

Gerbens-Leenes, P. W., Nonhebel, S., & Krol, M. S. (2010). Food consumption patterns and economic growth. Increasing affluence and the use of natural resources. *Appetite, 55*(3), 597–608. https://doi.org/10.1016/j.appet.2010.09.013

Gereffi, G. (1990). Chapter 1: Paths of industrialisation: An overview. In G. Gereffi, & D. L. Wyman (Eds.), *Manufacturing Miracles*. Princeton University Press.

Hawkes, C. (2006). Uneven dietary development: linking the policies and processes of globalization with the nutrition transition, obesity and diet-related chronic diseases. *Globalization and Health,* 2(4), 1–18. https://doi.org/10.1186/1744-8603-2-4

International Food Policy Research Institute. (2016). *Global nutrition report 2016: From promise to impact: Ending malnutrition by 2030.* https://www.ifpri.org/publication/global-nutrition-report-2016-promise-impact-ending-malnutrition-2030.

Koengkan, M., & Fuinhas, J. A. (2021). Does the overweight epidemic cause energy consumption? A piece of empirical evidence from the European region. *Energy, 236*(1), 119297. https://doi.org/10.1016/j.energy.2020.119297

Koengkan, M., Fuinhas, J. A., & Fuinhas, C. (2021). *Does urbanisation process increase the overweight epidemic? The case of Latin America and the Caribbean region.* SSRN. https://doi.org/10.2139/ssrn.3826196

Levin, A., Lin, C.-F., & Chu, C.-S. J. (2002). Unit root tests in panel data: Asymptotic and finite-sample properties. *Journal of Econometrics, 108*, 1–24. https://doi.org/10.1016/S0304-4076(01)00098-7

Manap, N. M. A., & Ismail, N. W. (2019). Food security and economic growth. *International Journal of Modern Trends in Social Sciences,* 2(8), 108–118. https://doi.org/10.35631/IJMTSS.280011

Marques, A. C., Fuinhas, J. A., & Pais, D. F. (2018). Economic growth, sustainable development and food consumption: Evidence across different income groups of countries. *Journal of Cleaner Production, 196*, 245–258.

NCD risk factor collaboration (NCD-RisC). (2016). Trends in the adult body-mass index in 200 countries from 1975 to 2014: A pooled analysis of 1698 population-based measurement studies with 19·2 million participants. *Lancet, 387*, 377–396. https://www.thelancet.com/journals/lancet/article/PIIS0140-6736(16)30054-X/fulltext.

Ng, M., Fleming, T., Robinson, M., Thomsom, B., & Graetz, N. (2014). Global, regional, and national prevalence of overweight and obesity in children and adults during 1980–2013: A systematic analysis for the global burden of disease study 2013. *Lancet, 384*, 766–781. https://doi.org/10.1016/S0140-6736(14)60460-8

Our World in Data. (2021). *Obesity*. https://ourworldindata.org/obesity.

Pan American Health Organization (PAHO). (2019). *Ultra-processed food and drink products in Latin America: Sales, sources, nutrient profiles, and policy implications* (p. 2019). PAHO.

Pesaran, M. H. (2004). *General diagnostic tests for cross-section dependence in panels*. The University of Cambridge, Faculty of Economics. https://doi.org/10.17863/CAM.5113. Cambridge Working Papers in Economics, n. 0435.

Pesaran, M. H. (2007). A simple panel unit root test in the presence of cross-section dependence. *Journal of Applied Econometrics, 22*(2), 256–312. https://doi.org/10.1002/jae.951

Popkin, B. M. (2002). An overview on the nutrition transition and its health implications: The bellagio meeting. *Public Health Nutrition, 5*, 93–103.

Popkin, B. M., & Gordon-Larsen, P. (2004). The nutrition transition: Worldwide obesity dynamics. *International Journal of Obesity, 28*, S2–S9.

Rask, K. J., & Rask, N. (2011). Economic development and food production-consumption balance: A growing global challenge. *Food Policy, 36*, 186e196. https://doi.org/10.1016/j.foodpol.2010.11.015

Roskam, A. J., Kunst, A. E., Van Oyen, H., Demarest, S., Klumbiene, J., Regidor, E., Helmert, U., Jusot, F., Dzurova, D., & Mackenbach, J. P. (2010). Comparative appraisal of educational inequalities in overweight and obesity among adults in 19 European countries. *International Journal of Epidemiology, 39*(2), 392–404. https://doi.org/10.1093/ije/dyp329

Skoufias, E., Di Maro, V., Gonzalez-Cossío, T., & Ramirez, S. R. (2011). Food quality, calories and household income. *Applied Economics, 43*, 4331–e4342. https://doi.org/10.1080/00036846.2010.491454

Stryker, R. E. (1979). The World Bank and agricultural development: Food production and rural poverty. *World Development, 7*(3), 325–336. https://doi.org/10.1016/0305-750x(79)90060-3

Sullivan, M. C., Hawes, K., Winchester, S. B., & Miller, R. (2008). Developmental origins theory from prematurity to adult disease. *Journal of Obstetric, Gynecologic, & Neonatal Nursing, 37*(2), 158–164. https://doi.org/10.1111/j.1552-6909.2008.00216.x

World Health Organization (WHO). (2020). *Obesity and overweight. Key facts*. http://www.who.int/mediacentre/factsheets/fs311/en/#.

World Bank open data. (2021). http://www.worldbank.org/.

9 Does the obesity epidemic increase the consumption of fossil fuels in Latin America and Caribbean countries?

9.1 Introduction

Fast economic growth has been experienced since the 1980–90 in Latin American and Caribbean (LAC) countries. The gross domestic product (GDP) per capita grew 12 times between 1970 and 2016 (see Chapter 4). The exponential increase in overweight and obesity prevalence has paralleled economic growth, and adult obesity has tripled since 1975 (UN, 2019). Additionally, to accompany the economic growth, there is the need to increase the consumption of energy, either fossil fuel energy or renewable energy (Humphrey & Stanislaw, 1979).

Economic growth in LAC countries accelerated after the structural and stabilisation programmes imposed by the International Monetary Fund (IMF). These adjustment programmes are neoliberal policies that consisted mainly of the complete opening of their economies to international trade and capital, deregulation of the economy, privatisation, reduction of public expenditures, creation of appropriate conditions for foreign investment and the reduction of the role of the state in the economy. Moreover, the 'commodities boom' that occurred between the beginning of the 2000s and end of 2014 also accelerated the process of openness as well as the economic growth in the region (e.g. Fuinhas et al., 2021, pp. 19–234; Koengkan, Fuinhas, & Marques, 2020; and Koengkan & Fuinhas, 2020). Fig. 9.1 shows the GDP per capita for the period 1990–2018 in LAC countries (excluding high-income countries), which registered a growth rate of 47.5% between 1990 and 2018.

This rapid economic growth between 1990 and 2014 impacted energy use. In 1990, the energy use (kg of oil equivalent per capita) was 1053,23 and in 2014 reached a value of 1358,20 as shown in Fig. 9.2 below.

Energy use in the LAC region, according to Balza et al. (2016, pp. 1–39), is 220% higher than in the early 1970s and represented an average annual growth rate of 2.8%, where the total energy used increased from 190 million tonnes of oil equivalent (MTOE) in 1971 to 610 MTOE in 2013. Of this increase, the industrial and transport sector contributed together with more than 302 MTOE. To get a sense of this, according to the explanations from Balza et al. (2016, pp. 1–39), since the 1970s, the transport sector had a yearly increase of around 3.5% in energy use, while the industrial sector had a rise of 3% in the same period.

Indeed, 89% of this energy use in the LAC region in 1970 emanates from fossil fuel energy sources (e.g. oil, coal and gas), and 11% comes from renewable energy sources (e.g. hydroelectric). In 2014, the consumption of fossil fuels reached a value of 74% of

Obesity Epidemic and the Environment. https://doi.org/10.1016/B978-0-323-99339-5.00004-2

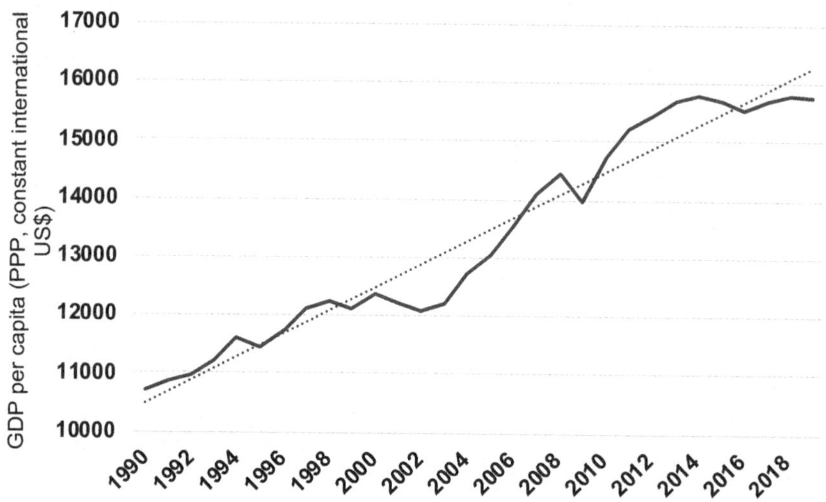

Figure 9.1 Gross domestic product (GDP) per capita (PPP, constant international $) for the Latin American and Caribbean (LAC) region, between 1990 and 2018.
The authors created this figure with the World Bank Open Data (2021). http://www.worldbank.org/.

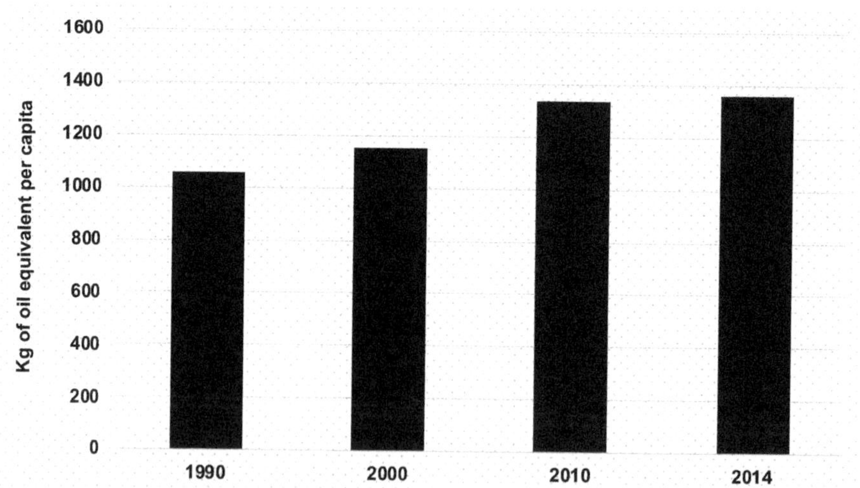

Figure 9.2 Energy use (kg of oil equivalent per capita) in the Latin American and Caribbean (LAC) region, between 1990 and 2014.
The authors created this figure with the World Bank Open Data. (2021). http://www.worldbank.org/.

total energy use in the region. This reduction is related to the increased share of the renewable energy sources (e.g. hydropower, nuclear, geothermal, biomass, waste, solar and wind) that had a participation of 25.04% in the total energy use (Fuinhas et al., 2021, pp. 19–234).

The high share of fossil fuel energy sources in total energy use in the LAC region clarifies the importance of initiatives in the energy sector to reduce non-renewable energy and emissions. Therefore, energy planning must consider a scenario of climate change, where additional efforts directed to limiting the emissions from the energy sector are necessary, especially in developing countries such as the LAC countries, where there are expectations of an increase in the energy demand in the coming decades (Fuinhas et al., 2021, pp. 19–234) as it is estimated that energy use in the LAC region will continue to grow steadily. This growth will be accompanying economic growth and a rise in the middle classes in the countries from the region. Therefore, the total energy use is projected to expand by more than 81.2% through 2040 at an average annual rate of 2.2%, reaching over 1538 MTOE by the end of the outlook period (see Table 9.1 below).

In the LAC region, the process of energy transition bloomed mainly from the 1970s or precisely from 1973 in Brazil and Paraguay with the development of the Itaípu Treaty that resulted in the construction of the large hydropower Itaípu dam during 1974–84 (Fuinhas et al., 2021, pp. 19–234). This dam sits on the Parana River, approximately 14 km from the international bridge connecting Ciudad del Este in Paraguay and Foz de Iguaçu in Brazil. The project proposal included 14 units of 765 MW generators to 10.7 gigawatt (GW)—which has now been expanded to 14 GW. Moreover, the cost of the project was estimated at 100 million USD, which will be owned equally by Centrais Elétricas Brasileiras (ELECTROBRAS) and Administración Nacional de Electricidad (ANDE) (IEA, 2020).

Their construction was made to attend to the great energy demand caused by the Brazilian Miracle (Brazilian Portuguese: *milagre econômico Brasileiro*), a period of

Table 9.1 Energy use forecast to 2040.

Country and region	2013	2040	Growth	CAGR
Argentina	81	123	52.6%	1.6%
Brazil	294	577	96.6%	2.5%
Chile	39	99	154.7%	3.5%
Colombia	32	67	110.3%	2.8%
Mexico	191	400	109.2%	2.8%
Venezuela	69	104	50.7%	1.5%
Other countries	144	169	17.3%	0.6%
LAC region	849	1538	81.2%	2.2%

Energy use (MTOE) forecast for 2040 in the LAC region. CAGR denotes 'compound annual growth rate'.
This figure was created by the authors and was based on the data from Fuinhas, J.A., Koengkan, M., & Santiago, R., (2021). Physical capital development and energy transition in Latin America and the Caribbean. Elsevier. https://doi.org/10.1016/C2020-0-01491-X and Balza, L.H., Espinasa, R., & Serebrisky, T., (2016). Lights on? Energy needs in Latin America and the Caribbean to 2040. Inter-American Development Bank. https://publications.iadb.org/en/publication/17053/lights-energy-needs-latin-america-and-caribbean-2040.

extraordinary economic growth and development in Brazil from 1968 to 1973. Indeed, during this time, the average annual GDP growth was close to 10% during the rule of the Brazilian military government (e.g. Fuinhas et al., 2021, pp. 19–234; Veloso et al., 2008).

Other energy transition initiatives arose in the region in the same decade, such as the Proalcool programme in 1975 in Brazil after the first oil shock in 1973. This programme is a combination of policy instruments that evolve and are mainly used to substitute for imported petroleum and to address the needs of both supply and demand sides (e.g. Fuinhas et al., 2021, pp. 19–234; Gielen et al., 2019; Koengkan & Fuinhas, 2020; and Solomon & Krishna, 2011). Additionally, this programme drives biomass-based ethanol demand, but the sector's long-term success continues to be impacted by economic cycles and changing government priorities (Gielen et al., 2019). Moreover, in that same year in Mexico, the Public Electricity Service Law (Ley del Servicio Público de Energía Eléctrica) was published, which deals with all aspects of renewable energy for public service, including generation, transmission, distribution, transformation and supply of wind, solar, solar photovoltaic, solar thermal and marine energy sources. This legislation established that only the public electricity service is an exclusive competence of the Mexican state, with the only the national electricity companies providing electricity on a least-cost basis (IEA, 2020). In the 1980s, these initiatives spread across the LAC region, where in 1985, Guatemala established the decree-Law 17 of 1985, regulated by the Governmental Accord 420-1985. This law aimed to increase the use of biofuels in the transport sector while reducing the imports of fossil fuels. Moreover, this legislation established a blending mandate for bioethanol of 5% (IEA, 2020).

However, in the 1990s, energy transition initiatives turned to the self-production of renewable energy. The first country that established legislation authorising the self-production of renewable energy was Costa Rica in 1990. The law authorising the self-production of renewable energy (*Ley que Autoriza la Generación eléctrica autónoma o paralela*) regulated the utility-scale private sector projects in Costa Rica (IEA, 2020). In the 2002, Uruguay established the law of national interest in bioenergy, which declared bioenergy to be of national interest for Uruguay. Indeed, this declaration was aligned with the National Energy Policy (IEA, 2020). In 2018, Argentina established the RenovAR3 auction round called 'MiniRen' for renewable power capacity from wind, solar and hydropower energy sources procurement. Around 400 megawatts (MW) of renewable energy capacity was opened for competition (IEA, 2020).

All these initiatives made the LAC region the only region in the world with a large share of renewable sources in the total energy consumption, where in 2014, the share of renewable energy (e.g. hydropower, nuclear, geothermal, biomass, waste, solar and wind) in the energy consumption in the LAC region was 25%. This is a sizable proportion when compared with the world average of 14.35% (Fuinhas et al., 2021, pp. 19–234). Indeed, from this 25% of renewable energy sources, 21% comes from hydroelectricity (e.g. small and large hydro dams). However, the share of energy consumption from hydropower has declined since the end of the 1990s due to the development of other energy sources from natural gas and new renewable energy sources (e.g. geothermal, biomass, waste, solar and wind) (Flavin et al., 2014). Exhibit 9.1 discusses the increase in renewable energy consumption and investment in the LAC region.

Exhibit 9.1 Renewable energy consumption and investment in the LAC region

Energy consumption from new renewable energy sources has undergone rapid growth since the end of the 1990s. However, in 2018, the consumption of this kind of energy comprised only 5.03% of total energy consumption in the LAC region, where the other renewables that include geothermal, biomass and waste had a share of 2.51%, wind 2.12% and solar 0.40% (see Fig. 9.3 below).

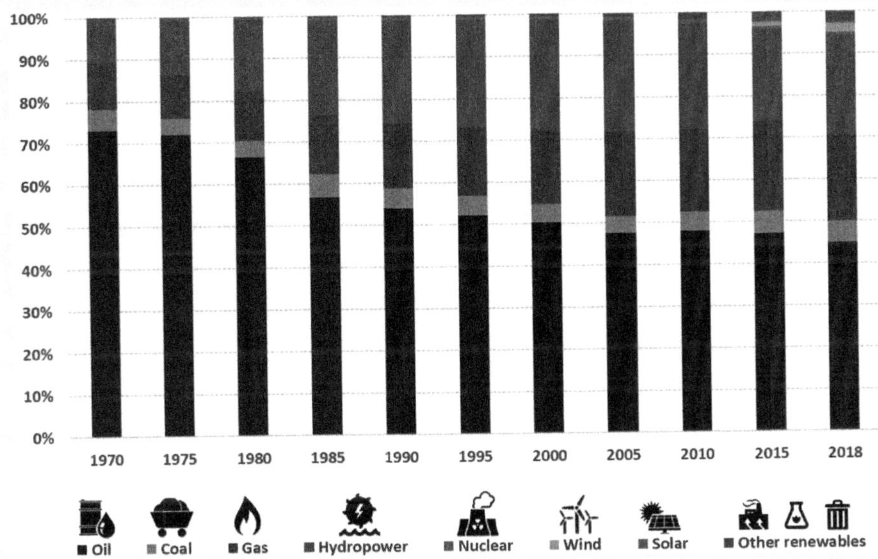

Figure 9.3 Energy consumption by source in Central and South America region between 1970 and 2018. Energy consumption is measured in terawatt-hours (TWh). Other renewables include geothermal, biomass and waste energy.
This figure was created by Fuinhas, J.A., Koengkan, M., & Santiago, R. (2021). Physical capital development and energy transition in Latin America and the Caribbean. Elsevier. https://doi.org/10.1016/C2020-0-01491-X and is based on the Our World in Data (2021). Obesity. https://ourworldindata.org/obesity database.

Continued

Several drivers have influenced the increase in energy consumption (e.g. economic growth, globalisation, trade, financial liberalisation, urbanisation, population growth, energy prices and so on). Nevertheless, the literature has missed a possible connection between increased energy consumption and health, such as obesity

Exhibit 9.1 Renewable energy consumption and investment in the LAC region—cont'd

This increase reflects investments in the installed capacity of renewable energy sources that occurred in the region. Investments in this kind of energy source more than doubled between 2006 and 2012, where in 2006, the installed capacity was 11.3 gigawatts (GW) and reached a value of 26.6 GW in 2012. The biomass, waste and wind energy sources make up most of this growth. Indeed, this increase results from high investments made in new renewable energy sources, with these investments adding a value of 2.4 billion United States Dollars (USD) in Brazil in 2005 and reaching a value of 3.5 billion USD in 2018 (see Fig. 9.4 below).

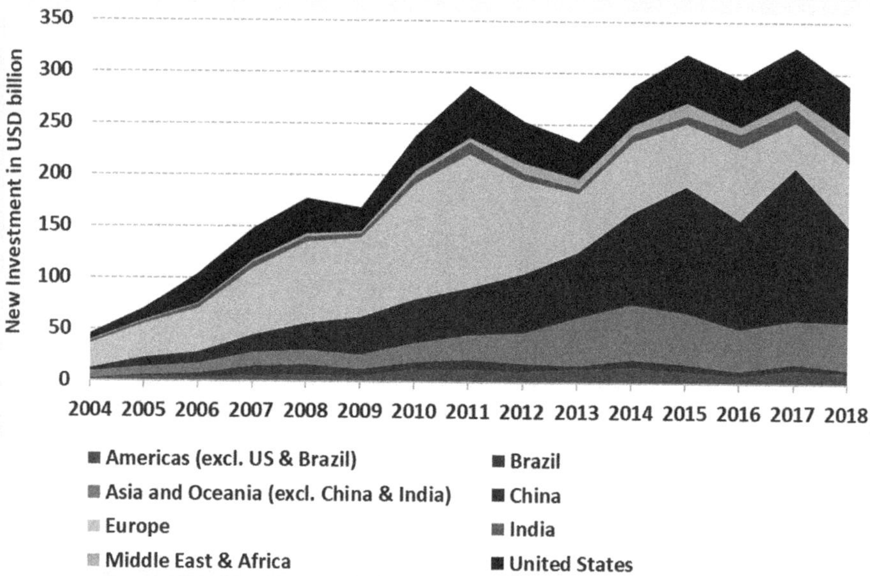

Figure 9.4 Global trends in renewable energy investment between 2004 and 2018. New investment in United States dollars (billion USD).
This figure was created by Fuinhas, J.A., Koengkan, M., & Santiago, R. (2021). Physical capital development and energy transition in Latin America and the Caribbean. Elsevier. https://doi.org/10.1016/C2020-0-01491-X, based on the International Energy Agency (IEA). (2020). Policies Database. https://www.iea.org/policies?region=Central%20&%20South%20America&page=1.

(Koengkan & Fuinhas, 2021). Globally, 13% of adults over 18 years were obese in 2016, while in 1990, this value was 6.8%. That means an astonishing increase of 91% between 1990 and 2016. In most high-income countries, such as the United States, 36% of adults were obese in 2016, while in the United Arab Emirates, this figure was 32%, in the United Kingdom, it was 28%, and in the European region, it was 23%. However, in upper-middle-income economies, such as China, obesity reached (6.2%) of the adult population in 2016, while in the Latin America (LA) region reached (19%) of the adult population. Moreover, in the lower- to middle-income economies, for example, India, the obesity problem reached (4%) of the adult population (see Fig. 9.5 below).

In the LA region, 19% of the adult population were obese in 2016, while this value was 9% in 1990 (see Fig. 9.6 below). That means, there was an increase of 108% between 1990 and 2016.

The increase of the share of adults that are obese is related to the rapid economic development, as mentioned before. When approaching the major LA economies (e.g. Argentina, Brazil, Chile, Colombia, Ecuador, Mexico, Peru and Venezuela), this trend of growth in the obesity can be a public health problem (see Fig. 9.7 below).

In 1990, the obesity problem reached 17% of the adult population in Argentina, while in Chile, this figure was 17%, Mexico 16%, Venezuela 15%, Colombia 12%, Peru 10%, Brazil 10% and Ecuador 9%. In 2016, the obesity problem reached the following values: Mexico 30%, Chile 28%, Argentina 28%, Venezuela 26%, Colombia 22%, Brazil 22%, Peru 20% and Ecuador 20%. In the LA region, the

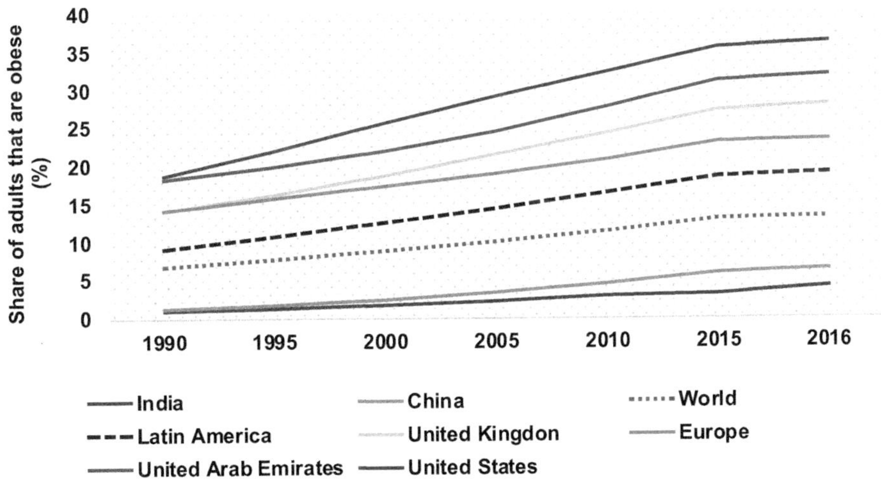

Figure 9.5 Share of adults that are obese (%) in the world, between 1990 and 2016. Obesity is defined as having a body mass index (BMI) equal to or greater than 30. BMI is a person's weight in kilograms divided by their height in meters squared.
The authors created this figure with the Our World in Data (2021). Obesity. https://ourworldindata.org/obesity.

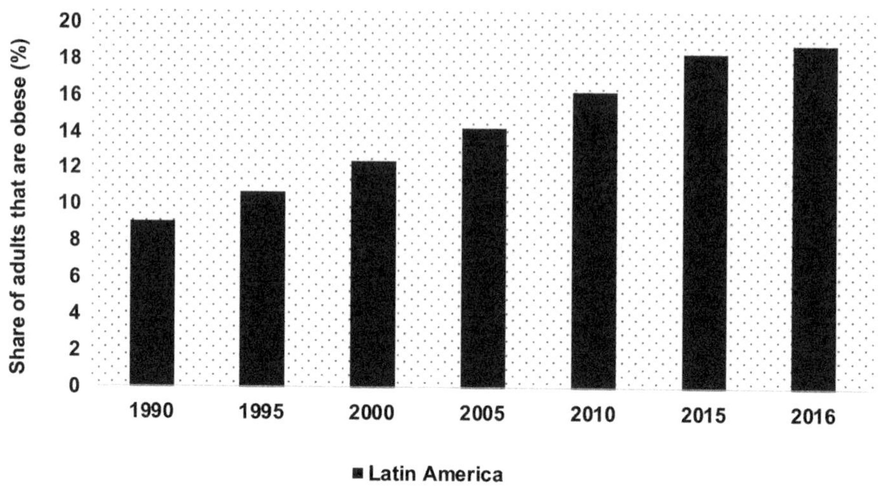

Figure 9.6 Share of adults that are obese (%) in the Latin American region, between 1990 and 2016. Being overweight is defined as having a body mass index (BMI) greater than or equal to 25. BMI is a person's weight in kilograms divided by their height in metres squared.
The authors created this figure with the Our World in Data (2021). Obesity. https://ourworldindata.org/obesity.

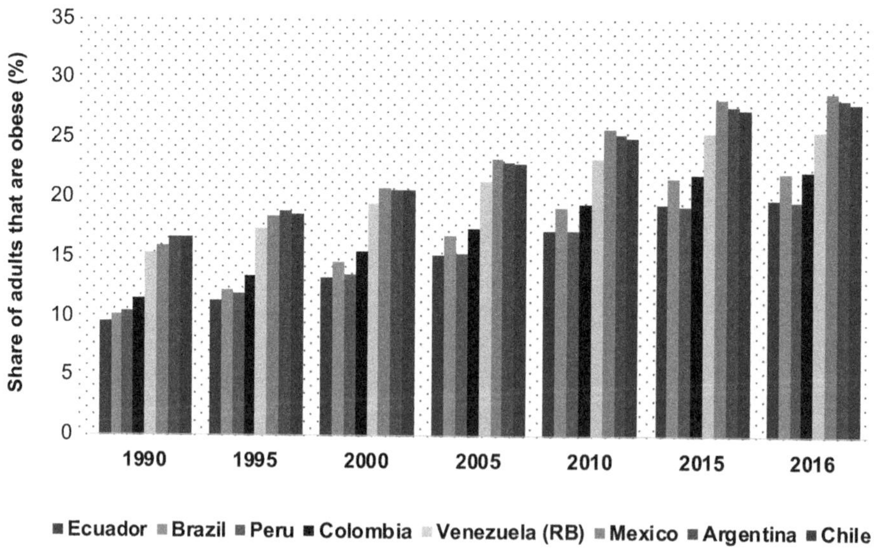

Figure 9.7 Share of obese (%) adults in the major Latin American economies between 1990 and 2016. Being overweight is defined as having a body mass index (BMI) greater than or equal to 25. Obesity is characterised by a BMI greater than or equal to 30. BMI is a person's weight in kilograms divided by their height in metres squared.
The authors created this figure with the Our World in Data (2021). Obesity. https://ourworldindata.org/obesity.

prevalence of obesity in women is similar to that of countries with higher per capita income (Kain et al., 2003).

Indeed, the obesity problem is caused by several factors, such as environmental, social, economic, genetic, physiological and political factors, that have interacted to varying degrees over time. Alongside other factors, the processes of urbanisation, globalisation and technological progress have promoted overweight and obesity. Indeed, the increase in weight caused by the factors mentioned above contributes to making people less physically active. Consequently, they demand more energy from motor vehicles and modern household appliances. Indeed, as overweight and obese people use motor vehicles as a mode of transportation to move around and have modern household appliances, this behaviour helps to reduce the physical effort that would otherwise be spent. It can be concluded that this behavior of using motor vehicles and modern household appliances contributes to weight gain due to individuals' lower caloric expenditure (e.g. Koengkan, Fuinhas, & Fuinhas, 2021, pp. 1–7; and Koengkan & Fuinhas, 2021).

This chapter will approach the impact of the obesity epidemic on the consumption of fossil fuels. The following research questions were formulated. Can the obesity epidemic increase fossil fuel energy consumption in the LAC region? A good starting point for analysing these phenomena is to choose a group of countries that have experienced fast economic, social and environmental transformations but were similar enough to be handled as a panel. Thus, this chapter will approach 20 countries from the LAC region, i.e., Argentina, Belize, Bolivia, Brazil, Chile, Colombia, Costa Rica, Dominican Republic, Ecuador, El Salvador, Guatemala, Guyana, Honduras, Jamaica, Mexico, Nicaragua, Panama, Paraguay, Peru and Uruguay, for the period between 1991 and 2016.

Moreover, this research will use three econometric approaches. First, a panel with fixed effects (least squares dummy variable) estimator will be used as a benchmark. Second, a panel quantile estimator with fixed effects will capture the non-linearities between explanatory variables and the explained ones. Finally, a majorise–minimise quantile regression estimator, which is a panel quantile regression robust to the absence, presence or multiple fixed effects, will be used to check the robustness of results.

9.2 Literature review

9.2.1 *Obesity and fossil fuel consumption*

Latin American countries face a severe obesity problem, as more than half its inhabitants were overweight in 2016, and about one in five were obese (Chapter 1). This problem is more acute in the region than in other countries with similar development levels and shows a worryingly rising trend (Chapter 1). At the onset of this epidemic are the changes in dietary habits towards highly processed fat- and sugar-rich food and more sedentary jobs and living styles (Popkin & Reardon, 2018).

The impact of the obesity epidemic goes far beyond its effects on the health and mortality rates of those directly affected. It drives countrywide transformations in health systems, economic growth (Chapter 5), food production (Chapter 8) and energy consumption.

The linkages between obesity and energy consumption are broad and often disregarded. The dietary patterns of obese people lean toward transfats and other highly caloric food types, which require a vast amount of energy to produce and rise the pressure on waste management systems (Michaelowa & Dransfeld, 2008). Energy consumption also increases in transportation due to the higher amounts of fuel needed to transport heavier people (Michaelowa & Dransfeld, 2008). Jacobson and King (2011) estimate that overweight and obesity are responsible for a 0.8% increase in automobile fuel consumption in the United States. Koengkan and Fuinhas (2021) assess the impact of the overweight epidemic on energy consumption in 31 European countries using quantile regression. They report that overweight generates an increase in energy consumption, which they attribute to a higher demand for processed foods and higher use of home appliances driven by a sedentary lifestyle.

Obesity is also responsible for an increase in greenhouse gas (GHG) emissions through higher energy consumption. According to Magkos et al. (2020), it accounts for a 1.6% increase in GHG emissions worldwide. Dietary choices related to the quantity and type of food ingested have vast repercussions on energy consumption and GHG emissions. Heller (2020), using data from the United States, shows that GHG emissions associated with the fifth quantile of the food impact factor are 7.9 times higher than in the first. The effect of dietary patterns on the energy consumption of food systems was studied by Canning et al. (2017). They show that the adherence to dietary guidelines for Americans would result in a 3%–74% decrease in the energy consumption related to food. The proliferation of fast-food chains related to the obesity epidemic also raises energy consumption, as food away from home has higher energy requirements per unit of mass than food at home (Mackie & Wemhoff, 2020).

9.2.2 Economic growth and fossil fuel consumption

Energy consumption in LAC has been inextricably linked to economic growth. This region experienced average annual GDP growth of 3.15% between 1990 and 2014, and, over the same period, energy use had an average annual increase of 1.07% (World Bank Open Data, 2021). Even though the energy intensity of these economies decreased, that does not imply that the pressure on fossil fuels consumption has eased, as the newly installed renewable energy capacity proved to be insufficient to satisfy the additional energy demand. Over this period, the share of fossil fuels in energy use increased from 70.76% to 87.88%, while the renewables share decreased from 32.44% to 27.6% (World Bank Open Data, 2021). A similar pattern is observable in electricity production, where the share of fossil fuels increased approximately 13 percentage points, and the renewables share saw a drop of similar size (World Bank Open Data, 2021).

There is a huge number of studies that focus on the causality nexus between economic growth and energy consumption, but their findings are mixed: some report a positive impact of economic growth on energy consumption (e.g. Fuinhas et al., 2021, pp. 19–234; Haque, 2021; Shahbaz et al., 2018), while others point to a bidirectional relationship (e.g. Koengkan, Losekann, & Fuinhas, 2019; and Marques et al., 2017a,b), and some argue there is no relation between them (Acheampong, 2018).

Research on the relation between economic growth and fossil fuels consumption is also abundant and covers a wide array of countries and periods. Many studies find a positive relationship running from economic growth to fossil fuel consumption in specific countries such as Bangladesh (Murshed & Alam, 2021), Kenya (Sarkodie & Adom, 2018) and regions such as Africa (e.g. Kolawole et al., 2017; and Mensah et al., 2019) and Latin America (Koengkan, Losekann, & Fuinhas, 2019). There is also evidence of bidirectional causality between these variables in Brazil (Pao & Fu, 2013), the American Southern Common Market (Koengkan, Fuinhas, & Santiago, 2020), the Commonwealth of Independent States (Rasoulinezhad & Saboori, 2018) and 53 different countries (Asafu-Adjaye et al., 2016).

Some authors focus on the linkages between economic growth and the consumption of specific fossil fuels. Apergis and Payne, 2010a find evidence of bidirectional causality between coal consumption and economic growth in 25 countries of the Organisation for Economic Co-operation and Development (OECD). Lotfalipour et al. (2010) report that growth causes natural gas consumption in Iran, while Apergis and Payne (2010b) and Ozturk and Al-Mulali (2015) find evidence of bidirectional causality between these variables, using panels comprising 67 countries and seven Gulf Cooperation countries, respectively. Finally, Bildirici and Bakirtas (2014) conduct a disaggregated analysis of the influence of economic growth on coal, oil and natural gas consumption in Brazil, Russia, India, China and South Africa (BRICS) plus Turkey. Their results are highly country and energy-source dependent and range from unilateral causality from growth to fossil fuel consumption, bidirectional causality and absence of a relationship.

9.2.3 Food production and fossil fuel consumption

The world faces the difficult challenge of increasing food production to fulfil the nutritional needs of an ever-growing population without degrading the environment. To achieve this goal, agricultural production must rise, which implies that productivity must soar, given the limited availability of arable land. During the twentieth century, the world and especially developed countries managed to increase agricultural productivity through its mechanisation and the use of fertilisers and other chemicals (Conforti & Giampietro, 1997). This process leads to a massive surge in energy use in agriculture, mostly stemming from fossil fuels (e.g. Guzman et al., 2018; and Harchaoui & Chatzimpiros, 2019).

Several authors have studied the evolution of agricultural energy efficiency in various countries and reach mixed conclusions: some find improvements (Bajan et al., 2020), while others report a degradation (e.g. Guzman et al., 2018; and Harchaoui & Chatzimpiros, 2019). However, they all agree that total energy use in agriculture has been rising steadily. Evaluating energy efficiency across countries and over time is a complex task, given its dependence on the land quality and goods produced (e.g. Dyer et al., 2010; Woods et al., 2010). Animal production has much higher energy requirements than cereal production. Thus, the change of dietary patterns driven by economic growth and obesity that foster the demand for meat products has a detrimental impact on agricultural energy efficiency (Arizpe et al., 2011).

It is essential to keep in mind that the impact of food production on energy consumption does not derive uniquely from the agricultural sector. Bajan et al. (2020) analysed the energy consumption in the whole food production systems of 14 countries with the highest agricultural production. They report that the agricultural sector accounts for the more significant share of food systems' energy consumption in developing countries, driven by its higher weight in food production. In contrast, food processing is responsible for most energy consumption from food production in developed countries due to more advanced technologies.

To the best of our knowledge, research on the causal relationship between food production and energy consumption is scarce. Using a non-linear autoregressive distributed lag (NARDL) model, Koondhar et al. (2021) show that food production Granger-causes energy consumption in Pakistan. Anser et al. (2020) report a positive impact of food production on fossil fuel energy consumption for the whole world.

9.2.4 Renewable energy and fossil fuel consumption

Renewable power is essential in a world eager for energy, as it plays a decisive role in avoiding the depletion of fossil fuels and the global warming their combustion generates. The cost of renewable energy decreased substantially over the last decade. This trend is expected to continue in the coming years (Bogdanov et al., 2021), which has driven sizable growth rates in renewable installed capacity and increased its energy generation share. However, the rising share of renewables in the energy mix does not necessarily mean that a transition from fossil fuels to renewables occurs, as both may be growing together in response to higher energy demand (York & Bell, 2019). Furthermore, the intermittency of new renewable energy forms, such as wind and solar, requires complementary energy storage systems or additional standby capacity of fossil fuels to ensure a steady power supply (Marques et al., 2018). Thus, the substitution/complementary relationship between renewable energy and fossil fuels in electricity generation is eminently a question that must be answered through empirical research.

The last decade witnessed an explosion in studies focusing on the relationship between renewables and fossil fuels. Many of those report that energy transition is taking place in various countries and regions, such as Kenya (Sarkodie & Adom, 2018), the European Union (Gökgöz & Güvercin, 2018), the Commonwealth of Independent States (Rasoulinezhad & Saboori, 2018) and the American Southern Common Market (Koengkan, Fuinhas, & Marques, 2020). Others find mixed evidence, such as Marques et al. (2018), who report that substitutability exists in 10 European countries for solar and hydropower, but not wind energy, and Sinha et al. (2018), who find mixed evidence depending on the method used. Pao and Fu (2013) argue that there is no relationship between Brazil's total renewable and non-renewable energy production.

9.3 Data and methods

This section will be divided into two parts. The first will approach the group of countries and data/variables used in the chapter, while the second will show the method.

9.3.1 Data

This chapter will use annual data from 1991 to 2016 of 20 countries from the LAC region, i.e. **Argentina**, **Belize**, **Bolivia**, **Brazil**, **Chile**, **Colombia**, **Costa Rica**, **Dominican Republic**, **Ecuador**, **El Salvador**, **Guatemala**, **Guyana**, **Honduras**, **Jamaica**, **Mexico**, **Nicaragua**, **Panama**, **Paraguay**, **Peru** and **Uruguay**. The use of time-series between 1991 and 2016 is due to data availability until 2016 for the variable OBESITY for all countries selected. The variables which were chosen to perform this investigation will be shown in Table 9.2 below.

'Log' denotes variables in natural logarithms, 'Obs.' denotes the number of observations in the model, 'Std.Dev' denotes the Standard Deviation, and 'Min and Max' denote minimum and maximum. All variables are in natural logarithms to standardise the interpretation of results and linearise the relationships between variables. These summary statistics were obtained from the command *sum* of **Stata 16.0**. The board below shows how to obtain the summary statistics of variables.

How to do:

```
**The summary statistics**
sum l_logfossil l_logobesity l_loggdp_ppp l_logfood_prod l_logrene
```

This section presents the group of countries from the LAC region on which this chapter will focus and the variables used. In the following subsection, we will present the method used to carry out the empirical investigation of this chapter.

9.3.2 Method

The empirical research was performed using three econometric approaches. First, a panel with fixed effects (least squares dummy variable) estimator was used as a benchmark. Second, a panel quantile estimator with fixed effects captured the non-linearities between explanatory variables and the explained ones. Finally, a majorise−minimise quantile regression estimator, which is a panel quantile regression robust to absence, presence or multiple fixed effects, was used to check the robustness of results.

The following equation specifies the relationship between the dependent variable and the covariates,

$$Y_{it} = \alpha_i + X'_{it}\beta + \mu_{it}, \tag{9.1}$$

where α_i, i = 1, …, N, are fixed unknown constants (fixed effects) that are estimated along with β; X_{it}, i = 1, …, N, and t = 1, …, T, is a k-dimensional vector of explanatory variables; and μ_{it} is the error term assumed to be i.i.d. over individuals and

Table 9.2 Variables' description and summary statistics.

Variables' description	
Variables	**Source**
Electric power consumption (kWh per capita) from fossil fuels (e.g. coal, oil, petroleum and natural gas products). In this investigation, we called this variable **'FOSSIL'.**	World Bank Open Data (2021)
Portion of adults that are obese (in percent). Obesity is defined as having a body mass index (BMI) equal to or greater than 30. BMI is a person's weight in kilograms divided by their height in metres squared. In this investigation, we called this variable **'OBESITY'.**	Our World in Data (2021)
Gross domestic product (GDP) per capita based on purchasing power parity (PPP). This variable is converted to international dollars using purchasing power parity rates. An international dollar has the same purchasing power over GDP as the US dollar has in the United States. GDP at purchaser's prices is the sum of gross value added by all resident producers in the country plus any product taxes and minus any subsidies not included in the value of the products. It is calculated without deductions for the depreciation of fabricated assets or depletion and degradation of natural resources. This variable is in the current 2017 international dollars. In this investigation, we called this variable **'GDP_PPP'.**	World Bank Open Data (2021)
Cereal production (metric tons) per capita. This variable measures the production data on cereals related to crops harvested for dry grain only. Cereal crops harvested for hay or harvested green for food, feed, or silage and used for grazing are excluded. In this investigation, we called this variable **'FOOD_PROD'.**	World Bank Open Data (2021)
Electricity consumption from renewable sources, excluding hydroelectric (kWh) per capita. This variable measures the electricity consumption from renewable sources, excluding hydroelectric, including geothermal, solar, tides, wind, biomass and biofuels. In this investigation, we called this variable **'RENE'.**	World Bank Open Data (2021)

Summary statistics

Variables	**Obs.**	**Mean**	**Std. Dev**	**Min**	**Max**
LogFOSSIL	452	4.1222	0.3390	3.0029	4.5114
LogOBESITY	520	2.7910	9.2778	2.0918	3.3428
LogGDP_PPP	520	8.8576	0.5317	7.5885	10.1393
LogFOOD_PROD	520	−1.9132	1.3903	−7.4333	0.2754
LogRENE	426	3.5427	1.4730	−1.8832	7.0419

(Log) denotes variables in the natural logarithms.

time. The fixed effects α_i seize all unobservable time-invariant divergences across individuals.

The quantile via moments approach will be used as the primary econometric technique. As stated in Chapters 4 and 5, this approach was proposed by Machado and Silva (2019) as an alternative for quantile regression with non-additive fixed effects formerly advanced by Powell (2016). This method also has the advantage of providing information on how the regressors affect the entire conditional distribution. Moreover, the method can also be adapted to do estimates in the presence of cross-sectional dependence with endogenous variables (Machado & Silva, 2019). In other words, this method is not based on the estimation of conditional means but on moment conditions that find conditional means under exogeneity. This characteristic is closely related to that of the Chernozhukov and Hansen (2008) model. Under suitable conditions, quantile via moments can identify the exact structural quantile function. So this method is suitable to be used in non-linear models and models with multiple endogenous variables. It makes this method adequate to be used for non-linear models and is much more straightforward, especially in models with multiple endogenous variables (Machado & Silva, 2019). This method can provide information on how the regressors affect the entire conditional distribution. Moreover, according to Machado and Silva (2019), the quantile via moments model, given data $\left\{\left(Y_{it}, X_{it}'\right)'\right\}$ from a panel of n individuals $i = 1, \ldots,$ n over T periods, is constructed around the following Eq. (9.2):

$$Y_{it} = a_i + X_{it}'\beta + \left(\delta_i + Z_{it}'\gamma\right)U_{it} \ , \tag{9.2}$$

with, $P\{\delta_i + Z_{it}'\gamma > 0\} = 1$. The parameters (α_1, δ_i), $i = 1, \ldots, n$, capture the individual i fixed effects and Z is a k-vector of known differentiable (with probability 1) transformations of the components of X with element l given by $Z_l = Z_l(X)$, $l = 1, \ldots, k$. In empirical research, we will use a special case of Eq. (9.1), a linear heteroskedasticity model, in which $Z = X$. The sequence $\{X_{it}\}$ is *i.i.d.* for any fixed i and independent across t. U_{it} are *i.i.d.* (across i and t), statistically independent of X_{it}, and normalised to satisfy the moment condition $E(U) = 0 \wedge E(|U|) = 1$ (Machado & Silva, 2019). Eq. (9.2) implies that the conditional quantile-τ is given by Eq. (9.3):

$$Q_Y(\tau|X_{it}) = (a_i + \delta_i q(\tau)) + X_{it}'\beta + Z_{it}'\gamma q(\tau) \ , \tag{9.3}$$

with $q(\tau) = F_U^{-1}(\tau)$, where F_U is the distribution function of U. The authors propose a recursive estimation method based on a set of moment conditions, which is computationally fast and straightforward and does not require the use of simulations that would render its widespread adoption difficult (Powell, 2016). They also prove that the resulting estimates are consistent and asymptotically normal.

After computing the estimates, the marginal effect of the explanatory variable l on quantile τ of the dependent variable can be retrieved from Eq. (9.4):

$$\beta_l(\tau|X) = \beta_l + \frac{\partial Z'}{\partial X_l}\gamma q(\tau), \tag{9.4}$$

As stated before, to check the robustness of results, the majorise–minimise quantile regression estimator, in the formalisation proposed by Rios-Avila (2020), was used. This estimator is in line with the method of quantiles via moments proposed by Machado and Silva (2019). This approach estimates quantile regressions both without fixed effects and as if multiple fixed effects were present. When no fixed effects are specified, this methodology allows us to compare the results with the ones from the estimator proposed by Machado and Silva (2019).

Some preliminary tests must be performed before the estimators can be considered adequate to handle the characteristics of data and the nature of relationships under analysis. Thus, the empirical analysis begins with verifying variables' properties, including checking for cross-sectional dependence, orders of integration, normality, multicollinearity and panel effects (see Table 9.3 below).

A battery of post-estimation tests was performed to confirm the appropriateness of estimations that were carried out (see Table 9.4 below). These tests are required to

Table 9.3 Preliminary tests.

Test	Finality
Cross-sectional dependence (CSD) test (Pesaran, 2004)	This test identifies the presence of cross-sectional dependence (CSD) in the panel's data. The null hypothesis of this test is the presence of cross-section independence CD $\sim$ N(0,1).
Panel unit root test (CIPS) test (Pesaran, 2007)	This test verifies the presence of unit roots in the variables. The panel unit root test (CIPS) null hypothesis is that all series have a unit root.
Shapiro–Wilk test (Royston, 1983)	These tests verify the normality of the model. The null hypothesis of these tests is the presence of normality.
Skewness and kurtosis test (D'Agostino et al., 1990)	This test checks the normality based on skewness and another based on kurtosis and then combines the two tests into an overall test statistic. The null hypothesis of this test is that the data is normally distributed.
Pairwise correlations	To assess the level of correlation between the variables in the panel data.
Variance inflation factor (VIF) test (Belsley et al., 1980)	This test verifies the presence of multicollinearity between the variables.
Hausman test	This test identifies heterogeneity, i.e., whether the panel has random effects (REs) or fixed effects (FEs).
Bias-corrected LM-based test (Born & Breitung, 2015)	This test checks the presence of serial correlation in fixed-effects panel models. The null hypothesis of this test is the non-presence of autocorrelation up to the second order.

Table 9.4 Post-estimation tests.

Test	Finality
The Wald test (Agresti, 1990)	This test verifies the global significance of the estimated models. The null hypothesis of the Wald test is that all the coefficients are equal to zero.

assess the global significance of the estimated models and the presence of heteroscedasticity. The empirical analysis was carried out using the econometric software **Stata 16.0.**

9.4 Empirical results

As mentioned before, this section will approach the empirical results of our investigation. We will show the results from preliminary tests, panel with fixed effects estimation, panel quantile regression with fixed effects estimation, majorise–minimise quantile regression estimator and post-estimation tests. Indeed, to find the level of correlation between the variables in our panel data, the pairwise correlations test was calculated. The term 'co-relation' was first proposed by Galton (1888). The product-moment correlation coefficient is often called the Pearson product-moment correlation coefficient because Pearson (1896) and Pearson and Filon (1898) were partially responsible for popularising its use. Table 9.5 below shows the outcomes from the pairwise correlations.

The results of the pairwise correlations indicate a positive correlation between LogOBESITY and LogFOSSIL, LogGDP_PPP and LogFOSSIL, LogFOOD_PROD, LogFOSSIL and a negative correlation between the variable LogRENE and LogFOSSIL. Moreover, we found a positive correlation between the variables LogGDP_PPP and LogOBESITY, LogFOOD_PROD and LogOBESITY, LogFOOD_PROD and LogGDP_PPP, LogRENE and LogOBESITY, and LogRENE and LogGDP_PPP. Nevertheless, we found the existence of a non-correlation between LogRENE and LogFOOD_PROD. The results from the pairwise correlations were obtained from the command *pwcorr* of **Stata 16.0**. The board below shows how to carry out and obtain the results from the pairwise correlations.

How to do:

```
**The pairwise correlations**
pwcorr l_logfossil l_logobesity l_loggdp_ppp l_logfood_prod l_logrene, sig
```

After computing the pairwise correlations, it is necessary to identify if the variables have a normal distribution. To this end, the skewness/kurtosis test (D'Agostino et al., 1990) and the Shapiro–Wilk test (Royston, 1983) were computed. Table 9.6 below shows the outcomes from the normal distribution tests.

Table 9.5 Pairwise correlations.

Variables	LogFOSSIL	LogOBESITY	LogGDP_PPP	LogFOOD_PROD	LogRENE
LogFOSSIL	**1.0000**				
LogOBESITY	0.3844	**1.0000**			
	0.0000				
LogGDP_PPP	0.4021	0.7334	**1.0000**		
	0.0000	0.0000			
LogFOOD_PROD	0.1544	0.1141	0.2159	**1.0000**	
	0.0010	0.0092	0.0000		
LogRENE	−0.2536	0.4457	0.2730	−0.0933	**1.0000**
	0.0000	0.0000	0.0000	0.1568	

(Log) denotes variables in the natural logarithms.

Table 9.6 Normal distribution tests.

Variables	Obs.	Skewness	Kurtosis	Skewness/ Kurtosis tests		Shapiro–Wilk test	
				Prob > Chi2		Prob > z	
LogFOSSIL	452	0.0000	0.5361	0.0000	***	0.0000	***
LogOBESITY	520	0.0338	0.0000	0.0000	***	0.0000	***
LogGDP_PPP	520	0.2857	0.0837	0.1266		0.0202	**
LogFOOD_PROD	520	0.0000	0.0000	0.0000	***	0.0000	***
LogRENE	426	0.0219	0.0445	0.0129	**	0.0000	***

*** and ** denote statistically significant at the 1% and 5% levels, respectively; (Log) denotes variables in the natural logarithms.

The skewness/kurtosis and Shapiro–Wilk tests results indicate that LogFOSSIL, LogOBESITY, LogFOOD_PROD and LogRENE are not normally distributed where the null hypotheses of both tests can be rejected. Moreover, the skewness/kurtosis test indicates that the variable LogGDP_PPP is normally distributed. The skewness/kurtosis and Shapiro–Wilk tests were obtained from the commands *sktest* and *swilk* of **Stata 16.0**. The board below shows how to carry out and obtain the skewness/kurtosis and Shapiro–Wilk tests results.

How to do:

```
** Skewness/kurtosis test**
sktest l_logfossil l_logobesity l_loggdp_ppp l_logfood_prod l_logrene
** Shapiro–Wilk test **
swilk l_logfossil l_logobesity l_loggdp_ppp l_logfood_prod l_logrene
```

After realising normal distribution tests, we can proceed to the identification of multicollinearity between the model's variables. To this end, the VIF test that informs on the presence of multicollinearity needs to be computed. Table 9.7 below shows the results from the VIF test.

The results from the VIF test show that the presence of multicollinearity is not a concern, given the low VIF and mean VIF values registered, which are lower than the usually accepted benchmark of 10, in the case of the VIF values, and six in the case of the mean VIF values (Koengkan, Fuinhas, & Marques, 2020). This test helps us understand the degree of multicollinearity present in our models, leading to problems in estimation (e.g. Koengkan & Fuinhas, 2021; Koengkan, Fuinhas, & Silva, 2021; and Santiago et al., 2020). The results of the VIF-test were obtained from the command *estat vif* in **Stata 16.0.** The board below shows how to carry out and obtain the results from the VIF test.

How to do:

```
** The variance inflation factor test**
reg l_logfossil l_logobesity l_loggdp_ppp l_logfood_prod l_logrene
estat vif
```

Table 9.7 VIF test.

Variables	VIF	1/VIF	Mean VIF
LogFOSSIL			
LogOBESITY	2.41	0.4149	1.72
LogGDP_PPP	2.16	0.4639	
LogFOOD_PROD	1.07	0.9303	
LogRENE	1.24	0.8050	

(Log) denotes variables in the natural logarithms.

After carrying out the VIF test, it is necessary to find cross-sectional dependence (CSD) in the panel data. The Pesaran CD test developed by Pesaran (2004) was used in this investigation. The null hypothesis of this test is the non-presence of cross-sectional dependence CD $\sim$ N (0,1) for $N \rightarrow \infty$, and that T is sufficiently large. Table 9.8 below shows the results from the Pesaran CD test.

The results from the CSD test show the presence of cross-sectional dependence in the variables LogFOSSIL, LogOBESITY, LogGDP_PPP and LogFOOD_PROD. The presence of cross-section dependence can signify that the countries selected in our study share the same characteristics and shocks, as Fuinhas, Marques, & Koengkan (2017) and Koengkan, Fuinhas, and Marques (2019) stated. The variable LogRENE could not be computed by CSD test because this test requires strongly balanced data. The results of the Pesaran CD test were obtained from the command *xtcd* in **Stata 16.0**. The board below shows how to carry out and obtain the results from the Pesaran CD test.

How to do:

```
** The Pesaran CD test**
xtcd l_logfossil l_logobesity l_loggdp_ppp l_logfood_prod l_logrene, resid
```

In the presence of cross-sectional dependence, it is necessary to verify the order of integration of variables (Koengkan & Fuinhas, 2020). To this end, the CIPS test that was developed by Pesaran (2007) was calculated. Table 9.9 below shows the results from the unit root test.

Table 9.8 Pesaran CD test.

Variables	CD-test	*P*-value		Corr	Abs (corr)
LogFOSSIL	20.93	0.000	***	0.338	0.494
LogOBESITY	61.81	0.000	***	0.999	0.999
LogGDP_PPP	60.55	0.000	***	0.979	0.979
LogFOOD_PROD	2.49	0.013	**	0.040	0.367
LogRENE	N.A				

*** and ** denote statistical significance at the 1% and 5% levels, respectively; (Log) denotes variables in the natural logarithms; NA denotes not available.

Table 9.9 Unit root test.

	2nd generation unit root test				
	Panel unit root test (CIPS) (Zt-bar)				
	Without trend			With trend	
Variables	**Lags**	**Zt-bar**		**Zt-bar**	
LogFOSSIL	1	−1.585	**	−1.578	**
LogOBESITY	1	−2.306	***	−2.736	***
LogGDP_PPP	1	−1.099	*	−1.619	**
LogFOOD_PROD	1	−2.243	**	−1.760	**
LogRENE	1	0.662		−1.493	**

***, ** and * denote statistically significant at the 1%, 5% and 10% levels, respectively; (Log) denotes variables in natural logarithms.

Therefore, the results of the CIPS test indicate that the variables LogFOSSIL, LogOBESITY, LogGDP_PPP and LogFOOD_PROD are stationary, while the variable LogRENE is on the borderline between I(0) and I(1) of the order of integration. The results of the CIPS-test were obtained from the command *multipurt* in **Stata 16.0**. The board below shows how to carry out and obtain the results from the CIPS test.

```
How to do:
  ** The CIPS-test**
  multipurt l_logfossil l_logobesity l_loggdp_ppp l_logfood_prod l_logrene, lags (1)
```

After finding the presence of the order of integration between the variables of our model, the next step of this investigation is to find the presence of individual effects in the model. To this end, the Hausman test, which compares the random (RE) and fixed-effects estimates (FE), was computed. The null hypothesis of this test is that the difference in coefficients is not systematic, where the random effects is the most suitable estimator (Fuinhas et al., 2017). The results of this test are presented in Table 9.10 below.

The results of this test show that the null hypothesis should be rejected (**chi2 (4) = 31.71*****, statistically significant at 1% level). That is, there is the presence of fixed effects in the model. The results of the Hausman test were obtained from the command *hausman* with *option sigmaless* in **Stata 16.0**. The board below shows how to carry out and obtain the results from the Hausman test.

```
How to do:
  **The Hausman test**
  qui:xtreg l_logfossil l_logobesity l_loggdp_ppp l l_logfood_prod l_logrene,fe
  estimates store fixed
  qui:xtreg l_logfossil l_logobesity l_loggdp_ppp l_logfood_prod l_logrene,
  estimates store random
  hausman fixed random, sigmaless
```

Table 9.10 Hausman test.

Variables	(b) Fixed	(B) Random	(b-B) Difference	Sqrt(diag(V_b-V-B)) S.E.
LogOBESITY	0.1921	0.1859	0.0062	0.0129
LogGDP_PPP	0.0160	0.0287	−0.0127	0.0096
LogFOOD_PROD	0.0346	0.0136	0.0209	0.0104
LogRENE	−0.0362	−0.0369	0.0007	0.0010
Chi2 (4)	31.71***			

*** denotes statistically significant at the 1% level; (Log) denotes variables in the natural logarithms.

After identifying the presence of fixed effects, it is necessary to check serial correlation in the fixed-effects panel model. To this end, the bias-corrected LM-based test developed by Born and Breitung (2015) was computed. The null hypothesis of this test is the non-presence of autocorrelation up to the second order. Table 9.11 below shows the results from the bias-corrected LM-based test.

The results from the bias-corrected LM-based test indicate the presence of autocorrelation up to the second order in the fixed-effects panel model, where the null hypothesis can be rejected. The results of the bias-corrected LM-based test were obtained from the command *xtqptest* in **Stata 16.0**. The board below shows how to carry out and obtain the results from this test.

How to do:

```
**The bias-corrected LM-based test**
xtqptest l_logfossil l_logobesity l_loggdp_ppp l_logfood_prod l_logrene, order(1)
```

After carrying out the preliminary tests, we can estimate the fixed effects model, quantile regression model with fixed effects and the majorise–minimise quantile regression model and the post-estimation test (e.g. the Wald test). In the fixed effects model, we opted to compute the following estimators (e.g. fixed effects (FE), FE robust standard errors (FE robust) and FE Driscoll and Kraay (FE D.-K.)). The Driscoll and Kraay (1998) estimator was applied (e.g. Fuinhas et al., 2017) to cope with the presence of heteroskedasticity, contemporaneous correlation, first-order autocorrelation and cross-sectional dependence (spatial dependence or spatial

Table 9.11 Bias-corrected LM-based test.

Variables	LM (k)-stat	
LogFOSSIL	3.23	***
LogOBESITY	11.75	***
LogGDP_PPP	8.33	***
LogFOOD_PROD	2.96	***
LogRENE	3.49	***

*** denotes statistical significance at the 1% level; under H0, LM(k) ~ N(0,1).

regimes). Moreover, this estimator is a matrix estimator that generates robust standard errors for several phenomena found in the sample errors. Table 9.12 shows the results from the model estimation and post-estimation test.

The results from the fixed-effects model indicate that a percentage point increase in the variable LogOBESITY, and LogFOOD_PROD increases the consumption of fossil fuels (LogFOSSIL) by 0.1922% and 0.0405%, respectively. In comparison, a percentage point increase in the variable LogRENE reduces the consumption of fossil fuels (LogFOSSIL) by 0.0379%. The quantile regression model results with fixed effects point out that the variable LogOBESITY increases the consumption of fossil fuels in the 25th, 50th and 75th quantiles, and the variable LogFOOD_-PROD in the 25th quantile. However, the variable LogRENE decreases the consumption of fossil fuels in the 25th, 50th and 75th quantiles. Moreover, the majorise–minimise quantile regression model with fixed effects indicates that the variable LogOBESITY increases the consumption of fossil fuels in the 25th, 50th and 75th quantiles, the variable LogGDP_PPP in the 75th quantile, and the variable LogFOOD_PROD in the 25th quantile. However, the variable LogRENE decreases the consumption of fossil fuels in the 25th, 50th and 75th quantiles. The results from the Wald test in all models point to the fact that time-fixed effects are needed, where the null hypotheses of the Wald test can be rejected. The fixed-effects model, quantile regression model with fixed effects and majorise–minimise quantile regression model were obtained from the commands *xtreg* with options *fe*, *fe robust* and *fe lag (1)*, *xtqreg* and *mmqreg* in **Stata 16.0**. The post-estimation test (e.g. The Wald test) was obtained from the command *xttest3* for the fixed effects model, the *testparm* for the quantile regression with fixed effects, and majorise–minimise quantile regression models. The board below shows how to carry out and obtain the models' regression and post-estimation test results.

How to do:

```
** Fixed effect model **
qui: xtreg l_logfossil l_logobesity l_loggdp_ppp l_logfood_prod l_logrene, fe
estimates store fe
qui: xtreg l_logfossil l_logobesity l_loggdp_ppp l_logfood_prod l_logrene, fe robust
estimates store fer
qui: xtscc l_logfossil l_logobesity l_loggdp_ppp l_logfood_prod l_logrene, fe lag(1)
estimates store dk
estimates table fe fer dk, star (.10 .05 .01) stats(N r2 r2_an F) b(%7.4f)
** The Wald test **
xtreg l_logfossil l_logobesity l_loggdp_ppp l_logfood_prod l_logrene, fe
xttest3
** Quantile regression model with fixed effects **
xtqreg l_logfossil l_logobesity l_loggdp_ppp l_logfood_prod l_logrene, i(country) quan-
tile(.25 .5 .75)
**The Wald test**
testparm l_logfossil l_logobesity l_loggdp_ppp l_logfood_prod l_logrene
** Majorise-minimise quantile regression model **
mmqreg l_logfossil l_logobesity l_loggdp_ppp l_logfood_prod l_logrene, abs(country) q(25
50 75)
```

Table 9.12 Estimations.

Independent variables	Dependent variable (LogFOSSIL)					
	Fixed-effects model					
	FE		FE Robust		FE D.K	
LogOBESITY	0.1922	***	***		**	
LogGDP_PPP	0.0160					
LogFOOD_PROD	0.0347	*	*		*	
LogRENE	−0.0363	***	***		***	
Constant	3.6630	***	***		***	
OBS	411					
Wald test (chi2) (8)	15297.30***					
Independent variables	**Dependent variable (LogFOSSIL)**					
	Quantile regression model with fixed effects					
	25th		**50th**		**75th**	
LogOBESITY	0.2641	***	0.1796	***	0.1235	*
LogGDP_PPP	−0.0223		0.0227		0.0527	
LogFOOD_PROD	0.0405	*	0.0336		0.0291	
LogRENE	−0.0379	***	−0.0359	***	−0.0347	***
Obs	411					
Wald test (chi2) (4)	64.34***					
	Majorise−minimise quantile regression model					
LogOBESITY	0.2641	***	0.1796	***	0.1235	**
LogGDP_PPP	−0.0223		0.0227		0.0527	*
LogFOOD_PROD	0.0405	*	0.0336		0.0291	
LogRENE	−0.0379	***	−0.0359	***	−0.0347	***
Constant	3.7729	***	3.6438	***	3.5581	***
OBS	411					
Wald test (chi2) (11)	105.17***					

***, ** and * denote statistically significant at the 1%, 5% and 10% levels, respectively; (Log) denotes variables in natural logarithms.

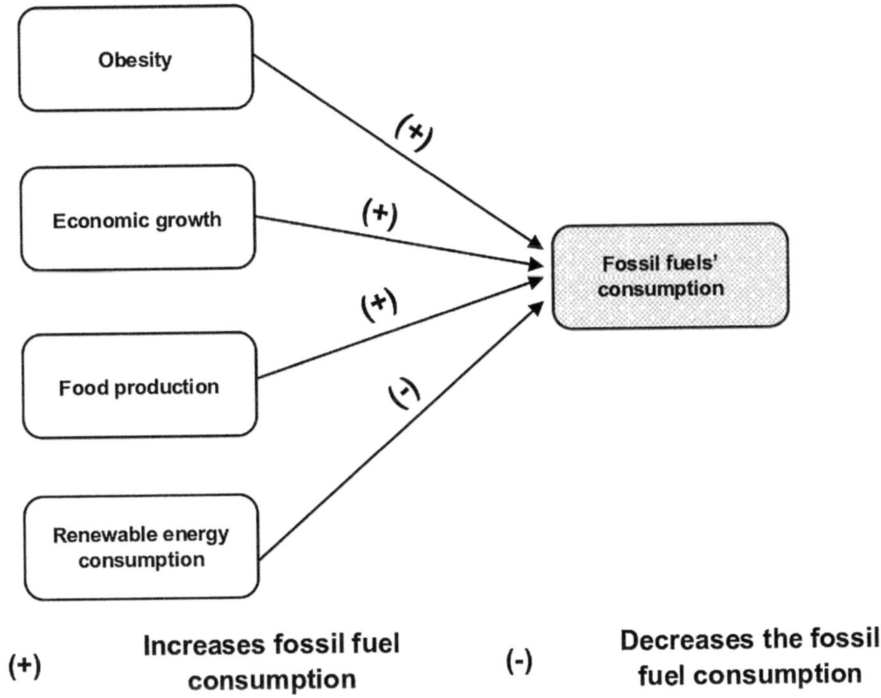

Figure 9.8 Summary of the impact of independent variables on dependent ones. This figure was created by the authors.

```
**The Wald test**
testparm l_logfossil l_logobesity l_loggdp_ppp l_logfood_prod l_logrene
```

Fig. 9.8 below summarises the impact of independent variables on dependent ones. This figure was based on the results from the majorise−minimise quantile regression model.

In this section, we showed the empirical results of this investigation. The following section will show the discussion of the empirical results.

9.5 Discussion

This section will provide some possible explanations for the results found in our empirical investigation. The positive impact of the obesity epidemic, economic growth and food production on energy consumption from fossil fuel sources could be related to economic development, urbanisation, poverty and globalisation, increasing the obesity epidemic, as found in Chapter 3. Indeed, the increase of the obesity epidemic due to these drivers encourages processed foods from multinational food corporations, fast-food chains and multinational supermarket chains. This increase impacts food production positively on farms. The evidence that the obesity epidemic increases food production in LAC countries was found in Chapter 8, where the obesity epidemic increases food production by 1.1501. Therefore, the increase in consumption and production of food positively impacts energy consumption to attend to the food demand, as mentioned in Chapters 5 and 8.

Moreover, the positive impact of economic growth on the consumption of fossil fuels could also be related to the increase in food production and the obesity epidemic, where the increase in food production positively affects economic activity. The evidence that food production and obesity increase economic activity in LAC countries was found in Chapter 5. Additionally, the increase in income caused by economic development also impacts food production and obesity positively. This evidence was found in Chapter 3, with economic growth increasing the obesity epidemic by 0.0041, and in Chapter 8, with economic growth increasing food production by 0.2867.

Another possible explanation for the positive impact of obesity on fossil fuel consumption is the lack of physical activities and outdoor activities because of overweight or obesity. This widespread physical inactivity is also reflected in the intensive use of home appliances, motorised transportation and screen-viewing leisure activities. All these affect the consumption of energy and economic growth positively.

Moreover, the negative impact of renewable energy consumption on non-renewable energy consumption could be related to the process of globalisation via financial openness. This increases capital stock and consequently reduces the cost of external financing, encouraging investment in renewable energy technologies. Thus, the reduction of fossil fuel consumption by renewable energy found in this analysis confirms the process of 'energy transition' in the LAC countries. According to Hauff et al. (2014), the term 'energy transition' indicates a growing trend of renewable energy sources to reduce the consumption of fossil fuels. The following section will show the main conclusions of this chapter.

9.6 Conclusion

This chapter approached the impact of the obesity epidemic on the consumption of fossil fuels. Twenty countries from the LAC region, i.e., Argentina, Belize, Bolivia, Brazil, Chile, Colombia, Costa Rica, Dominican Republic, Ecuador, El Salvador, Guatemala, Guyana, Honduras, Jamaica, Mexico, Nicaragua, Panama, Paraguay, Peru and Uruguay, for the period between 1991 and 2016, were used to carry out this investigation. To carry out this empirical analysis, this research used three econometric approaches. First, a panel with fixed effects (least squares dummy variable) estimator was used as a benchmark. Second, a panel quantile estimator with fixed effects captured the non-linearities between explanatory variables and the explained ones. Finally, a majorise–minimise quantile regression estimator, which is a panel quantile regression robust to absence, presence or multiple fixed effects, was used to check the robustness of results.

The preliminary tests, such as the pairwise correlations, indicated a positive correlation between LogFOSSIL, LogGDP_PPP, LogFOSSIL, LogFOOD_PROD and LogFOSSIL, and a negative correlation between the variable LogRENE and LogFOSSIL. Moreover, a positive correlation was found between LogGDP_PPP and LogOBESITY, LogFOOD_PROD and LogOBESITY, LogFOOD_PROD and LogGDP_PPP, LogRENE and LogOBESITY, and LogRENE and LogGDP_PPP. Nevertheless, the existence of a non-correlation between LogRENE and LogFOOD_PROD was found.

The skewness/kurtosis and Shapiro–Wilk tests indicated that the variables LogFOSSIL, LogOBESITY, LogFOOD_PROD and LogRENE are not normally distributed, where the null hypotheses of both tests can be rejected. However, the skewness/kurtosis test indicated that the variable LogGDP_PPP is normally distributed. The results from the VIF test showed that the presence of multicollinearity is not a concern, given the low VIF and mean VIF values registered, which are lower than the usually accepted benchmark of 10 in the case of the VIF values, and six in the case of the mean VIF values.

The results from the CSD test indicated the presence of cross-sectional dependence in the variables LogFOSSIL, LogOBESITY, LogGDP_PPP and LogFOOD_PROD. The presence of cross-sectional dependence can signify that the countries selected in our study share the same characteristics and shocks. Moreover, the variable LogRENE could not be computed by the CSD test because this test requires strongly balanced data. The results from the CIPS-test indicated that the variables LogFOSSIL, LogOBESITY, LogGDP_PPP and LogFOOD_PROD are stationary, while the variable LogRENE is on the borderline between I(0) and I(1) of the order of integration.

The results from the Hausman test indicated the presence of fixed effects in the model, where the results of this test showed that the null hypothesis should be rejected (chi2 (4) = 31.71***, statistically significant at the 1% level). Finally, the bias-correct LM-based test results indicated the presence of autocorrelation up to the second order in the fixed-effects panel model, where the null hypothesis can be rejected.

The results from the fixed-effects model indicated that the variable LogOBESITY increases the consumption of fossil fuels (LogFOSSIL) by 0.1922 and LogFOOD_PROD by 0.0405. In contrast, the variable LogRENE reduces the consumption of fossil fuels (LogFOSSIL) by −0.0379. The results from the quantile regression model with fixed effects pointed out that the variable LogOBESITY increases the consumption of fossil fuels in the 25th, 50th and 75th quantiles, and the variable LogFOOD_PROD in the 25th quantile. However, the variable LogRENE decreases the consumption of fossil fuels in the 25th, 50th and 75th quantiles. Moreover, the majorise–minimise quantile regression model with fixed effects indicated that the variable LogOBESITY increases the consumption of fossil fuels in the 25th, 50th and 75th quantiles, the variable LogGDP_PPP in the 75th quantile, and the variable LogFOOD_PROD in the 25th quantile. However, the variable LogRENE decreases the consumption of fossil fuels in the 25th, 50th and 75th quantiles. The results from the Wald test in all models indicate time-fixed effects are needed.

Therefore, the results confirmed that obesity increases the consumption of fossil-sourced electricity. However, this increase is also explained by economic growth and food production. Another result indicates a substitution relationship between renewable electric energy consumption and fossil electric energy consumption, reflecting the energy transition under which LAC countries are going.

As economic growth continues to take place in LAC, obesity and food production are likely to accompany this growth. However, unless some serious public health and industrial food measures are taken, energy consumption will continue to sustain all these socioeconomic evolutionary processes.

The increase in energy consumption is directed both to fossil and renewable energy sources. Under the energy transition, the share of renewable energy sources is likely to increase. This result may contribute to accomplishing the goals defined by international agreements on climate change, such as the Paris Agreement 2015. Nevertheless, electricity energy obtained from renewable sources also creates negative externalities, which should be accounted for. This negative effect arises from the intensive mineralisation in low- to middle-income countries to obtain, for instance, lithium needed to produce electric batteries. The future calls for a cost-benefit analysis and compensation rights to respond reasonably to the increasing use of renewable energy.

References

Acheampong, A. O. (2018). Economic growth, CO_2 emissions and energy consumption: What causes what and where? *Energy Economics, 74*, 677–692. https://doi.org/10.1016/j.eneco.2018.07.022

Agresti, A. (1990). *Categorical data analysis*. John Wiley and Sons. ISBN 0-471-36093-7.

Anser, M. K., Khan, M. A., Nassani, A. A., Aldakhil, A. M., Hinh Voo, X., & Zaman, K. (2020). Relationship of the environment with technological innovation, carbon pricing, renewable energy, and global food production. *Economics of Innovation and New Technology, 0*(0), 1–36. https://doi.org/10.1080/10438599.2020.1787000

Apergis, N., & Payne, J. E. (2010a). Coal consumption and economic growth: Evidence from a panel of OECD countries. *Energy Policy, 38*(3), 1353–1359. https://doi.org/10.1016/j.enpol.2009.11.016

Apergis, N., & Payne, J. E. (2010b). Natural gas consumption and economic growth: A panel investigation of 67 countries. *Applied Energy, 87*(8), 2759–2763. https://doi.org/10.1016/j.apenergy.2010.01.002

Arizpe, N., Giampietro, M., & Ramos-Martin, J. (2011). Food security and fossil energy dependence: An international comparison of the use of fossil energy in agriculture (1991–2003). *Critical Reviews in Plant Sciences, 30*(1–2), 45–63. https://doi.org/10.1080/07352689.2011.554352

Asafu-Adjaye, J., Byrne, D., & Alvarez, M. (2016). Economic growth, fossil fuel and non-fossil consumption: A pooled mean group analysis using proxies for capital. *Energy Economics, 60*, 345–356. https://doi.org/10.1016/j.eneco.2016.10.016

Bajan, B., Mrówczyńska-Kamińska, A., & Poczta, W. (2020). Economic energy efficiency of food production systems. *Energies, 13*(21), 5826. https://doi.org/10.3390/en13215826

Balza, L. H., Espinasa, R., & Serebrisky, T. (2016). *Lights on? Energy needs in Latin America and the Caribbean to 2040*. Inter-American Development Bank. https://publications.iadb.org/en/publication/17053/lights-energy-needs-latin-america-and-caribbean-2040.

Belsley, D. A., Kuh, E., & Welsch, R. E. (1980). *Regression diagnostics: Identifying influential data and sources of Collinearity*. Wiley. https://doi.org/10.1002/0471725153

Bildirici, M. E., & Bakirtas, T. (2014). The relationship between oil, natural gas and coal consumption and economic growth in BRICTS (Brazil, Russian, India, China, Turkey and South Africa) countries. *Energy, 65*, 134–144. https://doi.org/10.1016/j.energy.2013.12.006

Bogdanov, D., Ram, M., Aghahosseini, A., Gulagi, A., Oyewo, A. S., Child, M., Caldera, U., Sadovskaia, K., Farfan, J., De Souza Noel Simas Barbosa, L., Fasihi, M., Khalili, S., Traber, T., & Breyer, C. (2021). Low-cost renewable electricity as the key driver of the global energy transition towards sustainability. *Energy, 227*, 120467. https://doi.org/10.1016/j.energy.2021.120467

Born, B., & Breitung, J. (2015). Testing for serial correlation in fixed-effects panel data models. *Econometric Reviews, 35*(7), 1290–1316. https://doi.org/10.1080/07474938.2014.976524

Canning, P., Rehkamp, S., & Waters, A. (2017). *The role of fossil fuels in the US food system and the American diet*. Economic Research Report Number 224, January 2017 https://www.ers.usda.gov/webdocs/publications/82194/err-224.pdf.

Chernozhukov, V., & Hansen, C. (2008). Instrumental variable quantile regression: A robust inference approach. *Journal of Econometrics, 142*, 379–398. https://doi.org/10.1016/j.jeconom.2007.06.005

Conforti, P., & Giampietro, M. (1997). Fossil energy use in agriculture: An international comparison. *Agriculture, Ecosystems and Environment, 65*(3), 231–243. https://doi.org/10.1016/S0167-8809(97)00048-0

D'Agostino, R. B., Belanger, A. J., & D'Agostino, R. B., Jr. (1990). A suggestion for using powerful and informative. *The American Statistician, 44*(4), 316–321. https://doi.org/10.2307/2684359

Driscoll, J., & Kraay, A. C. (1998). Consistent covariance matrix estimation with spatially dependent data. *Review of Economics and Statistics, 80*(4), 549–560.

Dyer, J. A., Kulshreshtha, S. N., McConkey, B. G., & Desjardins, R. L. (2010). An assessment of fossil fuel energy use and CO_2 emissions from farm field operations using a regional level crop and land use database for Canada. *Energy, 35*(5), 2261–2269. https://doi.org/10.1016/j.energy.2010.02.013

Flavin, C., Gonzalez, M., Majano, A. M., Ochs, A., Rocha, M., & Tagwerker, P. (2014). Study on the development of the renewable energy Market in Latin America and the Caribbean. *IDB, 2*(14), 1–79. https://publications.iadb.org/publications/english/document/Study-on-the-Development-of-the-Renewable-Energy-Market-in-Latin-America-and-the-Caribbean.pdf.

Fuinhas, J. A., Koengkan, M., & Santiago, R. (2021). *Physical capital development and energy transition in Latin America and the Caribbean*. Elsevier. https://doi.org/10.1016/C2020-0-01491-X

Fuinhas, J. A., Marques, A. C., & Koengkan, M. (2017). Are renewable energy policies upsetting carbon dioxide emissions? The case of Latin America countries. *Environmental Science and Pollution Research, 24*(17), 15044–15054. https://doi.org/10.1007/s11356-017-9109-z

Galton, F. (1888). Co-relations and their measurement, chiefly from anthropometric data. *Proceedings of the Royal Society of London, 45*, 135–145.

Gielen, D., Boshell, F., Saygin, D., Bazilian, M. D., Wagner, N., & Gorini, R. (2019). The role of renewable energy in the global energy transformation. *Energy Strategy Reviews, 24*, 38–50. https://doi.org/10.1016/j.esr.2019.01.006

Gökgöz, F., & Güvercin, M. T. (2018). Energy security and renewable energy efficiency in EU. *Renewable and Sustainable Energy Reviews, 96*(June), 226–239. https://doi.org/10.1016/j.rser.2018.07.046

Guzmán, G. I., González de Molina, M., Soto Fernández, D., Infante-Amate, J., & Aguilera, E. (2018). Spanish agriculture from 1900 to 2008: A long-term perspective on agroecosystem energy from an agroecological approach. *Regional Environmental Change, 18*(4), 995–1008. https://doi.org/10.1007/s10113-017-1136-2

Haque, M. I. (2021). Oil price shocks and energy consumption in GCC countries: A system-GMM approach. *Environment, Development and Sustainability, 23*(6), 9336–9351. https://doi.org/10.1007/s10668-020-01027-y

Harchaoui, S., & Chatzimpiros, P. (2019). Energy, nitrogen, and farm surplus transitions in agriculture from historical data modeling. France, 1882–2013. *Journal of Industrial Ecology, 23*(2), 412–425. https://doi.org/10.1111/jiec.12760

Hauff, J., Bode, A., Neumann, D., & Haslauer, F. (2014). Global Energy Transitions a comparative analysis of key countries and implications for the international energy debate. *World Energy Council*, 1–30. https://www.extractiveshub.org/resource/view/id/13542.

Humphrey, W., & Stanislaw, J. (1979). Economic growth and energy consumption in the UK, 1700–1975. *Energy Policy, 7*(1), 29–42.

International Energy Agency (IEA). (2020). *Policies Database*. https://www.iea.org/policies?region=Central%20&%20South%20America&page=1.

Jacobson, S. H., & King, D. M. (2011). *A societal cost of obesity in terms of automobile fuel consumption*. Wiley Encyclopedia of Operations Research and Management Science. https://doi.org/10.1002/9780470400531.eorms0003

Kain, J., Vio, F., & Albala, C. (2003). Obesity trends and determinant factors in Latin America. *Cadernos de Saúde Pública, 19*(1), 1–10. https://doi.org/10.1590/S0102-311X2003000700009

Koengkan, M., & Fuinhas, J. A. (2020). The interactions between renewable energy consumption and economic growth in the Mercosur countries. *International Journal of Sustainable Energy, 39*(6). https://doi.org/10.1080/14786451.2020.1732978

Koengkan, M., & Fuinhas, J. A. (2021). Does the overweight epidemic cause energy consumption? A piece of empirical evidence from the European region. *Energy, 236*(1), 119297. https://doi.org/10.1016/j.energy.2020.119297

Koengkan, M., Fuinhas, J. A., & Fuinhas, C. (2021). *Does urbanisation process increase the overweight epidemic? The case of Latin America and the Caribbean region* (pp. 1–7). SSRN. https://doi.org/10.2139/ssrn.3826196

Koengkan, M., Fuinhas, J. A., & Marques, A. C. (2019). Chapter seven - the relationship between financial openness, renewable and non-renewable energy consumption, CO_2 emissions, and economic growth in the Latin American countries: An approach with a panel vector auto regression model. *The Extended Energy-Growth Nexus: Theory and Empirical Applications*, 199–229. https://doi.org/10.1016/B978-0-12-815719-0.00007-3

Koengkan, M., Fuinhas, J. A., & Marques, A. C. (2020). The role of financial openness and China's income on fossil fuels consumption: Fresh evidence from Latin American countries. *GeoJournal, 85*, 407–421. https://doi.org/10.1007/s10708-019-09969-1

Koengkan, M., Fuinhas, J. A., & Santiago, R. (2020). The relationship between CO_2 emissions, renewable and non-renewable energy consumption, economic growth, and urbanisation in the Southern common market. *Journal of Environmental Economics and Policy, 9*(4), 383–401.

Koengkan, M., Fuinhas, J. A., & Silva, N. (2021). Exploring the capacity of renewable energy consumption to reduce outdoor air pollution death rate in Latin America and the Caribbean region. *Environmental Science and Pollution Research, 28*, 1656–1674. https://doi.org/10.1007/s11356-020-10503-x

Koengkan, M., Losekann, L. D., & Fuinhas, J. A. (2019). The relationship between economic growth, consumption of energy, and environmental degradation: Renewed evidence from andean community nations. *Environment Systems and Decisions, 39*(1), 95–107. https://doi.org/10.1007/s10669-018-9698-1

Kolawole, A., Adesola, S., & De Vita, G. (2017). A disaggregated analysis of energy demand in sub-saharan Africa. *International Journal of Energy Economics and Policy, 7*(2), 224–235.

Koondhar, M. A., Udemba, E. N., Cheng, Y., Khan, Z. A., Koondhar, M. A., Batool, M., & Kong, R. (2021). Asymmetric causality among carbon emission from agriculture, energy consumption, fertiliser, and cereal food production – a non-linear analysis for Pakistan. *Sustainable Energy Technologies and Assessments, 45*(January), 101099. https://doi.org/10.1016/j.seta.2021.101099

Lotfalipour, M. R., Falahi, M. A., & Ashena, M. (2010). Economic growth, CO_2 emissions, and fossil fuels consumption in Iran. *Energy, 35*(12), 5115–5120. https://doi.org/10.1016/j.energy.2010.08.004

Machado, J. A. F., & Silva, J. M. C. S. (2019). Quantiles via moments. *Journal of Econometrics, 213*(1), 145–173. https://doi.org/10.1016/j.jeconom.2019.04.009Get

Mackie, C., & Wemhoff, A. P. (2020). Comparing greenhouse gas emissions associated with food away from home versus food at home in the United States. *Journal of Cleaner Production, 260*, 120930. https://doi.org/10.1016/j.jclepro.2020.120930

Magkos, F., Tetens, I., Bügel, S. G., Felby, C., Schacht, S. R., Hill, J. O., Ravussin, E., & Astrup, A. (2020). The environmental foodprint of obesity. *Obesity, 28*(1), 73–79. https://doi.org/10.1002/oby.22657

Marques, L. M., Fuinhas, J. A., & Marques, A. C. (2017a). Augmented energy-growth nexus: Economic, political and social globalisation impacts. *Energy Procedia, 136*, 97–101. https://doi.org/10.1016/j.egypro.2017.10.293

Marques, L. M., Fuinhas, J. A., & Marques, A. C. (2017b). On the dynamics of energy-growth nexus: Evidence from a world divided into four regions. *International Journal of Energy Economics and Policy, 7*(3), 208–215.

Marques, A. C., Fuinhas, J. A., & Pereira, D. A. (2018). Have fossil fuels been substituted by renewables? An empirical assessment for 10 European countries. *Energy Policy, 116*(November 2017), 257–265. https://doi.org/10.1016/j.enpol.2018.02.021

Mensah, I. A., Sun, M., Gao, C., Omari-Sasu, A. Y., Zhu, D., Ampimah, B. C., & Quarcoo, A. (2019). Analysis on the nexus of economic growth, fossil fuel energy consumption, CO_2 emissions and oil price in Africa based on a PMG panel ARDL approach. *Journal of Cleaner Production, 228*, 161–174. https://doi.org/10.1016/j.jclepro.2019.04.281

Michaelowa, A., & Dransfeld, B. (2008). Greenhouse gas benefits of fighting obesity. *Ecological Economics, 66*(2–3), 298–308. https://doi.org/10.1016/j.ecolecon.2007.09.004

Murshed, M., & Alam, M. S. (2021). Estimating the macroeconomic determinants of total, renewable, and non-renewable energy demands in Bangladesh: The role of technological innovations. *Environmental Science and Pollution Research, 2024*, 30176–30196. https://doi.org/10.1007/s11356-021-12516-6

Our World in Data. (2021). *Obesity*. https://ourworldindata.org/obesity.

Ozturk, I., & Al-Mulali, U. (2015). Natural gas consumption and economic growth nexus: Panel data analysis for GCC countries. *Renewable and Sustainable Energy Reviews, 51*, 998–1003. https://doi.org/10.1016/j.rser.2015.07.005

Pao, H. T., & Fu, H. C. (2013). The causal relationship between energy resources and economic growth in Brazil. *Energy Policy, 61*, 793–801. https://doi.org/10.1016/j.enpol.2013.06.113

Pearson, K. (1896). Mathematical contributions to the theory of evolution – III. Regression, heredity, and panmixia. *Philosophical Transactions of the Royal Society of London, Series A, 187*, 253–318.

Pearson, K., & Filon, L. N. G. (1898). Mathematical contributions to the theory of evolution. IV. On the probable errors of frequency constants and the influence of random selection on variation and correlation. *Philosophical Transactions of the Royal Society of London, Series A, 191*, 229–311.

Pesaran, M. H. (2004). *General diagnostic tests for cross-section dependence in panels*. The University of Cambridge, Faculty of Economics. https://doi.org/10.17863/CAM.5113. Cambridge Working Papers in Economics, n. 0435.

Pesaran, M. H. (2007). A simple panel unit root test in the presence of cross-section dependence. *Journal of Applied Econometrics, 22*(2), 256–312. https://doi.org/10.1002/jae.951

Popkin, B. M., & Reardon, T. (2018). Obesity and the food system transformation in Latin America. *Public Health/Nutrition, 19*(8), 1–23. https://doi.org/10.1111/obr.12694

Powell, D. (2016). *Quantile regression with nonadditive fixed effects*. RAND Working Paper http://works.bepress.com/david_powell/14.

Rasoulinezhad, E., & Saboori, B. (2018). Panel estimation for renewable and non-renewable energy consumption, economic growth, CO_2 emissions, the composite trade intensity, and financial openness of the commonwealth of independent states. *Environmental Science and Pollution Research, 25*(18), 17354–17370. https://doi.org/10.1007/s11356-018-1827-3

Rios-Avila, F. (2020). *Extending Quantile regressions via method of moments using multiple fixed effects*. MIMEO.

Royston, P. (1983). A simple method for evaluating the Shapiro-Francia W' test for non-normality. *Statistician, 32*, 297–300. https://doi.org/10.2307/2987935

Santiago, R., Koengkan, M., & Fuinhas, J. A. (2020). The relationship between public capital stock, private capital stock and economic growth in the Latin American and Caribbean countries. *International Review of Economics, 67*, 293–317. https://doi.org/10.1007/s12232-019-00340-x

Sarkodie, S. A., & Adom, P. K. (2018). Determinants of energy consumption in Kenya: A NIPALS approach. *Energy, 159*, 696–705. https://doi.org/10.1016/j.energy.2018.06.195

Shahbaz, M., Zakaria, M., Shahzad, S. J. H., & Mahalik, M. K. (2018). The energy consumption and economic growth nexus in top ten energy-consuming countries: Fresh evidence from the quantile-on-quantile approach. *Energy Economics, 71*, 282–301. https://doi.org/10.1016/j.eneco.2018.02.023

Sinha, A., Shahbaz, M., & Sengupta, T. (2018). Renewable energy policies and contradictions in causality: A case of next 11 countries. *Journal of Cleaner Production, 197*, 73–84. https://doi.org/10.1016/j.jclepro.2018.06.219

Solomon, B. D., & Krishna, K. (2011). The coming sustainable energy transition: History, strategies, and outlook. *Energy Policy, 39*(11), 7422–7431. https://doi.org/10.1016/j.enpol.2011.09.009

United Nations (UN). (2019). *UN spotlights' explosive' obesity rates, hunger in Latin America and Caribbean*. https://news.un.org/en/story/2019/11/1051211.

Veloso, F. A., Villela, A., & Giambiagi, F. (2008). Determinantes do "Milagre" econômico Brasileiro (1968–1973): Uma Análise Empírica. *Revista Brasileira de Economia, 62*(2), 221–246. https://doi.org/10.1590/S0034-71402008000200006

Woods, J., Williams, A., Hughes, J. K., Black, M., & Murphy, R. (2010). Energy and the food system. *Philosophical Transactions of the Royal Society B: Biological Sciences, 365*(1554), 2991–3006. https://doi.org/10.1098/rstb.2010.0172

World Bank Open Data. (2021). http://www.worldbank.org/.

York, R., & Bell, S. E. (2019). Energy transitions or additions?: Why a transition from fossil fuels requires more than the growth of renewable energy. *Energy Research and Social Science, 51*(November 2018), 40–43. https://doi.org/10.1016/j.erss.2019.01.008

The increase of CO_2 emissions by obesity epidemic in Latin American and Caribbean countries

10

10.1 Introduction

Latin American and Caribbean (LAC) countries have registered accelerated economic growth. For example, in the LAC region, the gross domestic product (GDP) per capita (current USD) in 1970 was 612.40 USD, which in 2014, reached a value of 10,407.80 USD (see Fig. 10.1 below).

The LAC's GDP per capita had an average annual growth rate of approximately 3.75% in this period. This increase is related to the structural and stabilisation programmes imposed on Latin American countries by the International Monetary Fund (IMF). These adjustment programmes were neoliberal policies that consisted mainly of the complete opening of their economies to international trade and capital, deregulation of the economy, privatisation, reduction of public expenditures, creation of

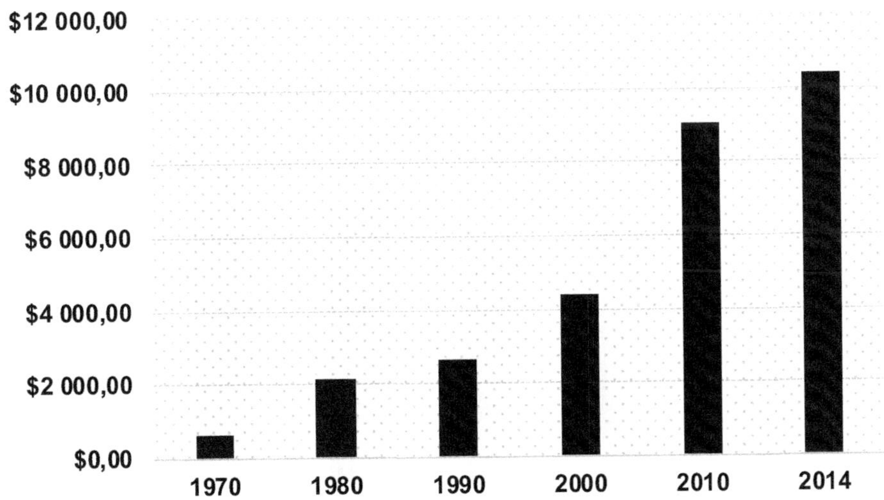

Figure 10.1 Gross domestic product (GDP) per capita (current USD) in the Latin American and Caribbean (LAC) region, between 1970 and 2014.
This figure was created by Fuinhas, J.A., Koengkan, M., & Santiago, R. (2021). Physical capital development and energy transition in Latin America and the Caribbean. Elsevier. https://doi.org/10.1016/C2020-0-01491-X and was based on World Bank Open Data (2021). http://www.worldbank.org/.

Obesity Epidemic and the Environment. https://doi.org/10.1016/B978-0-323-99339-5.00006-6

appropriate conditions for foreign investment, and the reduction of the role of the state in the economy. Moreover, the 'commodities boom' that occurred between the beginning of the 2000s and the end of 2014 also accelerated the process of openness and economic growth in the region (e.g. Fuinhas et al., 2021, pp. 19–234 and Koengkan et al., 2021).

The process of economic growth caused by economic and trade liberalisation influenced the increase of the obesity epidemic (see Chapters 3 and 4), food production (see Chapter 8), energy consumption from fossil fuels (see Chapter 9) and consequently environmental degradation (in this chapter). Therefore, in the LAC region, CO_2 emissions in 1970 were 1.7831 metric tons per capita and in 2014 reached a value of 3.1016 metric tons per capita (see Fig. 10.2 below).

Moreover, Fuinhas et al. (2021, pp. 19–234) add that during the period between 1990 and 2014, GHG emissions had an increase of 0.7%. In 1990, a value of 3.414 $MtCO_2eq$ (metric tons per capita) was registered, and in 2014, the emissions reached a value of 4.020 $MtCO_2eq$ 70% of this increase is related to the consumption of energy, 35%, and agriculture, forestry and other land use (AFOLU), 35%. In this energy consumption in the LAC region, liquid fuels account for 60.8%, while coal is only a modest contributor, with 7.6% in 2013. Regarding the structure of GHGs in the LAC region, electricity and heat production had a participation of 29% in 1990, AFOLU 66%, industry 2% and waste 3%. However, in 2014, the electricity and heat production had the participation of 48%, AFOLU 23%, industry 4% and waste 6% (see Fig. 10.3 below).

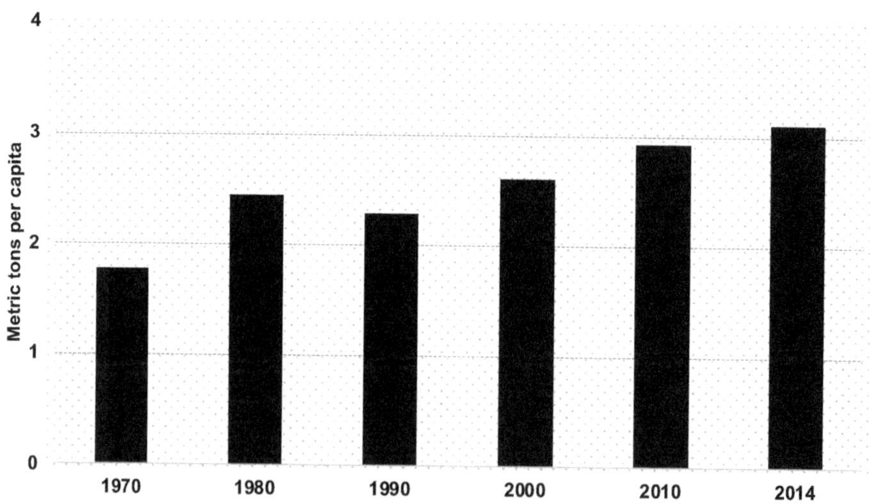

Figure 10.2 CO_2 emissions (metric tons per capita) in the Latin American and Caribbean (LAC) region between 1970 and 2014.
This figure was created by Fuinhas, J.A., Koengkan, M., & Santiago, R. (2021). Physical capital development and energy transition in Latin America and the Caribbean. Elsevier. https://doi.org/10.1016/C2020-0-01491-X and was based on World Bank Open Data (2021). http://www.worldbank.org/.

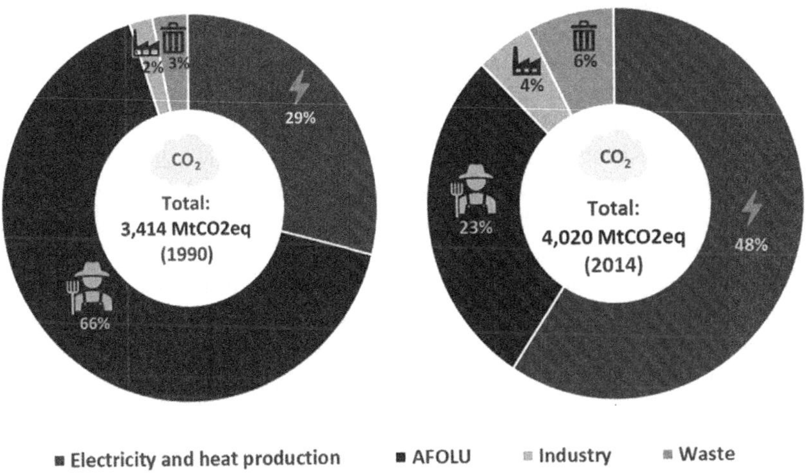

Figure 10.3 Direct greenhouse gases (GHGs) in the Latin American and Caribbean (LAC) region per sector in 1990 and 2014. Total GHGs in $MtCO_2eq$/year.
This figure was created by Fuinhas, J.A., Koengkan, M., & Santiago, R. (2021). Physical capital development and energy transition in Latin America and the Caribbean. Elsevier. https://doi.org/10.1016/C2020-0-01491-X and based on Bárcena, A., Samaniego, J., Galindo, L.M., Carbonell, J.F., Alatorre, J.E., Stockins, P., Reyes, O., Sánchez, L., & Mostacedo, J. (2019). A economia da mudança climática na América Latina e no Caribe (pp. 1−61). CEPAL. https://repositorio.cepal.org/bitstream/handle/11362/44486/1/S1801217_pt.pdf.

Despite the growth in emissions of 0.7% between 1990 and 2014, the region is a minor contributor per capita to the world's GHG, accounting for about 11% of total global emissions (e.g. Fuinhas et al., 2017, 2021, pp. 19−234). Indeed, the increase of CO_2 emissions in the LAC region is directly related to energy consumption.

This chapter will approach the impact of the obesity epidemic on CO_2 emissions. The following research question was formulated. Can the obesity epidemic increase CO_2 emissions in the LAC region? A good starting point for analysing these phenomena is to choose a group of countries that have experienced fast economic, social and environmental transformations but were similar enough to be handled as a panel. Thus, this chapter will approach 20 countries from the LAC region, i.e. Argentina, Belize, Bolivia, Brazil, Chile, Colombia, Costa Rica, Dominican Republic, Ecuador, El Salvador, Guatemala, Guyana, Honduras, Jamaica, Mexico, Nicaragua, Panama, Paraguay, Peru and Uruguay, for the period between 1991 and 2016. Moreover, this research will use the fixed-effects model to carry out this empirical investigation.

10.2 Literature review

Carbon dioxide emissions increased steadily in the LAC region over the past decades, from 270 million tonnes in 1990 to 463 million tonnes in 2019 (Friedlingstein et al., 2020). In this section, we will review the research on some of the main drivers behind this growth.

10.2.1 Obesity

The obesity epidemic causes an increase in CO_2 emissions through several channels: it increases the energy needed to transport heavier people (Jacobson & King, 2011); the demand for food production raises, in particular for highly intensive energy food types, such as transfats (Koengkan & Fuinhas, 2021 and Michaelowa & Dransfeld, 2008); waste management demand increases (Michaelowa & Dransfeld, 2008); and it reduces outdoor activities and fosters the intensive use of home appliances and motor transportation (Koengkan & Fuinhas, 2021).

Several authors have attempted to quantify the impact of the obesity epidemic on CO_2 and other greenhouse gas (GHG) emissions. Squalli (2017) finds that a 10% in obesity implies a 0.7% increase in CO_2 emissions in the United States, while Magkos et al. (2020) report that obesity generates a 1.6% increase in GHG emissions worldwide. In a study involving OECD countries, Michaelowa and Dransfeld (2008) find that a 5 kg increase in the average weight generates 10 million tonnes CO_2eq of additional GHG emissions from the transportation sector and 400 million tonnes of life-cycle emissions from the production of obesity-causing food products. Toti et al. (2019) report that metabolic food waste (the amount of food leading to excess body fat) is responsible for 239 million tonnes CO_2eq of GHG emissions worldwide and 34 million tonnes CO_2eq in Latin America.

10.2.2 Economic growth

In the early stages of their development process, countries need to use increasingly higher quantities of resources and energy to grow, driving CO_2 emissions. Later, as populations reach an acceptable level of material well-being, concerns about environmental preservation start weighing on governmental policies, and emissions may decrease. This hump-shaped relation between economic growth and CO_2 emissions, popularised by Grossman and Krueger (1991), has been thoroughly tested in Latin America and other regions.

Many studies focusing on different countries in the LAC region report evidence of a positive impact of economic growth on CO_2 emissions, such as Koengkan et al. (2019), (2020a), (2021), Jebli et al. (2019) and Valencia-Herrera et al. (2020). In addition, Román-Collado and Morales-Carrión (2018) estimate that 53% of the increase in CO_2 emissions in 20 Latin American countries between 1990 and 2013 was driven by economic growth.

The environmental Kuznets curve hypothesis supports Latin America in the research conducted by Hanif (2017) and Sapkota and Bastola (2017), which raises hope that countries in the region will eventually reach a development stage when growth no longer causes environmental degradation.

10.2.3 Fossil fuel

Liquid and gaseous fossil fuels accounted for the bulk of Latin America CO_2 emissions. In 1990, the emissions shares of liquid and gaseous fossil fuels were 76.38%

and 19.36%, respectively, while in 2014, the share of liquids decreased to 61.50%, and the share of gas increased to 26.36% (World Bank Open Data, 2021). The reason for this persistently huge weight of fossil fuels in Latin American CO_2 emissions resides in the increasing energy demand in the region and the countries' inability to promote a sufficiently swift transition to renewable energy sources (see Chapter 9). According to Roman-Collado (2018), the increase in the weight of fossil fuels in the total primary energy sources is responsible for 7% of the emissions increase in Latin America between 1990 and 2013, while the higher emissions per unit of fossil fuel, due to the change of energy mix, account for 4% of this increase.

Several authors have studied the effect of fossil fuel consumption on CO_2 emissions in Latin America and the Caribbean. Koengkan et al. (2021), using quantile regression for 19 LAC countries, show fossil fuel consumption drives emissions in the 25th and 50th quantiles, but not in the 75th. Koengkan et al. (2020b) estimate a long-run elasticity between fossil fuel consumption and emissions close to 0.036. Using a panel of 16 developing countries, Anser et al. (2020) also conclude that fossil fuels degrade the environment through higher emissions. A different approach was taken by Hanif (2017), who focused on the fossil fuel energy share to assess its effect on emissions. His results corroborate those of the previous authors.

10.2.4 Renewable energy

The LAC region is endowed with exceptional conditions for producing energy from renewable sources, such as wind, solar and hydroelectric (Aghahosseini et al., 2019 and Barbosa et al., 2017). However, the progress in expanding energy production from renewable sources was slow, and its share in total energy consumption has been decreasing in recent years (Washburn & Pablo-Romero, 2019).

Renewable energy will, unequivocally, play a decisive role in replacing fossil fuel sources and mitigating CO_2 emissions in the coming years. There is already some evidence of its effect on reducing emissions in the region. Among the research that supports the negative relationship between renewable energy consumption and carbon dioxide emissions, Koengkan et al. (2020b, 2021) and Anser et al. (2020) focus on groups of countries from LAC, while Robalino-Lopez et al. (2015) consider a single one (Venezuela). Furthermore, Koengkan and Fuinhas (2020) show that an increase in the ratio between renewable and fossil fuel energy consumption mitigates emissions. On the other hand, some authors reach the opposite conclusion, such as Zaman & Moemen (2017), who find a positive association between electricity production from renewable sources and carbon dioxide emissions.

10.2.5 Land use

Agriculture, forest and other land use (AFOLU) accounts for 42% of Latin America and the Caribbean's CO_2 emissions (Saravia-Matus et al., 2019). This figure is much higher than in any other world region (Lamb et al., 2021). Latin America has the world's largest reserve of agricultural land. A large part of these AFOLU emissions is related to the rapid expansion of agricultural land and production during the 21st century at the expense of forest land (Hong et al., 2021 and Houghton & Nassikas, 2017). Between 2001 and 2013, 17% of new cropland and 57% of new pastureland replaced forests (Graesser et al., 2015). The increase in emissions was also driven by a surge in livestock farming, both directly (grazing) and indirectly, due to land clearing for producing feed crops (Bennetzen et al., 2016).

The LAC region's environmental performance of the agricultural sector has improved in the last decades, as it managed to increase production faster than emissions (Bennetzen et al., 2016). Remarkably, the Caribbean region achieved strong decoupling, as it raised production while reducing emissions from this sector (Saravia-Matus et al., 2019).

Latin America is currently facing a conundrum, as it is in an enviable position to satisfy the upcoming surge in the world demand for food products. However, it must not do so at the expense of its forests and GHG emissions (Vosti et al., 2011). Thus, it must find a balance between production growth and environmental preservation through better practices that increase productivity and a rational expansion of agricultural land (Bataille et al., 2020).

10.2.6 Food production

The world food and feed demand is predicted to increase between 50% and 85% from 2009 to 2030 (Vosti et al., 2011), and a sizable part of this growth will likely be satisfied by Latin American farmers, given the region's vast reserves of agricultural land (Graesser et al., 2015). Thus, it is crucial to improve the productivity of LAC agriculture and reduce its GHG emissions, to foster production growth without compromising environmental sustainability.

Fossil fuel energy consumption and the associated CO_2 emissions in agriculture stem from its direct use and the production of fertilisers, pesticides and machinery, which are essential for modern agricultural practices. In the LAC region, the recent trend of CO_2 emissions growth was driven by a surge in livestock farming and the reduction of forest land driven by agricultural land expansion (Bennetzen et al., 2016). However, there are some signs of hope as the region managed to achieve a weak decoupling between agricultural production and emissions. That is, the growth rate of emissions has slowed down relative to the growth rate of production (Saravia-Matus et al., 2019).

It is essential to acknowledge that emissions in the food production chain do not stem exclusively from agriculture. Bajan et al. (2020) compare energy consumption in agriculture and food processing in 14 countries. They conclude that a more significant share of energy in the food production chain is devoted to agriculture in emerging countries. In contrast, in developed countries, most of the energy is used in food processing. This issue should be a warning sign for countries in the LAC region. As they transition to a higher development stage and the obesity epidemic worsens, demand for energy- and emissions-intensive processed food will probably rise.

10.3 Data and method

This section will be divided into two parts. The first will approach the group of countries and data/variables used in the chapter, while the second will show the method.

10.3.1 Data

This chapter will use annual data collected from 1991 to 2016 on 20 countries from the LAC region, i.e. **Argentina, Belize, Bolivia, Brazil, Chile, Colombia, Costa Rica, Dominican Republic, Ecuador, El Salvador, Guatemala, Guyana, Honduras, Jamaica, Mexico, Nicaragua, Panama, Paraguay, Peru** and **Uruguay**. The use of time-series between 1991 and 2016 is due to data availability until 2016 for the variable OBESITY for all countries selected. The variables that were chosen to perform this investigation will be shown in Table 10.1 below.

'Log' denotes variables in natural logarithms, 'Obs.' denotes the number of observations in the model, 'Std. Dev' denotes the standard deviation, and 'Min' and 'Max' denote minimum and maximum. All variables are in natural logarithms to standardise the interpretation of results and linearise the relationships between variables. These summary statistics were obtained from the command *sum* of **Stata 16.0**. The board below shows how to obtain the summary statistics of variables.

How to do:

The summary statistics

```
sum l_logco2 l_logobesity l_loggdp_pp l_logfossil l_logrene l_logland_use
l_logfood_prod
```

This section presents the group of countries from the LAC region on which this chapter will focus and the variables used. In the following subsection, we will present the method used to carry out the empirical investigation of this chapter.

10.3.2 Method

The empirical research was performed using a panel with fixed effects (least squares dummy variable) estimator. The panel specification is that expressed in Eq. (10.1)

$$Y_{it} = \alpha_i + X'_{it}\beta + \mu_{it}, \tag{10.1}$$

where α_i, $i = 1, \ldots,$ N, are fixed unknown constants (fixed effects) that are estimated along with β; X_{it}, $i = 1, \ldots,$ N, and $t = 1, \ldots,$ T, is a k-dimensional vector of explanatory variables; and μ_{it} is the error term assumed to be i.i.d. over individuals and time. The fixed effects α_i capture all unobservable time-invariant divergences across individuals.

Some preliminary tests must be performed before the estimators can be considered adequate to handle the characteristics of data and the nature of relationships under analysis.

Table 10.1 Description of variables and summary statistics.

Description of variables	
Variables	**Source**
Carbon dioxide emissions (metric tons per capita) are those stemming from the burning of fossil fuels and the manufacture of cement. In addition, they include carbon dioxide produced during the consumption of solid, liquid and gas fuels and gas flaring. In this investigation, we called this variable **'CO_2'**.	World Bank Open Data (2021)
Share of adults that are obese in (percent). Obesity is defined as having a body mass index (BMI) equal to or greater than 30. BMI is a person's weight in kilograms divided by his or her height in meters squared. In this investigation, we called this variable **'OBESITY'.**	Our World in Data (2021)
Gross domestic product (GDP) per capita based on purchasing power parity (PPP). This variable is converted to international dollars using purchasing power parity rates. An international dollar has the same purchasing power over GDP as the US dollar has in the United States. GDP at purchaser's prices is the sum of gross value added by all resident producers in the country plus any product taxes and minus any subsidies not included in the value of the products. It is calculated without deductions for the depreciation of fabricated assets or depletion and degradation of natural resources. Moreover, this variable is in the current 2017 international dollars. In this investigation, we called this variable **'GDP_PPP'.**	World Bank Open Data (2021)
Electric power consumption (kWh per capita) from fossil fuels (e.g. coal, oil, petroleum and natural gas products). In this investigation, we called this variable **'FOSSIL'.**	World Bank Open Data (2021)
Electricity consumption from renewable sources, excluding hydroelectric (kWh) per capita. This variable measures the electricity consumption from renewable sources, excluding hydroelectric, which includes geothermal, solar, tides, wind, biomass and biofuels. In this investigation, we called this variable **'RENE'**.	World Bank Open Data (2021)
Agricultural land (% of land area) refers to the share of arable land under permanent crops and permanent pastures. Arable land includes land defined by the FAO as land under temporary crops (double-cropped areas are counted once), temporary meadows for mowing or pasture, land under market or kitchen gardens, and temporarily fallow land. Land abandoned as a result of shifting cultivation is excluded. Land under permanent crops is land cultivated with crops that occupy the land for long periods and need not be replanted after each harvest, such as cocoa, coffee and rubber. This category includes land under flowering shrubs, fruit trees, nut trees and vines but excludes land under trees grown for wood or timber. Permanent pasture is land used for five or more years for forage, including natural and cultivated crops. In this investigation, we called this variable **'LAND_USE'**.	World Bank Open Data (2021)

Table 10.1 Description of variables and summary statistics.—cont'd

Description of variables	
Variables	**Source**
Cereal production (metric tons) per capita. This variable measures the production data on cereals related to crops harvested for dry grain only. Cereal crops harvested for hay or harvested green for food, feed or silage and those used for grazing are excluded. In this investigation, we called this variable **'FOOD_PROD'**.	World Bank Open Data (2021)

Summary statistics

Variables	Obs.	Mean	Std. Dev	Min	Max
$LogCO_2$	520	0.4975	0.5692	−0.7549	1.5552
LogOBESITY	520	2.7910	0.2778	2.0918	3.3428
LogGDP_PPP	520	8.8576	0.5317	7.5885	10.1393
LogFOSSIL	452	4.1222	0.3390	3.0029	4.5114
LogRENE	426	3.5427	1.4730	−1.8832	7.0419
LogLAND_USE	520	3.4999	0.6152	1.7090	4.4483
LogFOOD_PROD	520	−1.9132	1.3903	−7.4333	0.2754

(Log) denotes variables in the natural logarithms.

Thus, the empirical analysis begins with verifying variables' properties, including assessing the time-variability of variables, variable patterns, detecting potential outliers and potential structural breaks, and checking for cross-sectional dependence, order of integrations, normality, multicollinearity and panel effects (see preliminary tests in Table 10.2 below). Next, a battery of post-estimation tests was performed to confirm the appropriateness of estimations that were realised (see post-estimation tests in Table 10.2 below). Indeed, these tests are required to assess the global significance of the estimated models, the presence of heteroscedasticity, serial correlation and cross-sectional independence in the residuals. The empirical analysis was carried out using the econometric software **Stata 16.0**.

10.4 Empirical results

This section will approach the empirical results of our investigation. Therefore, we will show the results from preliminary tests, main model regression and post-estimation tests. The first step that we will undertake is carrying out the preliminary tests, such as (1) VIF test; (2) Pesaran CD test; (3) unit root tests; (4) Hausman test; and (5) bias-corrected LM-based test.

To identify the presence of multicollinearity between the variables, the variance inflation factor (VIF) test that informs on the presence of multicollinearity will be computed. Table 10.3 below shows the results of the VIF-test.

The results from the VIF test show that the presence of multicollinearity is not a concern, given the low VIF and mean VIF values registered, which are lower than the usually accepted benchmark of 10, in the case of the VIF values, and six in the case of the mean VIF values (Koengkan et al., 2020). This test helps us understand the degree of

Table 10.2 Preliminary and post-estimation tests.

Preliminary tests	
Test	**Finality**
Descriptive statistics analysis	Confirm the time-variability of variables, i.e. if they are time-variant, time-quasi-variant or time-invariant and detect potential outliers.
Cross-sectional dependence (CSD) test (Pesaran, 2004)	This test identifies the presence of cross-sectional dependence (CSD) in the panel's data. The null hypothesis of this test is the presence of cross-sectional independence CD ~ N(0,1).
Levin-Lin-Chu unit-root test (LLC-test) (Levin et al., 2002)	This test verifies the presence of unit roots in the variables. The null hypothesis of the LLC-test is that all the panels contain a unit root.
Panel unit root test (CIPS) test (Pesaran, 2007)	This test verifies the presence of unit roots in the variables. The panel unit root test (CIPS) null hypothesis is that all series have a unit root.
Variance inflation factor (VIF) test (Belsley et al., 1980)	This test verifies the presence of multicollinearity between the variables.
Hausman test	This test identifies heterogeneity, i.e. whether the panel has random effects (RE) or fixed effects (FE).
Bias-corrected LM-based test (Born & Breitung, 2015)	This test checks the presence of serial correlation in fixed-effects panel models. The null hypothesis of this test is the non-presence of autocorrelation up to the second order.
Post-estimation tests	
Test	**Finality**
Wooldridge test (Wooldridge, 2002)	Test for serial correlation in panel-data models. The null hypothesis of the Wooldridge test is no first-order autocorrelation.
Modified Wald test (Greene, 2000)	Test for cross-sectional independence in the residuals of a fixed effect regression model. The null hypothesis of this test is cross-sectional independence in the residuals. Please note that this Breusch-Pagan test is not that commonly employed to test for heteroscedasticity.

Table 10.3 VIF test.

Variables	VIF	1/VIF	Mean VIF
	$LogCO_2$		
LogOBESITY	3.03	0.3295	1.93
LogGDP_PPP	2.34	0.4282	
LogFOSSIL	2.03	0.4924	
LogRENE	1.83	0.5471	
LogLAND_USE	1.10	0.9108	
LogFOOD_PROD	1.24	0.8045	

(Log) denotes variables in the natural logarithms.

multicollinearity that may be present in our models, which can lead to problems in estimation (e.g. Koengkan & Fuinhas, 2021; Koengkan et al., 2021 and Santiago et al., 2020). The results of the VIF test were obtained from the command *estat vif* in **Stata 16.0.** The board below shows how to carry out and obtain the results from the VIF test.

How to do:

```
** The variance inflation factor test**

reg l_logco2 l_logobesity l_loggdp_pp l_logfossil l_logrene l_logland_use
l_logfood_prod

estat vif
```

After checking the existence of multicollinearity between the variables, we need to identify cross-sectional dependence (CSD) in the panel data. Therefore, the Pesaran CD test developed by Pesaran (2004) was used in this investigation. The null hypothesis of this test is the non-presence of cross-section dependence CD ~ N (0,1) for $N \rightarrow \infty$, and that T is sufficiently large. Table 10.4 below shows the results from the Pesaran CD test.

The results from the CSD test show the presence of cross-section dependence in variables LogCO_2, LogOBESITY, LogGDP_PPP, LogFOSSIL and LogFOOD_PROD. The presence of cross-section dependence can signify that the countries selected in our study share the same characteristics and shocks mentioned by Fuinhas et al. (2017) and Koengkan et al. (2019). The results also indicate the non-presence of cross-sectional dependence for the variable LogLAND_USE. The variable LogRENE could not be computed by the CSD test because this test requires strongly balanced data. The results of the Pesaran CD test were obtained from the command *xtcd* in **Stata 16.0**. The board below shows how to carry out and obtain the results from the Pesaran CD test.

How to do:

```
**The Pesaran CD-test**

xtcd l_logco2 l_logobesity l_loggdp_pp l_logfossil l_logrene l_logland_use
l_logfood_prod
```

Table 10.4 Pesaran CD test.

Variables	CD-test	*P*-value		Corr	Abs (corr)
LogCO_2	29.31	0.000	***	0.417	0.589
LogOBESITY	70.23	0.000	***	0.999	0.999
LogGDP_PPP	68.35	0.000	***	0.972	0.972
LogFOSSIL	20.93	0.000	***	0.338	0.494
LogRENE	N.A				
LogLAND_USE	−0.68	0.498		−0.011	0.581
LogFOOD_PROD	2.49	0.013	**	0.040	0.367

*** and ** denote statistical significance at 1% and 5% levels, respectively; (Log) denotes variables in the natural logarithms; NA denotes not available.

After the CSD test, it is necessary to verify the order of integration of the variables. To this end, the panel unit root tests, such as the LLC-test developed by Levin et al. (2002) and the CIPS-test developed by Pesaran (2007), were calculated. Table 10.5 below shows the results from the unit root tests.

The results of the LLC-test indicate that the variables LogCO_2, LogOBESITY, LogLAND_USE and LogFOOD_PROD are stationary, while the variable LogGDP_PPP is on the borderline between I(0) and I(1) of the order of integration. However, the variables LogFOSSIL and LogRENE could not be computed by the LLC-test because the test requires strongly balanced data. Moreover, the results from the CIPS-test indicate that the variables LogOBESITY, LogFOSSIL and LogFOOD_-PROD are stationary, while the variables LogCO_2, LogGDP_PPP and LogRENE are on the borderline between I(0) and I(1) of the order of integration. However, the variable LogLAND_USE indicates non-stationarity. The results of the LLC-test and CIPS-test were obtained from the commands *xtunitroot* and *multipurt* in **Stata 16.0**. The board below shows how to carry out and obtain the results from the LLC and CIPS tests.

How to do:

```
** The Levin-Lin-Chu unit-root test**

xtunitroot llc l_logco2
xtunitroot llc l_logco2, trend

xtunitroot llc l_logobesity
xtunitroot llc l_logobesity, trend

xtunitroot llc l_loggdp_ppp
xtunitroot llc l_loggdp_ppp, trend

xtunitroot llc l_logfossil
xtunitroot llc l_logfossil, trend

xtunitroot llc l_logrene
xtunitroot llc l_logrene, trend

xtunitroot llc l_logland_use
xtunitroot llc l_logland_use, trend

xtunitroot llc l_logfood_prod
xtunitroot llc l_logfood_prod, trend

**The CIPS-test**

multipurt l_logco2 l_logobesity l_loggdp_pp l_logfossil l_logrene l_logland_use
l_logfood_prod,lags(1)
```

After identifying the unit-roots of variables, we need to find individual effects in the model. To this end, the Hausman test, which compares the random (RE) and fixed effects (FE), was computed. The null hypothesis of this test is that the difference in coefficients is not systematic, where the random effects are the most suitable estimator (Fuinhas et al., 2017). The results of this test are presented in Table 10.6 below.

Table 10.5 Unit root tests.

Variables	Levin-Lin-Chu unit-root test (LLC-test)					Panel unit root test (CIPS) (Zt-bar)				
	Without trend			With trend		Without trend			With trend	
	Lags	Adjusted t		Adjusted t		Lags	Zt-bar		Zt-bar	
LogCO_2	1	−4.3015	***	−2.2363	**	1	−2.046	**	−0.884	
LogOBESITY	1	−19.4790	***	−3.1963	***	1	−2.306	**	−2.736	***
LogGDP_PPP	1	0.1277		−2.9921	**	1	−1.099		−1.619	**
LogFOSSIL	1	NA				1	−1.585	**	−1.578	**
LogRENE	1	NA				1	0.662		−1.493	**
LogLAND_USE	1	−3.1546	***	−2.6453	**	1	−0.303		−0.762	
LogFOOD_PROD	1	−2.6409	**	−1.4412	*	1	−2.243	**	−1.760	**

***, ** and * denote statistically significant at the 1%, 5% and 10% levels, respectively; (Log) denotes variables in the natural logarithms; NA denotes not available.

Table 10.6 Hausman test.

Variables	(b) Fixed	(B) Random	(b-B) Difference	Sqrt(diag(V_b-V-B)) S.E.
Trend	−0.0381	−0.0323	−0.0057	0.0024
LogOBESITY	0.6435	0.4879	0.1556	0.0611
LogGDP_PPP	0.7274	0.6926	0.0348	0.0228
LogFOSSIL	0.7321	0.8312	−0.0990	0.0414
LogRENE	0.0382	0.0418	−0.0036	0.0017
LogLAND_USE	0.2142	0.0462	0.1679	0.0890
LogFOOD_PROD	0.0522	0.0119	0.0402	0.0161
Chi2 (7)	15.37***			

*** denotes statistically significant at the 1% level; (Log) denotes variables in the natural logarithms.

The results of this test show that the null hypothesis should be rejected (**chi2 (7) = 15.37*****, statistically significant at the 1% level). That is, there is the presence of fixed effects in the model. The results of the Hausman test were obtained from the command *hausman* with option *sigmaless* in **Stata 16.0**. The board below shows how to carry out and obtain the results from the Hausman test.

How to do:

```
**The Hausman test**

qui:xtreg l_logco2 trend l_logobesity l_loggdp_pp l_logfossil l_logrene
l_logland_use l_logfood_prod,fe
estimates store fixed
qui:xtreg l_logco2 trend l_logobesity l_loggdp_pp l_logfossil l_logrene
l_logland_use l_logfood_prod,re
estimates store random
hausman fixed random, sigmaless
```

After identifying fixed effects, it is necessary to check serial correlation in the fixed-effects panel model. To this end, the bias-corrected LM-based test developed by Born and Breitung (2015) was computed. The null hypothesis of this test is the non-presence of autocorrelation up to the second order. Table 10.7 below shows the results from the bias-corrected LM-based test.

Table 10.7 Bias-corrected LM-based test.

Variables	LM (k)-stat	
LogCO_2	7.90	***
LogOBESITY	12.72	***
LogGDP_PPP	8.95	***
LogFOSSIL	3.23	***
LogRENE	3.49	***
LogLAND_USE	4.37	***
LogFOOD_PROD	2.96	***

*** denotes statistically significant at the 1% level; (Log) denotes variables in the natural logarithms.

The results from the bias-correct LM-based test indicate the presence of autocorrelation up to the second order in the fixed-effects panel model, where the null hypothesis can be rejected. The results of the bias-corrected LM-based test were obtained from the command *xtqptest* in **Stata 16.0**. The board below shows how to carry out and obtain the results from the bias-corrected LM-based test.

How to do:

**** The Bias-corrected LM-based test ****

```
xtqptest l_logco2 l_logobesity l_loggdp_pp l_logfossil l_logrene l_logland_use
l_logfood_prod, order(1)
```

After the preliminary tests, we can take the second step, which is the estimation of fixed effects model regression to assess the impact of obesity on environmental degradation (CO_2 emissions) in LAC countries, as well as the third step, which is performing post-estimation tests (e.g. Wooldridge test and modified Wald test). Indeed, in the fixed effects model, we opted to compute the following estimators: fixed effects (FE), FE robust standard errors (FE Robust) and FE Driscoll and Kraay (FE D.-K.). The Driscoll and Kraay (1998) estimator was applied (e.g. Fuinhas et al., 2017) to cope with the presence of heteroscedasticity, contemporaneous correlation, first-order autocorrelation and cross-sectional dependence (spatial dependence or spatial regimes). Moreover, this estimator is a matrix estimator that generates robust standard errors for several phenomena found in the sample errors. Table 10.8 below provides the results of fixed-effects model regression and the results of the post-estimation tests.

Table 10.8 Fixed-effects model and post-estimation tests.

	Dependent variable (LogCO2)			
Independent variables	**FE**		**FE Robust**	**FE D.K**
Trend	−0.0381	***	***	***
LogOBESITY	0.6436	***	*	***
LogGDP_PPP	0.7275	***	***	***
LogFOSSIL	0.7322	***	***	***
LogRENE	0.0383	***	***	***
LogLAND_USE	0.2142	*		*
LogFOOD_PROD	0.0522	**		*
Constant	−11.0479	***	***	***
Obs	411		411	411

Post-estimation tests for fixed-effects model

Statistics	Wooldridge test	Modified Wald test
	F(1,17) = 42.133***	chi2 (18) = 1013.43***

***, ** and * denote statistically significant at the 1%, 5% and 10% levels, respectively; (Log) denotes variables in the natural logarithms.

The results from the fixed-effects model regression indicate that the variable LogOBESITY increases environmental degradation (LogCO_2) by 0.6436, while the other variable LogGDP_PPP increase it by 0.7275, LogFOSSIL by 0.7322, LogRENE by 0.0383, LogLAND_USE by 0.2142 and LogFOOD_PROD by 0.0522. Moreover, the results from the Wooldridge test indicate that the null hypothesis can be rejected, indicating the presence of the first-order autocorrelation between the variables of the model. In contrast, the modified Wald test indicates that the null hypotheses can be rejected, indicating the presence of cross-sectional dependence in the residuals of model regression.

The fixed effects model regression results were obtained from the command *xtreg* with options, *fe*, *fe robust* and *fe lag(1)* in **Stata 16.0**. Moreover, the results from the post-estimation test, such as the Wooldridge test and modified Wald test, were obtained from the commands *xtserial* and *xttest2* in **Stata 16.0**. The board below shows how to carry out and obtain the results from the fixed-effects model regression and the results from the post-estimation tests.

How to do:

****The fixed effects model (FE) ****

```
xtreg l_logco2 trend l_logobesity l_loggdp_pp l_logfossil l_logrene
l_logland_use l_logfood_prod,fe
```

****The fixed effects model (FE Robust) ****

```
xtreg l_logco2 trend l_logobesity l_loggdp_pp l_logfossil l_logrene
l_logland_use l_logfood_prod,fe robust
```

****The fixed effects model (FE Driscoll-Kraay) ****

```
xtreg l_logco2 trend l_logobesity l_loggdp_pp l_logfossil l_logrene
l_logland_use l_logfood_prod, fe lag(1)
```

Post-estimation tests

****Wooldridge test****

```
xtserial l_logco2 trend l_logobesity l_loggdp_pp l_logfossil l_logrene
l_logland_use l_logfood_prod
```

****Modified Wald test****

```
xtreg l_logco2 trend l_logobesity l_loggdp_pp l_logfossil l_logrene
l_logland_use l_logfood_prod,fe
xttest2
```

Fig. 10.4 below summarises the impact of independent variables on dependent ones. This figure was based on the results from the fixed-effects model regression.

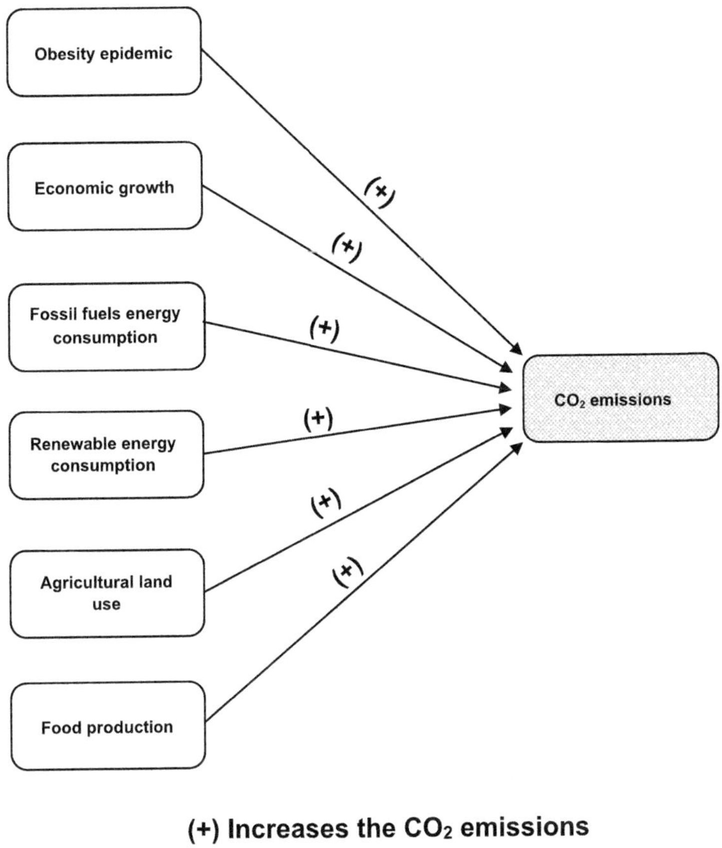

Figure 10.4 Summary of the impact of independent variables on dependent ones. This figure created by the authors.

In this section, we showed the empirical results of this investigation. The following section provides a discussion of the empirical results.

10.5 Discussion

As mentioned before, this section will approach the possible explanations for the results found in our empirical investigation. After identifying that the obesity epidemic encourages environmental degradation by increasing CO_2 emissions, the next step is to answer the following question: **What is the possible explanation for this phenomenon?** One possible way of explaining this effect is that the obesity epidemic is caused by an increase in the consumption of processed foods from multinational food corporations, fast-food chains and multinational supermarket chains, as well as the food production on farms, as indicated by some authors (e.g. Fox et al., 2019; Gerbens-Leenes et al., 2010; Koengkan & Fuinhas, 2021 and Popkin, 1998). The increased consumption of processed foods from multinational food corporations and farms will positively impact energy consumption from non-renewable energy sources. The evidence that the obesity epidemic increases food production and consequently fossil fuel consumption was found in Chapters 8 and 9 of this book.

Another explanation for this phenomenon is related to the reduction of outdoor activities by the obesity problem. This reduction will consequently encourage the use of intensive motorised transportation, screen-viewing leisure activities and home appliances, as indicated by some authors (e.g. Bell et al., 2002; Koengkan & Fuinhas, 2021 and Sobal, 2001). Koengkan and Fuinhas (2021) also add that the increase in the use of home appliances and motorised transportation has implications for energy demand from fossil fuel energy sources, where the consumption of this kind of sources increases considerably.

Additionally, the positive impact of the obesity epidemic on CO_2 emissions is also indirectly related to economic growth, globalisation and urbanisation. Therefore, the levels of obesity are related to the increasing economic activity, where the development caused by economic growth has effects on dietary changes (e.g. Springmann et al., 2016). Indeed, the income transition from lower to high income caused by this process tends to cause the consumption of fatty and energy-dense animal food sources. Consequently, income contributes to the rise in obesity levels, except for countries where home-produced food is prevalent (e.g. Gerbens-Leenes et al., 2010 and Roskam et al., 2010).

The globalisation process also causes an increase in obesity due to dietary changes. According to Popkin (1998), Fox et al. (2019) and Koengkan and Fuinhas (2021), the process of globalisation will contribute to the extension of food chains. As mentioned by the authors above, this extension will enable economies of scale in food production processes. Consequently, the economies of scale in food production processes will enable a diet rich in energy-caloric foods. Food with high sugar and salt contents is less expensive and more accessible to lower-income classes in developed countries. Moreover, the supply of processed foods that are unhealthy is related to the increase in multinational food companies, multinational supermarkets and fast-food chains caused by the globalisation process.

The process of urbanisation also plays a role in the increase in the obesity problem. That is, the process of urbanisation allows better accessibility to food due to supermarkets, multinational supermarkets and fast-food chains offering a ready supply of processed foods. That consequently causes the decline of farm stands and open markets with healthier foods (Reardon et al., 2003). This same process also exposes people to mass media marketing of food and beverages that influence traditional diets (Hawkes, 2006). Moreover, urbanisation increases car use and reduces walking or biking for transportation or leisure and consequently contributes to obesity, and obesity increases car use.

All these explanations align with the findings of Koengkan and Fuinhas (2021). They found that economic growth, globalisation and urbanisation positively affect the overweight problem in the European region. They consequently encourage energy consumption from non-renewable energy sources and subsequently increase CO_2 emissions/environmental degradation. Furthermore, the explanations above align with the results from the complementary analysis that was carried out in this investigation.

10.6 Conclusion

This chapter approached the impact of the obesity epidemic on CO_2 emissions in 20 countries from the LAC region between 1991 and 2016. This research used the fixed-effects model as a method for the approach.

The results from the preliminary tests, such as the VIF test, showed that the presence of multicollinearity is not a concern, given the low VIF and mean VIF values registered, which are lower than the usually accepted benchmark of 10, in the case of the VIF values, and six in the case of the mean VIF values. On the other hand, the results from the CSD test showed the presence of cross-section dependence in variables LogCO_2, LogOBESITY, LogGDP_PPP, LogFOSSIL and LogFOOD_PROD. The presence of cross-section dependence can signify that the countries selected in our study share the same characteristics and shocks. Moreover, the results also indicate the non-presence of cross-section dependence for the variable LogLAND_USE. The variable LogRENE could not be computed by the CSD test because this test requires strongly balanced data.

The results of the LLC-test indicated that the variables LogCO_2, LogOBESITY, LogLAN_USE and LogFOOD_PROD are stationary, while the variable LogGDP_PPP is on the borderline between I(0) and I(1) of the order of integration. However, the variables LogFOSSIL and LogRENE could not be computed by the LLC test because the test requires strongly balanced data. Moreover, the results from the CIPS test indicated that the variables LogOBESITY, LogFOSSIL and LogFOOD_PROD are stationary, while the variables LogCO_2, LogGDP_PPP and LogRENE are on the borderline between I(0) and I(1) of the order of integration. However, the variable LogLAND_USE indicates non-stationarity. The results from the Hausman test indicated the presence of fixed effects, where the null hypothesis should be rejected (chi2 (7) = 15.37***, statistically significant at the 1% level). The results from the bias-corrected LM-based test indicate the non-presence of autocorrelation up to the second order in the fixed-effects panel model, where the null hypothesis cannot be rejected

The results from the fixed-effects model regression indicated that the variable LogOBESITY increases environmental degradation (LogCO_2) by 0.6436, while the other variable LogGDP_PPP increases it by 0.7275, LogFOSSIL by 0.7322, LogRENE by 0.0383, LogLAND_USE by 0.2142 and LogFOOD_PROD by 0.0522. Moreover, the results from the Wooldridge test indicate that the null hypothesis can be rejected, indicating the presence of the first-order autocorrelation between the variables of the model. In contrast, the Modified Wald test indicates that the null hypotheses can be rejected, indicating the presence of cross-sectional dependence in the residuals of model regression.

The aim of the analysis performed in this chapter was to find the main determinants of the increasing levels of CO_2 emissions in LAC countries. The determinants considered were grouped into four major groups: public health, economic growth, energy consumption and food industry growth. The public health determinant accounted for the obesity pandemic that LAC countries have experienced since the 1990s; the increasing GDP per capita measures economic growth; the energy consumption includes the consumption of energy based both on fossil sources and in renewable sources; and finally, the food industry considers the share of land use for food crops and the food production indicator.

The most relevant results are obesity's increasing effect on CO_2 emissions and the increasing effect of land use and food production on CO_2 emissions. In Chapter 8, it was shown that obesity leverages food production, so the effects are complementary and reinforce each other. In this way, policy measures aiming at reducing the prevalence of obesity in LAC countries may have direct and indirect mitigating effects on CO_2 emissions. One possible explanation for this may be a shift in the food industry, favoring more environmentally friendly production processes as consumers change their preferences to less processed food.

Finally, the increasing use of renewable energy does not end with CO_2 emissions, and there is a positive correlation between those emissions and energy consumption, no matter the source. However, the elasticity of CO_2 emissions concerning renewable energy is low. So the impact of increasing the use of renewable energy on CO_2 emissions is low, meaning that it is an environmentally friendly source of energy. In Chapter 9, we found that the consumption of fossil fuel energy expanded as food production and obesity prevalence also expanded. Therefore, policy measures to reduce obesity in LAC countries may contribute to the falling consumption of fossil energy directly and indirectly through lower food production. This fall in the use of fossil energy will cut back CO_2 emissions, contribute to the accomplishment of the international agreements on pollution and climate change, and contribute to the improvement of general standards of public health.

References

Aghahosseini, A., Bogdanov, D., Barbosa, L. S. N. S., & Breyer, C. (2019). Analysing the feasibility of powering the Americas with renewable energy and inter-regional grid interconnections by 2030. *Renewable and Sustainable Energy Reviews, 105*, 187–205. https://doi.org/10.1016/j.rser.2019.01.046

Anser, M. K., Hanif, I., Alharthi, M., & Chaudhry, I. S. (2020). Impact of fossil fuels, renewable energy consumption and industrial growth on carbon emissions in Latin American and Caribbean economies. *Atmosfera, 33*(3), 201–213. https://doi.org/10.20937/ATM.52732

Bajan, B., Mrówczyńska-Kamińska, A., & Poczta, W. (2020). Economic energy efficiency of food production systems. *Energies, 13*, 5826. https://doi.org/10.3390/en13215826

Bárcena, A., Samaniego, J., Galindo, L. M., Carbonell, J. F., Alatorre, J. E., Stockins, P., Reyes, O., Sánchez, L., & Mostacedo, J. (2019). *A economia da mudança climática na América Latina e no Caribe* (pp. 1–61). CEPAL. https://repositorio.cepal.org/bitstream/handle/11362/44486/1/S1801217_pt.pdf.

Bataille, C., Waisman, H., Briand, Y., Svensson, J., Vogt-Schilb, A., Jaramillo, M., Delgado, R., Arguello, R., Clarke, L., Wild, T., Lallana, F., Bravo, G., Nadal, G., Le Treut, G., Godinez, G., Quiros-Tortos, J., Pereira, E., Howells, M., Buira, D., … Imperio, M. (2020). Net-zero deep decarbonization pathways in Latin America: Challenges and opportunities. *Energy Strategy Reviews, 30*, 100510. https://doi.org/10.1016/j.esr.2020.100510

Bell, A. C., Ge, K., & Popkin, B. M. (2002). The road to obesity or the path to prevention: Motorized transportation and obesity in China. *Obesity Research, 10*, 277–283. https://doi.org/10.1038/oby.2002.38

Belsley, D. A., Kuh, E., & Welsch, R. E. (1980). *Regression diagnostics: Identifying influential data and sources of collinearity*. Wiley. https://doi.org/10.1002/0471725153

Ben Jebli, M., Ben Youssef, S., & Apergis, N. (2019). The dynamic linkage between renewable energy, tourism, CO_2 emissions, economic growth, foreign direct investment, and trade. *Latin American Economic Review, 28*(1), 2. https://doi.org/10.1186/s40503-019-0063-7

Bennetzen, E. H., Smith, P., & Porter, J. R. (2016). Agricultural production and greenhouse gas emissions from world regions-The major trends over 40 years. *Global Environmental Change, 37*, 43–55. https://doi.org/10.1016/j.gloenvcha.2015.12.004

Born, B., & Breitung, J. (2015). Testing for serial correlation in fixed-effects panel data models. *Econometric Reviews, 35*(7), 1290–1316. https://doi.org/10.1080/07474938.2014.976524

Breusch, T. S., & Pagan, A. R. (1979). A simple test for heteroscedasticity and random coefficient variation. *Econometrica, 47*, 1287–1294. https://doi.org/10.2307/1911963

De Barbosa, L. S. N. S., Bogdanov, D., Vainikka, P., & Breyer, C. (2017). Hydro, wind and solar power as a base for a 100% renewable energy supply for South and Central America. *PLoS ONE, 12*(3). https://doi.org/10.1371/journal.pone.0173820

Driscoll, J., & Kraay, A. C. (1998). Consistent covariance matrix estimation with spatially dependent data. *Review of Economics and Statistics, 80*(4), 549–560.

Fox, A., Feng, W., & Asal, V. (2019). What is driving global obesity trends? Globalisation or "modernisation"? *Globalisation and Health, 15*(32), 1–16. https://doi.org/10.1186/s12992-019-0457-y

Friedlingstein, P., O'Sullivan, M., Jones, M. W., Andrew, R. M., Hauck, J., Olsen, A., Peters, G. P., Peters, W., Pongratz, J., Sitch, S., Le Quéré, C., Canadell, J. G., Ciais, P., Jackson, R. B., Alin, S., Aragão, L. E. O. C., Arneth, A., Arora, V.Bates, N. R., … (2020). Global carbon budget 2020. *Earth System Science Data, 12*, 3269–3340. https://doi.org/10.5194/essd-12-3269-2020

Fuinhas, J. A., Koengkan, M., & Santiago, R. (2021). *Physical capital development and energy transition in Latin America and the caribbean*. Elsevier. https://doi.org/10.1016/C2020-0-01491-X

Fuinhas, J. A., Marques, A. C., & Koengkan, M. (2017). Are renewable energy policies upsetting carbon dioxide emissions? The case of Latin America countries. *Environmental Science and Pollution Research, 24*(17), 15044–15054. https://doi.org/10.1007/s11356-017-9109-z

Gerbens-Leenes, P. W., Nonhebel, S., & Krol, M. S. (2010). Food consumption patterns and economic growth. Increasing affluence and the use of natural resources. *Appetite, 55*(3), 597–608. https://doi.org/10.1016/j.appet.2010.09.013

Graesser, J., Aide, T. M., Grau, H. R., & Ramankutty, N. (2015). Cropland/pastureland dynamics and the slowdown of deforestation in Latin America. *Environmental Research Letters, 10*(3). https://doi.org/10.1088/1748-9326/10/3/034017

Greene, W. (2000). *Econometric analysis*. Prentice–Hall.

Grossman, G., & Krueger, A. (1991). *Environmental impacts of a North American free trade agreement*. National Bureau of Economic Research 3914. https://doi.org/10.3386/w3914

Hanif, I. (2017). Economics-energy-environment nexus in Latin America and the Caribbean. *Energy, 141*, 170–178. https://doi.org/10.1016/j.energy.2017.09.054

Hawkes, C. (2006). Uneven dietary development: Linking the policies and processes of globalisation with the nutrition transition, obesity and diet-related chronic diseases. *Globalisation and Health, 2*(4), 1–18. https://doi.org/10.1186/1744-8603-2-4

Hong, C., Burney, J. A., Pongratz, J., Nabel, J. E. M. S., Mueller, N. D., Jackson, R. B., & Davis, S. J. (2021). Global and regional drivers of land-use emissions in 1961-2017. *Nature, 589*(7843), 554–561. https://doi.org/10.1038/s41586-020-03138-y

Houghton, R. A., & Nassikas, A. A. (2017). Global and regional fluxes of carbon from land use and land cover change 1850–2015. *Global Biogeochemical Cycles, 31*, 456–472. https://doi.org/10.1002/2016GB005546

Jacobson, S. H., & King, D. M. (2011). *A societal Cost of obesity in terms of automobile fuel consumption.* Wiley Encyclopedia of Operations Research and Management Science. https://doi.org/10.1002/9780470400531.eorms0003

Koengkan, M., & Fuinhas, J. A. (2020). Exploring the effect of the renewable energy transition on CO_2 emissions of Latin American & Caribbean countries. *International Journal of Sustainable Energy, 39*(6), 515–538. https://doi.org/10.1080/14786451.2020.1731511

Koengkan, M., & Fuinhas, J. A. (2021). Does the overweight epidemic cause energy consumption? A piece of empirical evidence from the European region. *Energy, 236*(1), 119297. https://doi.org/10.1016/j.energy.2020.119297

Koengkan, M., Fuinhas, J. A., & Marques, A. C. (2019). Chapter seven - the relationship between financial openness, renewable and nonrenewable energy consumption, CO_2 emissions, and economic growth in the Latin American countries: An approach with a panel vector auto regression model. *The Extended Energy-Growth Nexus: Theory and Empirical Applications*, 199–229. https://doi.org/10.1016/B978-0-12-815719-0.00007-3

Koengkan, M., Fuinhas, J. A., & Marques, A. C. (2020). The role of financial openness and China's income on fossil fuels consumption: Fresh evidence from Latin American countries. *GeoJournal, 85*, 407–421. https://doi.org/10.1007/s10708-019-09969-1

Koengkan, M., Fuinhas, J. A., & Santiago, R. (2020a). The relationship between CO_2 emissions, renewable and nonrenewable energy consumption, economic growth, and urbanisation in the Southern Common Market. *Journal of Environmental Economics and Policy, 9*(4), 383–401. https://doi.org/10.1080/21606544.2019.1702902

Koengkan, M., Fuinhas, J. A., & Santiago, R. (2020b). Asymmetric impacts of globalisation on CO_2 emissions of countries in Latin America and the Caribbean. *Environment Systems and Decisions, 40*(1), 135–147. https://doi.org/10.1007/s10669-019-09752-0

Koengkan, M., Fuinhas, J. A., & Silva, N. (2021). Exploring the capacity of renewable energy consumption to reduce outdoor air pollution death rate in Latin America and the Caribbean region. *Environmental Science and Pollution Research, 28*, 1656–1674. https://doi.org/10.1007/s11356-020-10503-x

Lamb, W. F., Wiedmann, T., Pongratz, J., Andrew, R., Crippa, M., Olivier, J. G. J., Wiedenhofer, D., Mattioli, G., Khourdajie, A., House, J., Pachauri, S., Figueroa, M., Saheb, Y., Slade, R., Hubacek, K., Sun, L., Ribeiro, S. K., Khennas, S., De La Rue Du Can, S., … Minx, J. (2021). A review of trends and drivers of greenhouse gas emissions by sector from 1990 to 2018. *Environmental Research Letters, 16*(7). https://doi.org/10.1088/1748-9326/abee4e

Levin, A., Lin, C.-F., & Chu, C.-S. J. (2002). Unit root tests in panel data: Asymptotic and finite-sample properties. *Journal of Econometrics, 108*, 1–24. https://doi.org/10.1016/S0304-4076(01)00098-7

Magkos, F., Tetens, I., Bügel, S. G., Felby, C., Schacht, S. R., Hill, J. O., Ravussin, E., & Astrup, A. (2020). The environmental foodprint of obesity. *Obesity, 28*(1), 73–79. https://doi.org/10.1002/oby.22657

Michaelowa, A., & Dransfeld, B. (2008). Greenhouse gas benefits of fighting obesity. *Ecological Economics, 66*(2–3), 298–308. https://doi.org/10.1016/j.ecolecon.2007.09.004

Our World in Data. (2021). *Obesity*. https://ourworldindata.org/obesity.

Pesaran, M. H. (2004). *General diagnostic tests for cross-section dependence in panels.* The University of Cambridge, Faculty of Economics. https://doi.org/10.17863/CAM.5113. Cambridge Working Papers in Economics, n. 0435.

Pesaran, M. H. (2007). A simple panel unit root test in the presence of cross-section dependence. *Journal of Applied Econometrics, 22*(2), 256–312. https://doi.org/10.1002/jae.951

Popkin, B. M. (1998). The nutrition transition and its health implications in low-income countries. *Public Health Nutrition, 1*(1), 5–21. https://doi.org/10.1079/PHN19980004

Reardon, T., Timmer, C. P., Barrett, C. B., & Berdegu, J. (2003). The rise of supermarkets in Africa, Asia, and Latin America. *American Journal of Agricultural Economics, 85*(5), 1140–1146. https://www.jstor.org/stable/1244885.

Robalino-López, A., Mena-Nieto, Á., García-Ramos, J. E., & Golpe, A. A. (2015). Studying the relationship between economic growth, CO2 emissions, and the environmental Kuznets curve in Venezuela (1980–2025). *Renewable and Sustainable Energy Reviews, 41*, 602–614. https://doi.org/10.1016/j.rser.2014.08.081

Román-Collado, R., & Morales-Carrión, A. V. (2018). Towards a sustainable growth in Latin America: A multiregional spatial decomposition analysis of the driving forces behind CO_2 emissions changes. *Energy Policy, 115*(May 2017), 273–280. https://doi.org/10.1016/j.enpol.2018.01.019

Roskam, A. J., Kunst, A. E., Van, O. H., Demarest, S., Klumbiene, J., Regidor, E., Helmert, U., Just, F., Dzurova, D., & Mackenbach, J. P. (2010). Comparative appraisal of educational inequalities in overweight and obesity among adults in 19 European countries. *International Journal of Epidemiology, 39*(2), 392–404. https://doi.org/10.1093/ije/dyp329

Santiago, R., Koengkan, M., & Fuinhas, J. A. (2020). The relationship between public capital stock, private capital stock and economic growth in the Latin American and Caribbean countries. *International Review of Economics, 67*, 293–317. https://doi.org/10.1007/s12232-019-00340-x

Sapkota, P., & Bastola, U. (2017). Foreign direct investment, income, and environmental pollution in developing countries: Panel data analysis of Latin America. *Energy Economics, 64*, 206–212. https://doi.org/10.1016/j.eneco.2017.04.001

Saravia-Matus, S. L., Aguirre Hörmann, P., & Berdegué, J. A. (2019). Environmental efficiency in the agricultural sector of Latin America and the Caribbean 1990–2015: Are greenhouse gas emissions reducing while agricultural production is increasing? *Ecological Indicators, 102*, 338–348. https://doi.org/10.1016/j.ecolind.2019.02.050

Sobal, J. (2001). Commentary: Globalisation and the epidemiology of obesity. *International Journal of Epidemiology, 30*, 1136–1137. https://doi.org/10.1093/ije/30.5.1136

Springmann, M., Godfray, H. C. J., Rayner, M., & Scarborough, P. (2016). Analysis and valuation of the health and climate change cobenefits of dietary change. *PNAS, 113*(15), 4146–4151. https://doi.org/10.1073/pnas.1523119113

Squalli, J. (2017). The environmental impact of obesity: Longitudinal evidence from the United States. *Public Health, 149*, 89–98. https://doi.org/10.1016/j.puhe.2017.04.016

Toti, E., Di Mattia, C., & Serafini, M. (2019). Metabolic food waste and ecological impact of obesity in FAO world's region. *Frontiers in Nutrition, 6*. https://doi.org/10.3389/fnut.2019.00126

Valencia-Herrera, H., Santillán-Salgado, R. J., & Venegas-Martínez, F. (2020). On the interaction among economic growth, energy-electricity consumption, CO_2 emissions, and urbanization in Latin America. *Revista Mexicana de Economía y Finanzas, 15*(4), 745–767. https://doi.org/10.21919/remef.v15i4.553

Vosti, S., Msangi, S., Lima, E., Quiroga, R., Batka, M., & Zanocco, C. (2011). *Agricultural greenhouse gas emissions in Latin America and the caribbean: Current situation, future trends and one policy experiment*. Inter-American Development Bank, Infrastructure and Environment. Discussion Paper IDB-DP-167.

Washburn, C., & Pablo-Romero, M. (2019). Measures to promote renewable energies for electricity generation in Latin American countries. *Energy Policy, 128*, 212–222. https://doi.org/10.1016/j.enpol.2018.12.059

Wooldridge, J. M. (2002). *Econometric Analysis of Cross-Section and Panel Data* (pp. 1–737). Cambridge, Massachusetts: The MIT Press.

World Bank Open Data. (2021). http://www.worldbank.org/.

Zaman, K., & Abd-el Moemen, M. (2017). The influence of electricity production, permanent cropland, high technology exports, and health expenditures on air pollution in Latin America and the Caribbean Countries. *Renewable and Sustainable Energy Reviews, 76*, 1004–1010. https://doi.org/10.1016/j.rser.2017.03.103

Index

'*Note:* Page numbers followed by "f" indicate figures, "t" indicate tables and "b" indicate boxes.'

9780323993395